The Minerals, Metals & Materials Series

Murat Tiryakioğlu · William Griffiths ·
Mark Jolly
Editors

Shape Casting

7th International Symposium Celebrating Prof. John Campbell's 80th Birthday

TMS

Springer

Editors
Murat Tiryakioğlu
University of North Florida
Jacksonville, FL, USA

William Griffiths
University of Birmingham
Birmingham, UK

Mark Jolly
Cranfield University
Cranfield, UK

ISSN 2367-1181 ISSN 2367-1696 (electronic)
The Minerals, Metals & Materials Series
ISBN 978-3-030-06033-6 ISBN 978-3-030-06034-3 (eBook)
https://doi.org/10.1007/978-3-030-06034-3

Library of Congress Control Number: 2018964679

Cover illustration: From Cao, Xinjin, "Heat Treatment of Liquid Metal: Precipitation and Sedimentation Processing of Liquid Al-11.5Si-0.4Mg Cast Alloy", Ph.D. Thesis, University of Birmingham, 2001; Supervisor: Prof. John Campbell

This Springer imprint is published by the registered company Springer Nature Switzerland AG
The registered company address is: Gewerbestrasse 11, 6330 Cham, Switzerland

Preface

This is seventh in the series of International Shape Casting Symposia held at the TMS Annual Meeting. The 2019 symposium has been organized to celebrate the 80th birthday of Prof. John Campbell, who served on the organizing committee between 2005 and 2015. Professor Campbell has been a strong advocate for solving the challenges facing the casting industry through the scientific method and incorporating physical fundamentals into solutions. He has been not only the leading expert on castings worldwide, but also an advisor, a teacher, a mentor, and a friend to many of us working on castings. Overall, he has been an inspiration for those of us who try to make a contribution to the field. John, thank you for all you have done. Happy 80th birthday!

As in previous volumes, we continue to explore the leading edge technologies and the latest innovations in casting process design and quality improvements relative to shape casting through presentations by researchers from around the world. We have given special emphasis to the contributions of Prof. Campbell, especially bifilms in castings.

Murat Tiryakioğlu
Lead Organizer

Contents

Part II Casting Defects and Their Characterization

Part III Process Innovation and Modelling

About the Editors

Murat Tiryakioğlu is a Professor of Mechanical Engineering and Advanced Manufacturing and Materials Innovation at the University of North Florida. He received his B.Sc. in mechanical engineering from Boğaziçi University (1990); M.S. (1991) and Ph.D. (1993) in engineering management from the University of Missouri, Rolla; and another Ph.D. in metallurgy and materials (2002) from the University of Birmingham, England.

He grew up around his family's foundry, which continues to thrive in Istanbul, Turkey. This has led to research interests in process design for high-quality castings, aluminum heat treatment modeling and optimization, process–structure–property relationships in metals, and statistical modeling and quality and reliability improvement, on which he has written a chapter, published over 100 technical papers, and edited ten books.

He was awarded the inaugural John Campbell Medal, from the Institute of Cast Metals Engineers (UK), and the Eugene Merchant Outstanding Young Manufacturing Engineer, by the Society of Manufacturing Engineers. He was also selected a TMS Young Leader in Light Metals. He is a member of TMS, ASM International, and ASQ and is an ASQ Certified Quality Engineer.

William Griffiths worked for several years in an iron foundry and eventually obtained a Ph.D. in the quenching of steel at Sheffield Hallamshire University. This led to an appointment in Foseco plc, followed by a position as a Research Fellow at the University of Nottingham, carrying out research in electromagnetic stirring of Al alloy, and as Federal-Mogul Research Fellow at UMIST, investigating interfacial heat transfer during casting solidification in Al alloys. Following this, he moved to the University of Birmingham to work on double oxide film defects with Prof. John Campbell, as a Lecturer, in 1999, and subsequently as Senior Lecturer.

He has been involved in research into solidification processing for about 30 years. The theme running through this research is the identification and study of the mechanisms that control metallurgical processes, particularly in the case of casting. It is vital that a detailed understanding of the principal phenomena that control the process is achieved before successful modeling can be carried out. His research has been aimed at experimental investigation of the fundamental mechanisms that govern solidification processing, and has embraced several different areas:

- Interfacial heat transfer during casting solidification, studying the acceleration of cooling rate associated with solidification in helium.
- Double oxide film defects in Al alloys, investigating how the addition of transition metals (Mo and W) can reduce their deleterious effects.
- The lost-foam casting process, investigating the influence of molecular weight of the pattern on casting quality.
- Positron emission particle tracking (PEPT), employing radioactive oxide particles which can be tracked once inserted into a liquid metal, to allow the study of the movement of inclusions in castings.

Mark Jolly is Professor of Sustainable Manufacturing and head of the Sustainable Manufacturing Systems Centre at Cranfield University (seven faculties, eight staff, and approximately 30 Ph.D.s). He has over 35 years' experience in manufacturing. He spent 13 years working in industry in automotive and tier 1&2 suppliers into manufacturing both in the UK and abroad before moving into academia in 1995. In 2012, he took up his current position at Cranfield after 17 years at the University of Birmingham. He has managed over £8.5 M of research projects since 1999. He has over 350 publications, technical reports, and articles and has given over 60 invited lectures and seminars in the UK and globally. He has taught at undergraduate and postgraduate levels. He is a judge for the TMMX Sustainable Manufacturer of the Year Award sponsored by IMechE and the Manufacturer as well as Chair of judging panels in a number of categories for the EEF Best Manufacturer Awards. He has sat on the EPSRC Peer Review College since 2003 and sits on the UK Metals Council and is a reviewer for the European Space Agency and Enterprise Ireland.

He is currently Chair of the Materials Science and Technology Division of IOM3, Vice-Chair of the Sustainable Development Group of IOM3, and was Chair of the Solidification Committee of The Minerals, Metals & Materials Society (USA) for 2 years until 2018. He is a Liveryman of the Worshipful Company of Founders.

His main areas of current research are in circular economy and sustainability including resource-efficient manufacturing, process modeling, and novel casting processes. Resource-efficient manufacturing targets traditional processes and aims at reducing the materials usage and energy with the ultimate aim of lowering carbon and water footprints of final components. He has worked with many large names across a number of sectors including Rolls-Royce, DePuy Synthes (Johnson & Johnson), GKN, and Coca-Cola. He has also championed working with many small companies especially SMEs who supply to large companies.

Part I
Entrainment and Bifilms

Update on Bifilms—*The* Fundamental Defect in Metals

John Campbell

Abstract The concept of the bifilm is now nearly 20 years old. It has been clearly shown to control structural defects in castings such as porosity and hot tearing, plus, in solid wrought products, mechanical properties such as ductility and fatigue. It appears to explain for the first time the structures of Al–Si alloys and cast irons. Furthermore, although there are a number of proposed mechanisms for crack initiation in metals, it seems most probable that these are invalid. It follows that there is probably no intrinsic crack failure mechanism for most engineering metals; metals should never crack. The only crack-initiating feature appears to be the bifilm, introduced accidentally during casting. By appropriate casting techniques, the introduction of bifilms can be avoided. For this reason, failure by fatigue and creep may be eliminated, together with invasive corrosion such as pitting, stress corrosion cracking and possibly other failure modes.

Keywords Bifilm · Creep · Fatigue · Corrosion · Griffith crack · Casting

Introduction

During the pouring of liquid metals, the liquid surface can become folded into the liquid, and liquid drops and splashes impinge. These turbulent random actions entrain both air bubbles and oxide films into the liquid. The oxide entrainment necessarily involves the oxide on the liquid surface being impinged against its opposing oxide, 'dry side' to 'dry side' and so forming an unbonded interface between this pair of films. I call it a 'bifilm'. As a result of the necessary logic of the entrainment mechanism, all oxide films in liquid metals are double films and all act as cracks. Turbulent pouring of liquid metals fills the liquids with cracks (Fig. 1).

J. Campbell (✉)
Emeritus Professor of Casting Technology, Department of Metallurgy and Materials, University of Birmingham, Birmingham, UK
e-mail: jc@campbelltech.co.uk

M. Tiryakioğlu et al. (eds.), *Shape Casting*, The Minerals, Metals & Materials Series,
https://doi.org/10.1007/978-3-030-06034-3_1

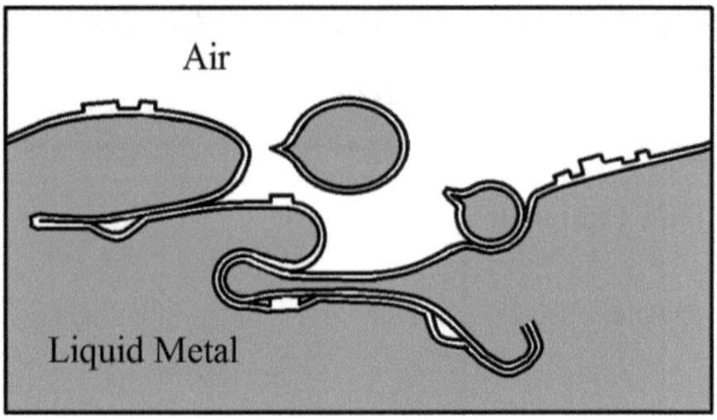

Fig. 1 Entrainment of bubbles and bifilms during the turbulence of pouring

The central interface of the bifilm contrasts with the outer interfaces in contact with the matrix. The outer interfaces have perfect atomic bonding with the matrix. The external interfaces act as a substrate for the formation and growth of many intermetallics and second phases. These twin creations of surface turbulence, *(i) cracks and (ii) substrates,* contribute significantly to the properties and structures of metals.

A further complication is the two forms the bifilm can take: (a) compact and convoluted as a result of bulk turbulence in the liquid; and (b) 'unfurled', straightened or flattened. The convoluted and compact bifilm is relatively harmless in the casting, but the unfurled, flattened bifilm constitutes an engineering crack, and so is an important and harmful defect. The bifilm can become straightened by a number of mechanisms, including (i) shrinkage, (ii) gas precipitation, (iii) dendrite pushing or (iv) as a result of attaching to, and being constrained to follow, the morphology of planar second phases during their growth [1].

Gas Porosity, Shrinkage Porosity, Hot Tearing

The trinity of major casting defects which occur during solidification, gas porosity, shrinkage porosity and hot tearing, all appear to be initiated by bifilms. Without the presence of bifilms, none of these defects could appear. This expectation is simply the consequence of all of these macroscopic features being initiated without difficulty by bifilms. In contrast, nucleation (as an atomic-sized event) is enormously difficult. Even classical heterogeneous nucleation of volume defects (considered as an atomic-sized event in the form of a cap bubble on a planar substrate) is still difficult. This is the consequence of the strength of interfaces which have grown in the melt and which are therefore in perfect atomic contact. We may generalize this observation, predicting that the only interfaces which can be separated are those which are actually unbonded. Such interfaces exist in the metal only as a result of entrainment from the outside. The bifilm as an entrained entity eliminates any significant initiation difficulty because its two adjacent films are unbonded. Its component films therefore easily separate with minimal gas pressure or minimal stress, making it the ideal initiation site for the growth of pores and cracks. These expectations are amply confirmed every day on the shop floors of foundries.

The hot tear is intriguing because often it is not clear whether (i) the tear has been initiated by a bifilm and the volume defect created in this way continues to grow into the solidifying metal under the action of continuing applied strain, or (ii) whether the tear itself *is* the bifilm. It seems probable that both varieties of tear exist.

In passing, we should note a relatively trivial but common misconception. The nucleation scenario presented in most textbooks depicts an embryonic bubble sitting at the bottom of a V-shaped depression in an inclusion. This feature is viewed as an easy initiation site (sometimes misleadingly called a 'nucleation site'). However, it is actually an *entrained* feature; not created by solidification; it had to be introduced from the outside. Thus, the adhering air bubble will be enclosed in the oxide skin entrained by its penetration of the surface of the liquid. All such entrained inclusions, such as sand inclusions, can be found to be shrouded in the surface oxide film. As such, they are rather like semi-rigid bifilms, in which one side of the defect is the rigid inclusion, the other side is the entrained oxide film, and between these two non-wetted surfaces is a layer of entrained air or other gas from the environment. As a rather rugged variety of bifilm, it is, of course, an excellent initiator of gas, shrinkage or tear defects.

Cracking—The Central Issue

We now turn to defects controlled by bifilms in the solid state.

The fact that metals deform and crack is familiar to the point that it appears to need no explanation; it has always happened, and we naturally assume that it will always happen; after sufficient deformation, sometimes very little, failure by cracking is to be expected. It is a familiar feature of the failure mode of metals.

Books have been written on fracture mechanics, the science of prediction whether a crack will propagate or not. However, relatively little has been studied on the initiation of cracks. The classical historical approach has been the concept of the initiation of a crack from a pre-existing micro-crack, as assumed by Griffith.

Significantly, there appears to have been no thought given to the origin of the pre-existing micro-cracks, nor much serious consideration of the problems of the initiation of a crack in a sound metal. This is the central issue to be discussed in this paper.

Crack Initiation

In brief, in a sound metal, it seems impossible to initiate a crack. All theories of cracking seem to agree on this point. As further corroboration, the problem of creating a volume defect such as a pore seems quite impossible in a 'sound' liquid or solid metal. This follows simply from the enormous strength of the metallic bond between atoms; it is really difficult to pull atoms apart. Such ultimate strengths are in the range of tens of GPa for metals.

Historically, the formation of cracks has been assumed to occur by a number of mechanisms [2]:

1. The condensation of vacancies: This has never been observed. Vacancies condense to fully collapsed features because of the enormous cohesive forces between atoms. The fully dense, collapsed features include dislocation rings or stacking fault tetrahedra, depending on the stacking fault energy. Vacancies in irradiated metals appear to condense to form bubbles on grain boundaries, but these boundaries are most probably the sites of bifilms which are, naturally, easily inflated.
2. Dislocation pile-ups: These features certainly occur, but have never been observed to result in the formation of cracks. More recently, Wang [3] has demonstrated by molecular dynamic simulations that the formation of cracks from pile-ups is not possible.
3. Formation in steels from so-called brittle carbides at grain boundaries: This misunderstanding has arisen because it was not known that carbides would form, crack-free and ductile, but sited on bifilms as preferred substrates, mainly located at grain boundaries. (The ductility of so-called brittle phases is beautifully demonstrated by common salt, NaCl, crystals when deformed under running water to dissolve away surface cracks and stress raisers. It bends easily [4]. The formation of intermetallics and second phases by either solidification or solid-state diffusion can only create atomically perfect interfaces. Without stress raisers of pre-existing micro-cracks, they will be unable to propagate failure and be forced to deform plastically when sufficiently stressed. The only unbonded interfaces will be those entrained from the outside world, particularly the oxide on the liquid metal surface.)
4. Formation from stresses around inclusions: Once again, the stresses as a result of applied thermal or mechanical sources are not expected to be sufficiently high to nucleate an atomic-sized embryonic pore at the inclusion/matrix interface. Interfaces too have enormous cohesive energy, typical of interatomic bonds. For instance [5], estimates of the tensile strength of the silicon/matrix interface in Al–Si alloys are in the region of 20 GPa which is of the same order as the strength of the matrix.

Clearly, none of the above four mechanisms is responsible for crack initiation.

(The fact that cracks are often seen to grow from inclusions is because the inclusion is the remnant of an entrainment defect such as a bifilm. The residual insoluble argon, from the 1% argon in the air, will constitute a permanent gas inclusion in the metal after the oxygen and nitrogen have been consumed by reaction. The argon trapped in the original bifilm will remain attached to the inclusion, such as an alumina inclusion in steel. The crack is thereby enabled to form from this pre-existing volume defect; an unbonded region, probably in the form of a cap bubble of argon, on the alumina inclusion. The widespread observations of cracks appearing to initiate in grain boundaries or even the centres of grains are commonly cited as evidence against bifilm initiation. However, all such evidence is unfortunately contaminated because of the universal poor casting technology for metals which effectively ensures a dense population of bifilm defects in existing research materials.)

This leaves the bifilm as, probably, the *only* potential candidate for crack formation. We shall expand on this possibility later in this account.

Crack Propagation

There is good theoretical work to establish that many of our engineering metals are intrinsically ductile. In the case of a crack subjected to a tensile opening stress, instead of the crack propagating, the crack tip merely emits dislocations. Macroscopically, this process develops into crack blunting. The crack opens, but does not propagate. The number of engineering metals behaving in this way is shown in Fig. 1.

Interestingly, from its universal acceptance of being the most malleable metal, this author had expected gold (Au) to appear on the extreme left of Fig. 1, consistent with its lack of oxides. The fact that lead (Pb) should, in theory, be more malleable is clearly countered by its content of entrained oxides.

Those metals which are intrinsically brittle, including W, Cr and Be, would have an interestingly more complex behaviour if sound (i.e. free from micro-cracks). If subjected to a tensile stress, they would be expected to deform plastically. However, in the event of the presence of a crack (introduced, for instance, as a hacksaw cut, or as part of the design or manufacture of the component) the crack tip does not put out dislocations, but immediately extends the crack.

Some intermediate metals, including the curiously heterogeneous group, Co, Fe, Mg, Mo and Nd, have mixed behaviour. Iron in particular is famously temperature-sensitive, exhibited in its 'ductile/brittle transition temperature' (DBTT). At this stage, from shop floor experience in foundries, it is known that Co alloys, steels and Mg alloys all gain in ductility and toughness as entrainment problems are reduced. There may be some reassurance therefore, at least at ambient temperatures and above, that in the absence of entrained defects acting as micro-cracks, the intermediate group of metals will also behave in a ductile mode. (Checks on the behaviour of steels below the DBTT have yet to be carried out, and will, of course, be of exceptional interest and significance.)

Crack Theory

Turning to the fundamentals of crack initiation, over the years the Griffith model has remained widely accepted. This is an energy concept, not reliant on any geometrical model. However, only relatively recently it has been shown to be inadequate, as a result of the stored elastic energy falling too quickly to provide sufficient energy for the propagation of the crack. The Griffith criterion therefore cannot account for the propagation of cracks [6].

Barenblatt was the first to come forward with a geometrical model, consisting of a planar region of the metal lattice which exhibited zero cohesion, surrounded by region of normal (high) cohesion [6]. This model fulfilled the conditions for both initiation and growth of cracks. At the time of its formulation, no one would have known how a planar region of a metal lattice could exhibit zero cohesion. The model, of course, happens to be a perfect fit with the structure of the bifilm. It therefore appears to provide the bifilm concept with a rigorous mathematical underpinning.

Application of the Bifilm Concept

1. Tensile Failure by Cracking

The relative freedom from bifilms enjoyed by many steels is seen in their elongation to failure results in the region of 50% and higher. The large populations of bifilms in aluminium alloys resist floating out because of their near-neutral buoyancy, so that for many foundries elongations are in the region of 1–3%. However, there is now much shop floor experience in foundries that cleaner castings exhibit greater elongation, so that by the use of better technology for reducing bifilm populations in metals, aluminum foundries can achieve up to 20% at this time. Higher elongations are predicted from further improvements in melting and casting techniques. The ductility should perhaps be measured as reduction of area (RA), with 100% RA as the ultimate target for all metals.

Essentially, the elimination of bifilms is predicted to eliminate failure by cracking in metals. Thus, a tensile test, irrespective of the strength of the alloy, should only ever result in plastic deformation to failure at 100% RA. Metals predicted to be ductile should never crack. Even those predicted to be brittle, on the right of Fig. 2, but without bifilms, may behave in a ductile mode (providing no 'designed' crack is introduced to the design of the component).

2. Creep

The inspirational work by Wilshire and colleagues at Swansea, UK, has transformed the interpretation of creep curves [7]. The intrinsic accuracy of the new interpretation is underpinned by the accuracy with which the curves can now be used to extrapolate limited data to estimate lifetimes of components many years ahead. Originally, creep curves were divided into three regions: an initial transient, a steady-state regime and a final transient. The new interpretation has assumed that only two regimes apply: an initial transient in which the creep rate falls progressively and a final transient in which creep accelerates to final rupture. The steady-state regime is seen to be merely an overlap of the two transient curves (Fig. 3). Many studies of the final transient show that the metal fails by the growth of cavities, which gradually link to cause final failure.

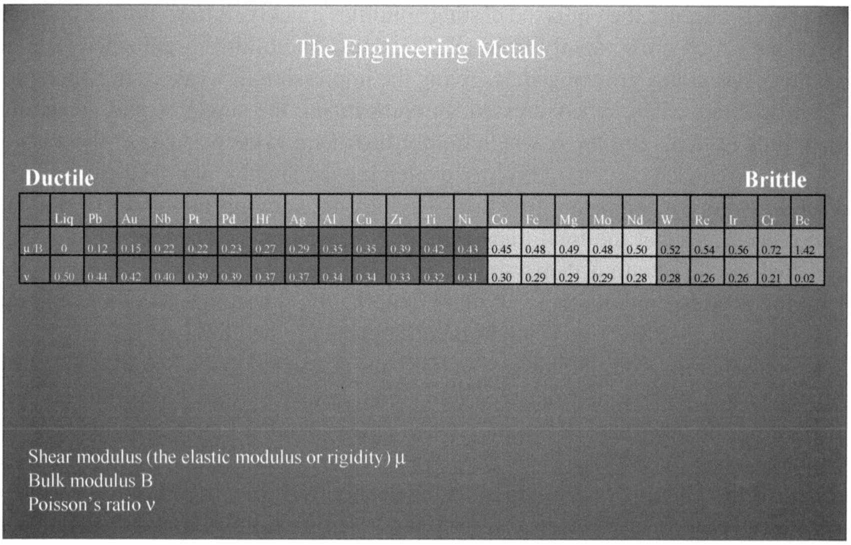

	Liq	Pb	Au	Nb	Pt	Pd	Hf	Ag	Al	Cu	Zr	Ti	Ni	Co	Fe	Mg	Mo	Nd	W	Re	Ir	Cr	Be
μ/B	0	0.12	0.15	0.22	0.22	0.23	0.27	0.29	0.35	0.35	0.39	0.42	0.43	0.45	0.48	0.49	0.48	0.50	0.52	0.54	0.56	0.72	1.42
ν	0.50	0.44	0.42	0.40	0.39	0.39	0.37	0.37	0.34	0.34	0.33	0.32	0.31	0.30	0.29	0.29	0.29	0.28	0.28	0.26	0.26	0.21	0.02

Fig. 2 A line-up of the engineering metals in terms of ductile or brittle behaviour with Au as the most malleable because of its lack of bifilms, compared to the predicted position of Pb

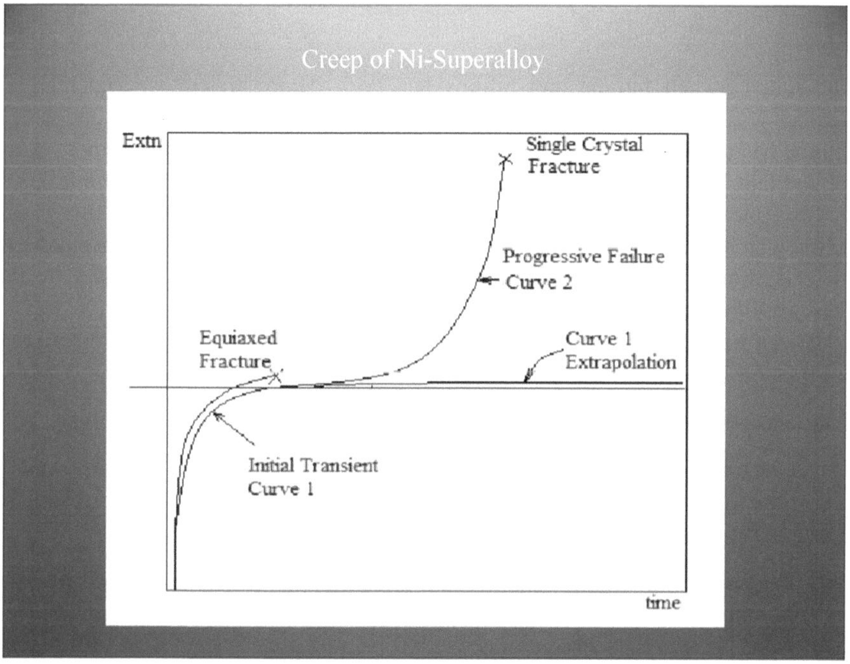

Fig. 3 Initial and final transient curves as the new description of creep behaviour. In the absence of bifilms, the final transient is predicted to be eliminated, eliminating creep strain

In Fig. 3 because the equiaxed casting solidifies quickly, it traps bifilms between the grains which grow rapidly from all directions. The bifilms are trapped at grain boundaries as grains grow together, giving the impression of weak grain boundaries which decohere easily when stressed. In comparison, the single crystal gives time for bifilms to float, and those which do not float tend to be pushed, so the casting has a much reduced content of bifilms, explaining the greatly improved creep life. If all bifilms are removed, the growth of cavities accelerating the strain to fracture cannot occur, so that the final transient curve is eliminated. The creep life is then predicted by the continued extrapolation of the initial transient. Effectively, the creep life becomes indefinite. Even more importantly, it becomes indefinite without exhibiting creep strain. Creep might therefore actually be eliminated.

In confirmatory work, Indian research [8] has illustrated that a Mg alloy cast by a counter-gravity technique, and therefore, relatively free from bifilm defects had only half of the creep rate of the alloy when gravity poured (and hence damaged by turbulence).

Following the logic of this paper, the nucleation and growth of cavities can only occur as a result of the presence of bifilms. It further follows that if bifilms were absent, the final transient behaviour would not exist. No cavities or cracks would open up. Thus, the creep behaviour now consists only of a continuation of the primary transient. Effectively, creep has been eliminated by eliminating defects. The importance of this prediction surely needs no emphasis. Confirmatory research is required as quickly as possible.

3. Fatigue

The link between fatigue and bifilms has been demonstrated experimentally a number of times (for instance [9]). The analogous relation between elongation and fatigue has also shown to be excellent [10]. In the experience of the author, every fatigue failure he has examined appears to have been initiated from a bifilm defect.

Nevertheless, it is conceivable that fatigue may be initiated from intrusive slip bands which could grow into an initiating crack. However, even in this situation, there is some evidence [11] that the slip bands themselves are initiated by the presence of entrained defects.

It seems possible, therefore, that most, if not all fatigue, might be eliminated by elimination of bifilms.

4. Corrosion

The bifilm is, unfortunately, an effective and efficient agent of invasive forms of corrosion [12]. The features of bifilms which aid them to enhance corrosion so damagingly are

(i) The ability of the corrodent to penetrate deeply into the bulk of the metal via the unbonded central interface, usually exhibiting patchy regions as an open crack.
(ii) The fact that the bifilm has often been the substrate for a number of intermetallics and second phases during solidification. Thus, these compounds sit

on the wetted outer surface of the bifilm, only a few atoms thickness away from the inflowing corrodent. The opportunity for bimetallic couples leading to enhanced corrosion is usually abundant.

Lin and co-workers [13] describe a clear example of the action on bifilms in exfoliation corrosion of Al alloys, in which the corrosion conditions generate pits, flaking or blisters, all of which are worsened by cold rolling and solution treatment, after which the alloy splits and splinters like wood. Apparently, this is a life-limiting degradation effect in airframes. Another work on 6061 Al alloy [14] finds corrosion associated with one or two corrosion pits to halve the fatigue life, but the effect was not understood. Corrosion pits seem to be always denoting the emergence of a bifilm at the surface of the metal.

More serious varieties of invasive corrosion include stress corrosion cracking (SCC). As an example of the probable role of bifilms in at least some SCC problems, Japanese research [15] revealed the rate of crack growth in a 690 steel subjected to embrittlement in the primary water of a pressurized water reactor. Rates varied from batch to batch in an inexplicable way. The clear micro-graphs showing transverse cracks across carbides arranged as 'beads on a string' along grain boundaries were typical of carbides which had grown on bifilms. Chinese work [16] on an austenitic stainless steel in water at 250–320 °C found that SCC preferred to initiate at the bottom of a corrosion pit and propagated along the phase boundary between ferrite and austenite. This again is typical of bifilm behaviour.

The phenomenon of hydrogen embrittlement may involve a key contribution from bifilms [12]. As an example, in hydrogen environment-assisted cracking of Monel K-500, the failures triggering the development of dimples in the ductile dimple fracture surfaces were always found to be carbides [17]. During the freezing of the alloy, carbides would have nucleated and grown on bifilms and so would be 'pre-cracked', opening easily to initiate the surrounding ductile shearing leading to the creation of a dimple. As an aside, adjacent ductile dimples meet at a razor edge, confirming the 100% RA of the matrix material.

The Elimination of Bifilms

All the above failure mechanisms appear to depend on the presence of bifilms. It is important to emphasize that these defects are not the result of a natural and necessary phenomenon such as solidification, which is, of course, merely a phase change, but are the mechanical accidents of entrainment of the surface oxide of the melt during melting and casting processes.

It follows that if we eliminate the mechanical phenomenon in which the oxide surface of the melt becomes submerged into the bulk melt, we eliminate failure by cracking. In practice, of course, it is also necessary to eliminate those populations of entrained defects which already exist in our metals as a result of their history of having been melted and cast perhaps many times previously.

Even if entrainment occurs, steels have great benefits resulting from their high density, floating out populations of entrained defects quickly, allowing them to enjoy their high tensile elongations. Furthermore, because of their high processing temperatures, some steels benefit from a liquid surface oxide. Impingement of drops and splashes in such steels is now liquid-to-liquid, resulting in liquid oxide droplets, not bifilms. In these conditions, bifilm formation is avoided. Even if the surface oxide is not melted, the resulting bifilm population may be transformed to more compact inclusions by coarsening-type processes, in which the film becomes an array of solid particles, reducing the crack-initiating potential of the defect. Even so, importantly, this change to its morphology does not completely eliminate the crack-forming potential; the expected retention of the argon bubble will be present as a non-wetted patch on the solid inclusion and so provides a micro-cavity which can form the initiation site for a crack.

In steelmaking, the pour from the furnace into the ladle is hugely damaging to the quality of the melt, but is not easily avoided for large tonnages. For limited sizes of products up to a few tons, the metal in the furnace could, in principle, be cast by transferring into the mould by counter-gravity, which would result in the most perfect liquid possible in the mould cavity. Steelmakers are recommended to consider such options seriously and urgently.

In those cases where the pour into the ladle cannot be avoided, significant valuable detraining of defects continues to occur by flotation prior to teeming the melt into moulds. The subsequent pours into ingot or other moulds can be greatly improved (probably eliminating in the region of 95% of entrained inclusions) if contact pouring is used. This is a low-cost, simple technique which, additionally, also improves safety in the foundry. There is no excuse for not pouring ingots by the contact pour technique. The alternative efforts to avoid air entrainment by the use of shrouds of various types, sometimes aided by inert gas, are all much less effective and are recommended to be abandoned.

Where the ladle empties into a continuous casting unit, every effort has to be made to eliminate *entrainment*. Curiously, and disappointingly, the major research in continuous casting is on the *detrainment* of inclusions from launders. This is largely a wasted effort because the dwell time in launders is usually insufficient to permit significant cleaning action. The re-direction of effort to prevent *entrainment* by the use of correctly tapered sprues and better-designed submerged entry nozzles (SENs) would be far more effective because the steel in the ladle, dwelling so long during the continuous cast process, is likely to be extremely clean. Transferring this superclean metal into a casting machine is the greatest risk to quality; the degradation process resulting from *entrainment* is the key issue to be solved.

For light alloys, the author has endeavoured to promote the improvement of metal quality by [18] a three-pronged attack on oxides

(i) eliminating the external oxide skins on charge materials by using a dry hearth furnace,

(ii) eliminating the internal oxides in the charge by filtration and/or sedimentation and/or enhanced by chemical addition, and
(iii) eliminating the re-introduction of bifilms by counter-gravity casting into moulds using a pump.

Although electromagnetic (EM) pumps have been used, and have now started to be supplanted by centrifugal pumps, both of these varieties of pumps suffer massive shearing stresses in their working volumes. Although this energetic turbulence occurs under the surface of the liquid metal, the forces involved are so high that entrainment of oxygen, sucked through or down the refractories, seems inescapable. It is not known at this time whether these varieties of pumps cause degradation of the metal. Pneumatic pumps, using simple displacement by a gas, seem to offer a method of counter-gravity in which the melt may be undamaged. Such pumps are in development at this time. The use of a pump seems essential for the controlled and quiescent filling of moulds to prevent the re-introduction of bifilms into clean metal. In addition, pumps greatly facilitate the automation and productionizing of the casting process [18].

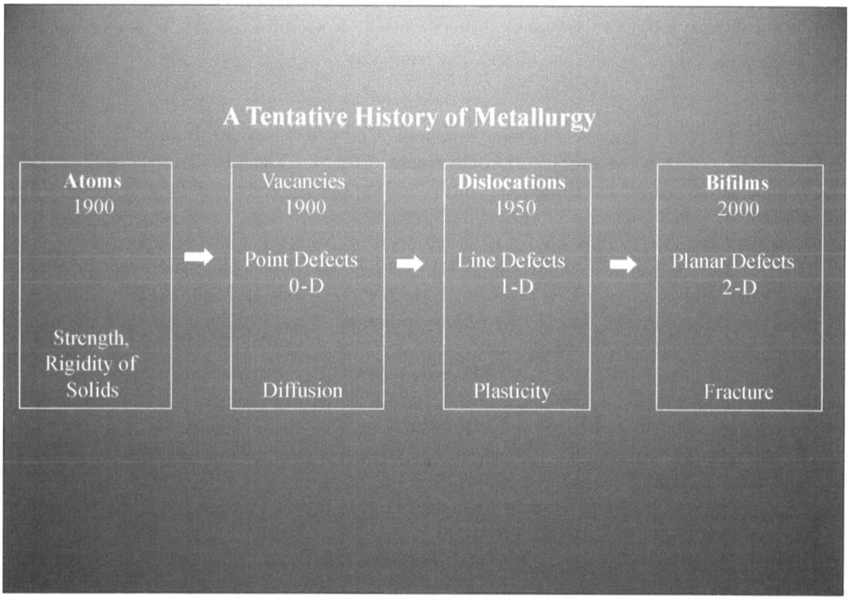

Fig. 4 A tentative suggested history of metallurgical concepts, noting the aspects of metallurgy which could be understood at each stage, with the recent central role of bifilms in the understanding of fracture

Conclusions

1. Bifilms are formed during the turbulence of melting and casting processes and provide unbonded interfaces which appear to be the only mechanism for the initiation of pores and cracks and cracks in metals (Fig. 4).
2. Current melting and casting processes deliver metals with highly variable bifilm populations, creating batch-to-batch variations in the quality of metal products.
3. The centrality of bifilms in metallurgy is expected from their predicted control of fracture (tensile failure, creep and fatigue) and invasive corrosion.
4. Bifilms in metals can be reduced and possibly eliminated only in the liquid state.
5. The control of the melting and casting process therefore has the capability to eliminate cracking, creep, fatigue and invasive corrosion and perhaps SCC and hydrogen embrittlement.
6. As dislocations are fundamental to plasticity, bifilms are fundamental to cracking.
7. The concept of the bifilm and its possible control represents a re-thinking of much metallurgy and the potential for engineering.

References

1. Campbell J (2015) Complete casting handbook, 2nd edn. Elsevier, Oxford
2. Campbell J (2011) The origin of Griffith cracks. Met Mat Trans B 42:1091–1097
3. Wang J (2015) Atomistic simulations of dislocation pileup: grain boundary interaction. JOM 67(7):1515–1525
4. Andrade EDC (1935) Plasticity of rock salt crystals. Nature 135: 310–310
5. Gall K, Horstemeyer MF, Schilfgaarde MV, Baskes MI (2000) Atomistic simulations of the tensile debonding of an Al–Si interface. J Mech Phys Solids 48:2183–2212
6. Charlotte M, Laverne J, Marigo J-J (2006) Initiation of cracks with cohesive force models. Eur J Mech A/Solids 25:649–669
7. Wilshire B, Sharning PJ (2008) Int Mater Rev 14:91–104
8. Srinivasan A, Pillai UTS, Swaminathan J, Pai BC (2006) Enhanced creep properties of low pressure die cast AZ91 Mg alloy. Int J Cast Metals Res 19(5):265–268
9. Tiryakioglu M, Eason PD, Campbell J (2013) Fatigue life of ablation-cast 6061-T6 components. Mater Sci Eng, A 559:447–452
10. Tiryakioglu M (2014) The relationship between elongation and fatigue life in A206 aluminum castings. Mater Sci Eng, A 601:116–122
11. Tiryakioğlu M (2011) Fracture surface facets and fatigue life potential of castings. Metall Mater Trans B 42(6) 1098–1103
12. Campbell J (2016) Consolidation and corrosion of metals: an overview of the role of bifilms in corrosion. Innov Corros Mater Sci 6:132–139. 000-000 1 2352-0949/16
13. Lin L, Liu Z, Li Y, Han X, Chen Xu (2012) Effects of severe cold rolling on exfoliation corrosion of Al–Zn–Mg–Cu–Cr alloy. J Mater Eng Perform 21(6):1070–1075
14. Almaraz GM, Aburto AD, Gomez EC (2014) Ultrasonic fatigue endurance of Al alloy 6061-T6 on pre- and non-corroded specimens. Metall Mater Trans 45:280–286
15. Yonezawa T, Watanate M, Hashimoto A (2015) Primary water SCC in 690 alloy. Metall Mater Trans 46:2768–2780

16. Lu YH, Chen ZR, Zhu XF, Shoji T (2014) SCC behaviour of austenitic stainless 23CN20-09 M in high temperature water. Mater Sci Technol 30(15):1944–1950
17. Harris ZD, Dolph JD, Pioszak GL, Rincon Troconis BC, Scully JR, Burns JT (2016) Hydrogen environment-assisted cracking of monel K-500. Metall Mater Trans 47:3488–3510
18. Campbell J (2015) Quality castings: a personal account of the development of the cosworth casting process. Aspect Design 2013 and American Foundry Society 2015

Entrainment Defects in Cast Iron

Zakareya Nashwan and W. D. Griffiths

Abstract Castings were made of two types of cast iron, flake grey iron (FGI) and spheroidal graphite iron (SGI), in test bar moulds having both turbulent and quiescent filling systems. The different filling system designs did not seem to affect the scatter of values of the ultimate tensile strength (UTS) in FGI. On the other hand, the turbulently filled moulds had increased scatter of UTS values in the case of the FGI more than in the case of the SGI. Slag particles and ternary eutectic phosphide defects were identified as responsible for the fracture of FGI test bars. Two types of defects, (classified as type I and II), symmetrical on either side of the fracture surfaces of tensile test bars, were found in the SGI test bars, but not the FGI test bars. Scanning electron microscopy and EDX analysis of defects suggested that type I was a thin film of magnesium silicate and type II consisted of folded-over carbon-rich films containing magnesium oxide. These defects are similar in form and structure to the double oxide film defects found in Al and Mg alloys.

Keywords Cast iron · Flake grey cast iron · Spheroidal graphite cast iron · Weibull · Entrainment defects · Bifilms · Oxide films

Introduction

Casting of metal is a near net shape economical process for mass production of mechanical parts, but variability in mechanical properties and reproducibility issues can limit their use, resulting in forging and fabrications being preferred by designers

Z. Nashwan (✉)
School of Metallurgy and Materials, College of Engineering
and Physical Sciences, University of Birmingham, Edgbaston,
Birmingham, UK
e-mail: zsn584@bham.ac.uk

W. D. Griffiths
College of Engineering and Physical Sciences,
University of Birmingham, Edgbaston, Birmingham, UK
e-mail: w.d.griffiths@bham.ac.uk

M. Tiryakioğlu et al. (eds.), *Shape Casting*, The Minerals, Metals & Materials Series,
https://doi.org/10.1007/978-3-030-06034-3_2

because of their reliability [1]. Liquid metal is subjected to various transfer operations such as pouring or other types of surface turbulence. Through these actions, any surface film is enfolded creating entrainment defects. There are three principle types (a) bifilms, doubled over films that act as cracks, (b) bubbles which create bubble trail defects and (c) exogenous inclusions. Recent research into the cause of failure for light alloys has identified double oxide films to be an important source of defects which leads the castings to fail unexpectedly below their expected properties [2]. When melting sufficiently reactive metals, an oxide film is formed on top of the melt which reforms as soon as it breaks. During metal transfer from furnace to ladle and casting, splashing and surface turbulent flow can occur. Double oxide film defects form when the oxidized surface layer is folded over onto itself and enters the bulk of the metal. The folded-over layers are unwetted and trap a layer of mould atmosphere between them. When a solidification front engulfs the double oxide film layer, they act as intrinsic cracks. Double oxide films in Al alloys are characterised by a thin layer of gas separating two unbonded layers of oxide, which could be simple oxides, such as MgO, or complex, such as spinel [2, 3]. When the layer of gas becomes big enough, it floats to the surface carrying the folded film with it, leaving behind a trail of collapsed oxide film which is identified as a bubble trail defect [4]. Thus, oxide bifilms and bubble trail defects are deleterious to the mechanical properties of cast light alloys [5, 6].

Although no major research has been done in entrainment defects in ferrous alloys, evidence of entrainment defects in cast iron can be found in the literature. Campbell [7] found that FGI can suffer from bubble trail defects from silicate and carbonaceous film, formed by oxidizing mould and core gases, caused leakage defects from the bubble trails. Moreover, Campbell attributed the plate fracture effect in SGI, occurring along large planer grain boundaries, to dendrite straightening of bifilms [8]. Folded magnesium silicate films with thickness 35–40 μm were found on the fracture surfaces of brake support brackets made of SGI [9]. These evidences, although few, suggest the resemblance of entrainment defect mechanisms and effects in light alloys to cast iron. The film nature and composition might vary, but the mechanism and the effect on mechanical properties might be the same. The work reported here was carried out to study the evidence and nature of entrainment defects in cast iron.

Experimental Procedure

Two grades of cast iron, FGI G220 and SGI 500/7, were cast according to the British Standard BS EN 1561:2011 [10] for FGI and BS EN 1563:2018 [11] for SGI. FGI G220 was melted in a clay-graphite crucible in an induction furnace, while the SGI 500/7 was melted directly in an induction furnace. Both alloys were cast into two types of test bar moulds using a resin-bonded sand, (NOVATHANE for FGI and ALPHASET for SGI), at a pouring temperature for both alloys of 1420 °C. The moulds had quiescent and surface turbulent running system design as shown in Fig. 1. The test bars had dimensions 100 mm in height and 11 mm in diameter. The two designs of the moulds reflected two different manners of running system design. The mould shown in Fig. 1a has a poorly designed running system (PDM) to produce surface turbulence, where the liquid iron flow from the pouring basin down through a rectangular downsprue into a thick runner bar, followed by bottom filling of the test bars. This filling manner would be accompanied with significant entrainment events and splashing. Figure 1b shows mould with a well-designed running system (WDM) which has a quiescent running system with a tapered downsprue, thin runner bar (to prevent rolling back waves during filling) and a 20 ppi SEDEX ceramic foam filter to reduce ingate velocity followed by bottom filling of the test bars.

The behaviour of the liquid iron filling the two moulds was verified in a 190-kV real-time X-ray. To examine the casting properties, test bars were machined from each type of moulds, 20 for FGI and 40 for SGI, but only form the BM design both according to ASTM E8/E8 M-15a [12]. The test bars had a diameter of 6.75 mm and gauge length of 30 mm. They were tested using a Zwick 1484 tensile machine at a strain rate of 1.0 mm/min, equipped with a contact extensometer. The fracture

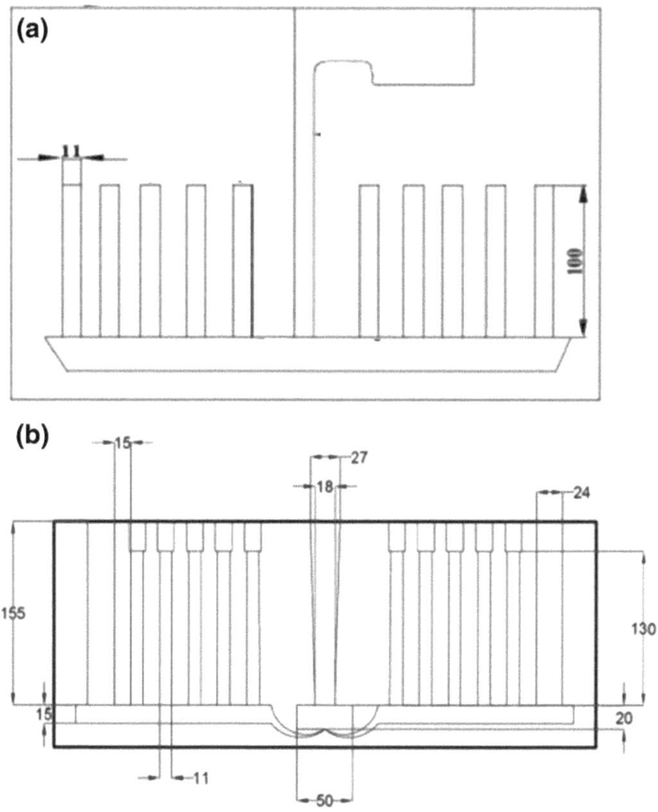

Fig. 1 **a** Design of the turbulent filling mould (PDM) and **b** design of the quiescent filling mould (WDM)

surfaces were examined with a PHILIPS XL30 FEG SEM equipped with an INCA Oxford electron dispersive X-ray spectrometer (EDX).

Weibull statistical analysis is widely used to study the distribution of mechanical data from flaw-containing materials such as castings [6, 13, 14]. The two-parameter Weibull distribution cumulative probability function is expressed as:

$$\mathrm{P} = 1 - \exp\left[-\left(\frac{\sigma}{\sigma^0}\right)^m\right] \tag{1}$$

where P is the probability of failure at a given stress or lower, σ is the ultimate tensile strength (UTS), σ^0 is the scale parameter, and m is the shape parameter also referred to as the Weibull modulus. Transforming Eq. (1) to a linear form by taking the natural logarithm twice and rearranging gives:

$$\ln[-\ln(1-\mathrm{P})] = m\ln(\sigma) - m\ln\left(\sigma^0\right) \tag{2}$$

When the left-hand side is plotted versus ln (σ), the Weibull modulus m can be determined from the slope of the plot and the shape parameter σ^0 from the intercept. P can be estimated based on the work by Tiryakioğlu and Hudak [15], who showed that the following estimator yielded unbiased estimates for both scale and shape parameters:

$$\mathrm{P} = \frac{i-a}{n+b} \tag{3}$$

Where i is rank in an ascending order, n is the sample size, and a and b are constants.

σ^0 and m are the parameters used to characterize the distribution of the UTS. The scale parameter σ^0 is a measure of mean strength and the minimum threshold value, and the Weibull modulus m represents a measure of the mean to standard deviation (signal to noise). Thus, the Weibull modulus is a single value that represents the scatter of properties, specifically in this paper, UTS. Higher values indicate a narrow spread of properties, fewer defects and more reliable and reproducible castings.

Results

Real-time X-ray inspection of the filling of the PDM showed the running system produced considerable splashing compared to the WDM where the filling was quiescent, as shown in Fig. 2, and hence, the castings would be expected to have significantly more entrainment defects.

The Weibull plots of the tensile properties obtained from the moulds FGI WDM and FGI PDM are shown in Fig. 3 for the UTS. Figure 4 shows the Weibull plots

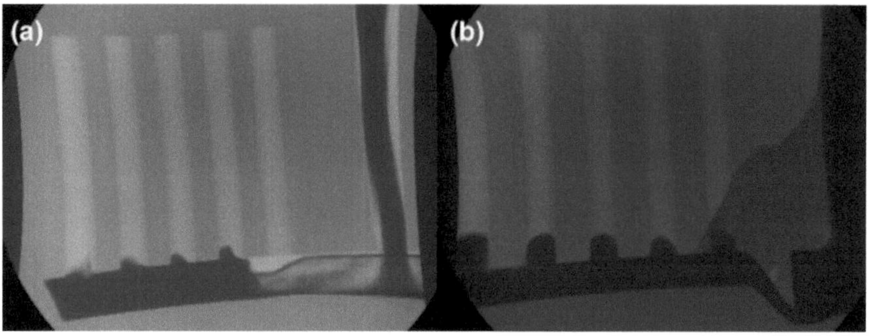

Fig. 2 Real-time X-ray image of the filling of **a** PDM showing surface turbulence and **b** WDM showing quiescent filling

Fig. 3 Weibull plot of UTS values from FGI cast into PDM and WDM

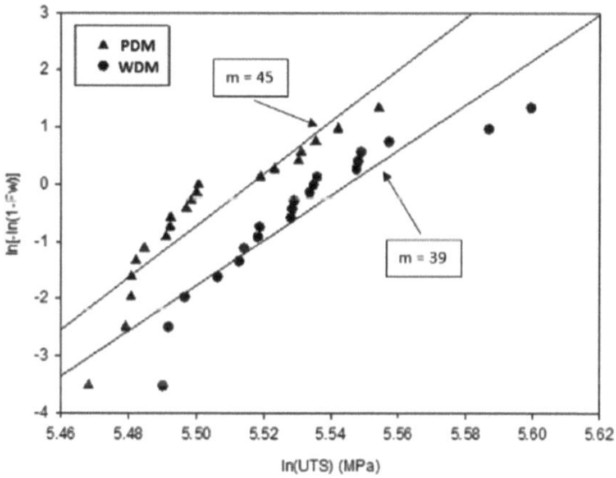

Fig. 4 Weibull plot of normalized UTS values from FGI and SGI cast into PDM

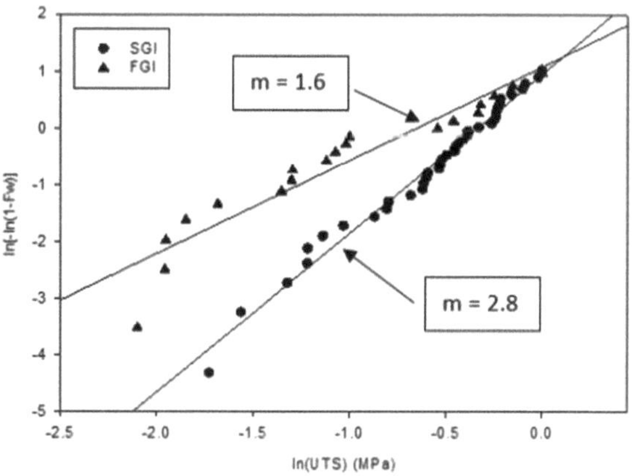

Table 1 Summary of the Weibull analysis of the UTS values resulting from the cast FGI test bars

	FGI PDM	FGI WDM
Mean value (MPa)	245	252
Equation of the Weibull plot	y = 45.53x − 251.15	y = 39.50x − 219.07
Weibull modulus	45	39
Position parameter (MPa)	249	256

Table 2 Summary of the Weibull analysis of the normalized UTS values from the cast FGI and SGI test bars

	Normalized FGI PDM	Normalized SGI PDM
Equation of the Weibull plot	y = 1.64x + 1.076	y = 2.80x + 0.95
Weibull modulus	1.6	2.8

obtained from the normalized UTS data of the FGI PDM and SGI PDM test bars. The Weibull parameters have been summarized in Tables 1 and 2.

For the FGI, the PDM casting produced a Weibull modulus for the UTS of 45, compared with a Weibull modulus of 39 associated with the WDM and its well-designed running system. The mean value and position parameters for the UTS were not significantly different either. The Weibull modulus for the normalized UTS values for FGI and SGI BM casting were 1.6 and 2.8, respectively.

Examination of the fracture surfaces of the test bars obtained from the SGI casting (see Fig. 5) revealed that many of them (24 out of 40) had two types of defects. Type I was an inter-dendritic shiny area, and close examination with SEM

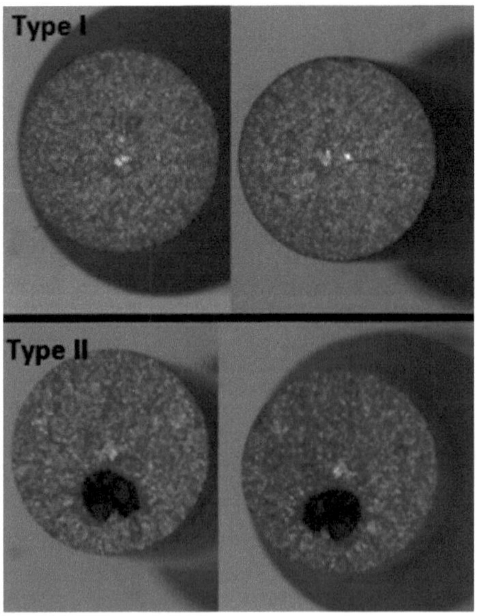

Fig. 5 Fracture surfaces from SGI PDM tensile test bars showing two types of defects (I & II) and their symmetrical nature. The test bar diameter was 6.75 mm

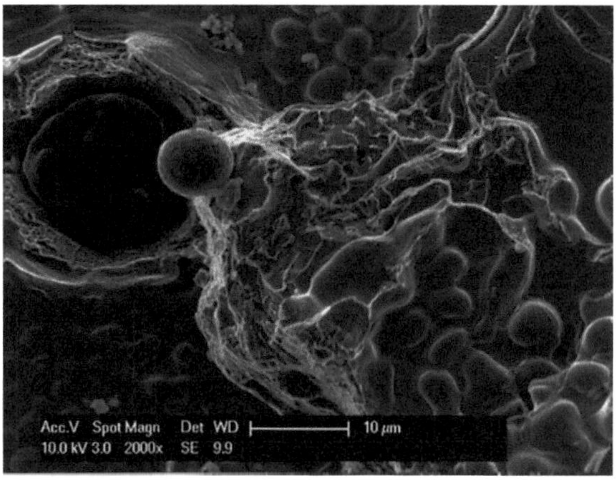

Fig. 6 SEM image of SGI type I defect shows the secondary dendrites with a graphite sphere breaking through a film-like layer

showed that it lacked graphite spheres, while type II defect was a thick black film. The defects areas were symmetrical on either side of the fracture surfaces of the test bars. Figure 6 shows an SEM image of a type I defect which shows small solidification shrinkage pores, as indicated by the secondary dendrite arms and a graphite sphere breaking through the film-like layer. EDX analysis revealed the film to contain oxygen, carbon, iron, magnesium and silicon (Fig. 7). A 5 kV acceleration volt for the SEM was used to show the features of the thin film (Fig. 8). Confetti-like fragments were also found between the secondary dendrite arms (see Fig. 9), and EDX analysis revealed them to contain oxygen and magnesium (Figs. 7 and 10). SEM imaging showed type II defects to consist of a thick rough film (Fig. 11). EDX analysis (Fig. 12) revealed the carbon, magnesium, silicon, aluminium and oxygen content of the film.

Examination of the fracture surfaces of the test bars obtained from the FGI castings revealed two sources of failure. Slag particles with high carbon content and eutectic phosphide defects are known as steadite. Figure 13 shows a SEM BSE

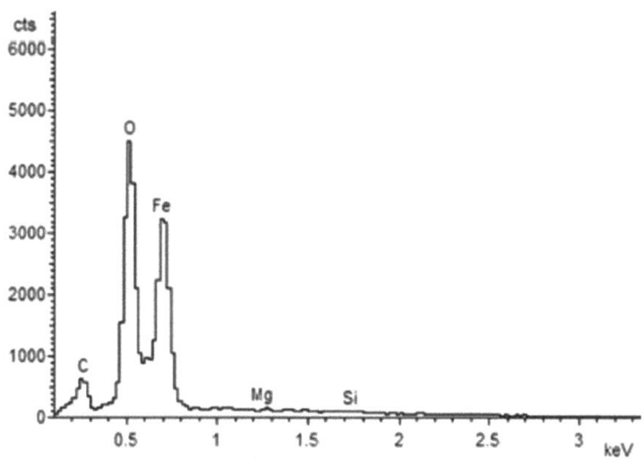

Fig. 7 EDX spectrum of SGI type I defect

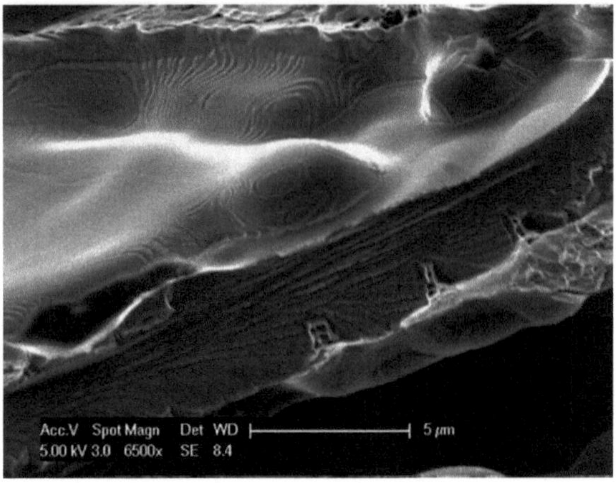

Fig. 8 A close-up SEM image of SGI type I defect shows the thin film nature covering the secondary dendrites

Fig. 9 SEM image of confetti-like fragment between the secondary dendrite arms in SGI type I defect

image of the steadite area illustrating the heavy element content of the defect, and EDX analysis of the area showed a clear phosphorous peak (Fig. 14). A preserved collapsed bubble with its tail was found by the author on the surface of a FGI engine block (Fig. 15), and it was on the transverse section of the casting facing upward. The bubble resembling the bubble trail found in Zn–Al die cast alloy [4].

Discussion

The WDM design would be associated with a reduction in the number of entrainment defects for two reasons. Firstly, the design of the running system, which was aimed at reducing surface entrainment events and maintaining quiescent filling of the test bars, and, secondly, the presence of a filter which would reduce the runner bar, ingate velocity and remove inclusions introduced during previous

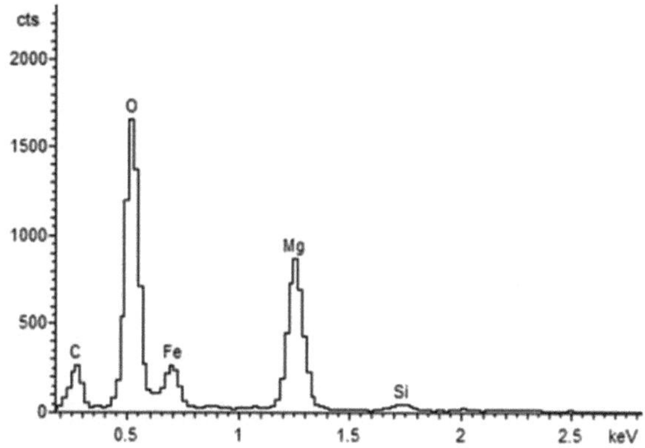

Fig. 10 EDX spectrum of the fragment shown in Fig. 9

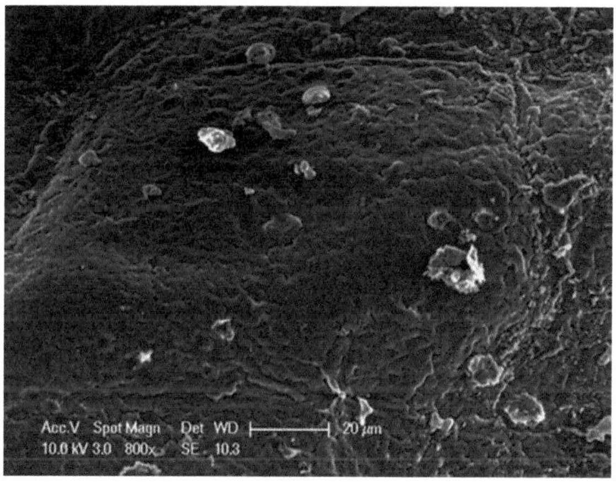

Fig. 11 SEM image for SGI type II film defect

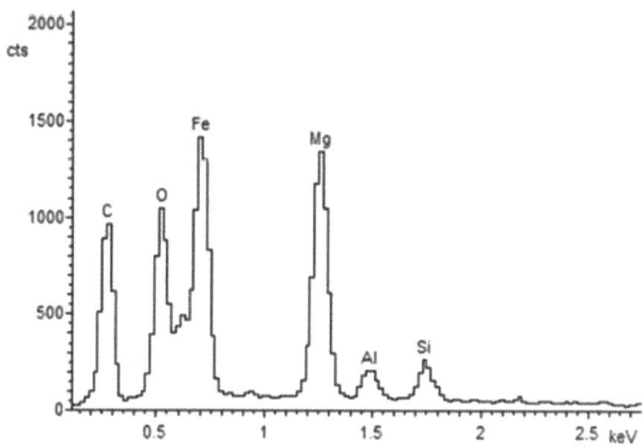

Fig. 12 EDX spectrum of the area in Fig. 11

Fig. 13 SEM BSE image of steadite defect on fracture surface of FGI test bar

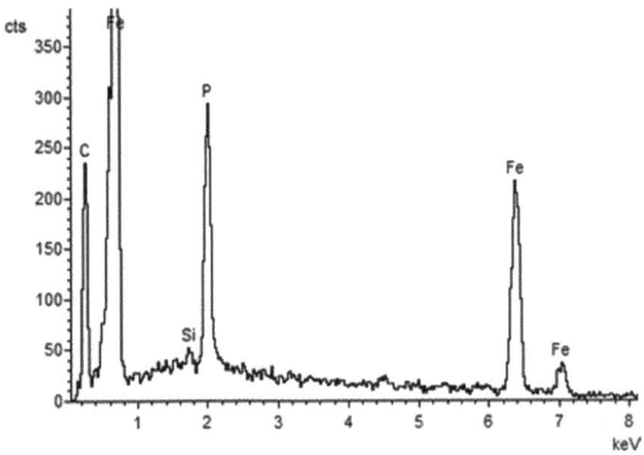

Fig. 14 EDX spectrum of the steadite defect

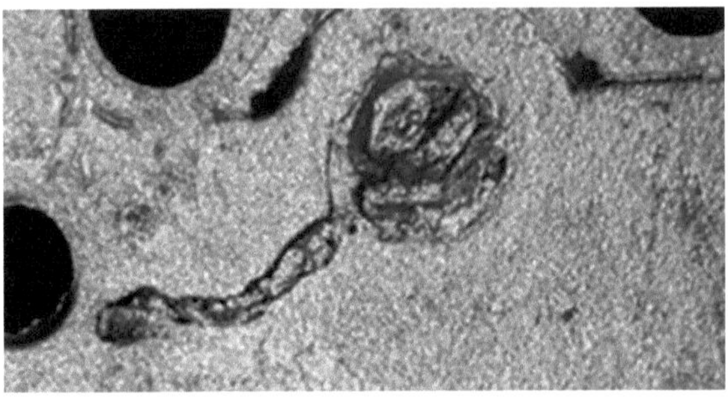

Fig. 15 A collapsed bubble with its tail attached (total length 10 cm) in a grey iron engine block casting

melting and casting operations. The effects were not reflected in the mechanical properties of FGI, in which the mean and Weibull modulus of the UTS of the WDM and the PDM castings were almost the same (45 for PDM and 39 for WDM). It is also worth mentioning that the Weibull modulus value for Al casting with bottom filled moulds is 38 [2], which is in line with both FGI values for bottom filled moulds. An effect of the surface turbulent filling system on the scatter of mechanical properties for FGI and SGI was apparent and was reflected in higher Weibull moduli of the normalised UTS for the SGI castings. The Weibull modulus of the SGI was 43% more than with the FGI, indicating that surface turbulent filling had a negative effect on the scatter of properties and reliability of FGI castings, more so than SGI castings.

Figure 5 through Fig. 12 show the examples of double oxide (type I and type II) film defects from the fracture surfaces of SGI castings. They were observed to have film-like structures (Figs. 6, 8 and 9), which were symmetrical on either side of the fracture surfaces (see Fig. 5), folded and wrinkled in appearance (see Figs. 8 and 9) and occurred frequently (on 24 out of 40 bars). The structure of the films represented in the symmetry on both sides of the fracture surfaces agreed well with the description of their formation proposed by Campbell [2]. Moreover, the appearance and formation of the film defects were similar to those found in light alloys.

A type I defect is a thin oxide film that was revealed by reducing the accelerating voltage of the SEM to 5 keV. The SEM beam interaction volume radius for iron decreases from 505 nm to 159 nm when the accelerating voltage is reduced from 10 keV to 5 keV [16]. The oxygen peak in the EDX analysis was increased in magnitude by the reduction in the interaction volume of the beam. The increase in the oxygen content and topographical features of the film with the reduction in accelerating voltage indicates the thin nature of the film. EDX analysis (Figs. 7 and 9) indicated that the film was a magnesium silicate.

Type II defects were thick black rough films with a crater-like shape on either side of the fracture surfaces (Figs. 5 and 11), indicating the bifilm nature of the defect and that it was not an inclusion. EDX analysis (Fig. 12) shows the oxide film to be rich in carbon and magnesium, and some silicon and aluminium were present. The black appearance of the film was due to the high carbon content. The chemical composition suggests the film to be MgO. The complex composition of the type II oxide film explains that it is thick and rough features, making it easy to identify visually.

Two types of defects were the reason behind failure in FGI test bars. Slag particles were found at the sites for crack propagation, and Fig. 13 reveals steadite defects on the fracture surfaces of the test bars backed by the EDX analysis result which shows a clear phosphors peak (Fig. 14). Localised spots with high phosphorus content (typically 10%) will result in a ternary eutectic phosphide reaction during solidification, which in turn will produce eutectic phosphide. Steadite solidification temperature is 200 °C lower than that of iron, resulting in nucleation of cracks and causing failure.

Conclusion

1. Evidence of bubble trail formation in FGI was found in a engine block casting.
2. There was no apparent effect of quiescent or surface turbulent filling on the strength of FGI.

3. The surface turbulently filled mould increased the scatter of UTS of FGI than SGI.
4. Two defects were found to contribute to the fracture of FGI test bars, slag particles and steadite defects; both were found on the fracture surface.
5. Two types of oxide bifilm defects (types I and II) were found on the fracture surfaces of SGI BM tensile test bars. Type I was a thin layer of magnesium silicate film causing inter-dendritic brittle fracture. Type II was a black thick rough bifilm easily seen on either side of the fracture surface optically, with a complex composition of carbon, oxygen, magnesium and aluminium.

Acknowledgements I would like to express my sincere thanks to Mr. Adrian Caden for his careful assistance with the experimental work.

I would also like to extend my gratitude to Grainger and Worrall foundry represented by Tim Garwood and Keith Dunham for their continuous help in producing the castings and allowing me to join them for industrial training.

References

1. Harding RA (2006) Towards more reliable investment castings. Int J Cast Met Res 19(5):289–301. https://doi.org/10.1179/136404606X153885
2. Campbell J (2015) Complete casting handbook: metal casting processes, metallurgy, techniques and design, 2nd edn. Butterworth-Heinemann, Amesterdam
3. Griffiths WD, Lai NW (2007) Double oxide film defects in cast magnesium alloy. Metall Mater Trans A 38:190–196. https://doi.org/10.1007/s11661-006-9048-7
4. Divandari M, Campbell J (2001) Mechanisms of bubble trail formation in castings. Transactions of American Foundry Society 109:433–442
5. Mirak AR, Divandari M, Boutorabi SMA, Taylor JA (2012) Effect of oxide film defects generated during mould filling on mechanical strength and reliability of magnesium alloy castings (AZ91). Int J Cast Met Res 25(3):188–194. https://doi.org/10.1179/1743133611Y.0000000037
6. Green NR, Campbell J (1993) Statistical distributions of fracture strengths of cast Al7SiMg alloy. Mater Sci Eng, A 173(1–2):261–266. https://doi.org/10.1016/0921-5093(93)90226-5
7. Campbell J (2007) Leakage defects via bubble trails in grey iron castings. Int J Metalcast 1 (1):7–16
8. Campbell J (2006) Entrainment defects. Mater Sci Technol 22(2):127–145. https://doi.org/10.1179/174328406X74248
9. Hsu F-Y, Wang K-N, Li C-L (2017) Bifilm defects in ductile-iron support bracket castings. Int J Cast Met Res 30(3):148–158. https://doi.org/10.1080/13640461.2016.1273621
10. BSI (2011) BS EN 1561:2011 founding. Grey cast irons, London
11. BSI (2018) BS EN 1563:2018 founding. Spheroidal graphite cast irons, London
12. ASTM International (2016) Standard test methods for tension testing of metallic materials
13. Tiryakioğlu M, Hudak D (2011) Guidelines for two-parameter weibull analysis for flaw-containing materials. Metall Mater Trans B 42(6):1130–1135. https://doi.org/10.1007/s11663-011-9556-8
14. Cox M, Harding RA, Campbell J (2003) Optimised running system design for bottom filled aluminium alloy 2L99 investment castings. Mater Sci Technol 19(5):613–625. https://doi.org/10.1179/026708303225002028

15. Tiryakioğlu M, Hudak D (2008) Unbiased estimates of the Weibull parameters by the linear regression method. J Mater Sci 43(6):1914–1919. https://doi.org/10.1007/s10853-008-2457-9
16. Goldstein J et al (2003) Scanning electron microscopy and x-ray microanalysis, 3rd edn. Springer, New York

Measurement of Air Entrainment During Pouring of an Aluminum Alloy

Lucas Archer, F. V. Guerra and Christoph Beckermann

Abstract Air entrainment during mold filling leads to oxide inclusions that can severely diminish the mechanical properties of castings. Despite this, the volume of air that is entrained during pouring of liquid metals has never been quantified. The present research focuses on air entrainment for a liquid aluminum alloy jet plunging into a pool. Experimental measurements with argon and air atmospheres are performed to determine the total volume of gas entrained during pouring and the volume of oxygen consumed. They are used to validate a previously developed model that predicts air entrainment as part of a casting simulation. The measured entrainment ratios from the experiments conducted in an argon atmosphere are in approximate agreement with the previously developed model. A large difference between entrainment ratios for argon and air atmospheres is found.

Keywords Casting · Pouring · Air entrainment · Oxide inclusions · Aluminum

Introduction

Oxide inclusions are a commonly reported defect in ferrous and non-ferrous castings. These inclusions often form when the liquid metal is exposed to oxygen during mold filling. Oxide inclusions in aluminum alloy castings consist of thin solid aluminum oxide films. They can usually not be removed post-casting and decrease the quality of cast aluminum parts.

Air entrainment during mold filling is a leading source of the oxygen that is consumed during the formation of inclusions. Air entrainment occurs at discontinuities in free surface flows and manifests itself in the form of bubbles. A common air entraining flow in metal casting is the plunging jet shown in Fig. 1. Air entrainment by a plunging jet has been investigated extensively for water as the

L. Archer · F. V. Guerra · C. Beckermann (✉)
Department of Mechanical Engineering, University of Iowa,
Iowa City, IA 52242, USA
e-mail: becker@engineering.uiowa.edu

M. Tiryakioğlu et al. (eds.), *Shape Casting*, The Minerals, Metals & Materials Series,
https://doi.org/10.1007/978-3-030-06034-3_3

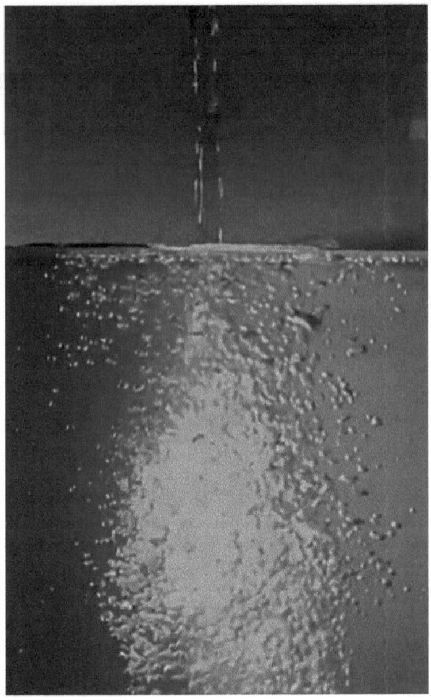

Fig. 1 Air entrainment by a jet of water plunging into a quiescent pool of water; the bubbles form at the intersecting perimeter of the jet with the pool

working fluid [1–7]. To the authors' best knowledge, air entrainment has never been measured for molten metals.

In this study, a technique is developed to quantitatively measure the volume of air entrained by a jet of liquid metal emanating from a bottom-pour ladle through a nozzle and plunging into a pool. The experimental setup is first tested with water as the working fluid. Simulations of the water experiments are performed using an available casting simulation software [8] to further validate a previously developed air entrainment model [9, 10]. Next, air entrainment via a plunging jet is measured for liquid A356 aluminum alloy. Experiments are conducted in both argon and air atmospheres. The jet velocity is varied by choosing different nozzle heights. The volume of entrained gas in the aluminum experiments is corrected using temperature measurements.

Experimental Procedure

To create the plunging liquid jet, A356 aluminum alloy in a bottom-pour ladle is emptied through a nozzle into a steel mold. The A356 aluminum alloy has a liquidus temperature of 616 °C. The bottom-pour ladle and mold shown in Fig. 2a were constructed to withstand the heat of the molten metal. The ladle features a refractory cement lining, a graphite stopper rod, and a nozzle protruding from the bottom. The nozzle was cast from refractory cement.

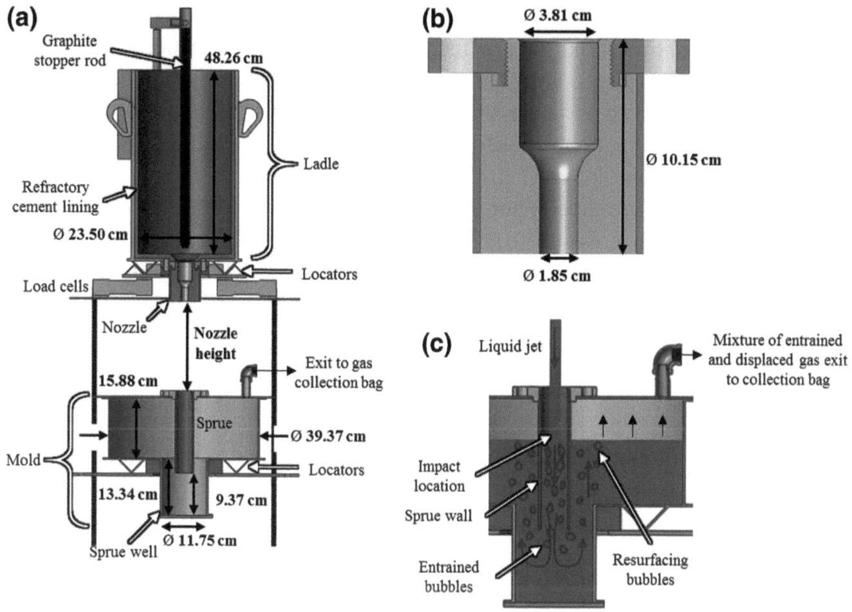

Fig. 2 CAD drawings of the (**a**) ladle and mold in the stand used for the water and aluminum experiments, and **b** close-up view of the nozzle geometry, with (**c**) an illustration of air entrainment via jet plunging into a mold with a sprue. All drawings are cut midway through the vertical plane

The geometry of the nozzle cavity shown in Fig. 2b achieves a quick but relatively smooth transition to a fully developed flow before the nozzle exit. The ladle rests above an array of four load cells that are used to measure the flow rate of the jet as it exits the nozzle. Locators below the ladle and mold ensure that the liquid jet impacts at the center of the sprue. The legs of the ladle and mold stand can be adjusted to change the nozzle height, which is depicted in Fig. 2a. Several cylindrical molds were welded together with 10-gauge steel. A pipe welded to the top of the mold leads to the gas collection system. A steel sprue with an inner diameter of 4.15 cm protrudes into the mold so that the sprue exit is 9.37 cm from the bottom of the sprue well. The sprue well is used to reduce the time and volume of liquid required to submerge the sprue.

Figure 2c shows an illustration of the entrained air as it resurfaces outside of the sprue and flows toward a gas collection bag. A similar measurement technique was used to measure air entrained by a plunging water jet in studies performed by Wanstall et al. [6] and Bates et al. [7]. As the liquid jet impacts in the pool, the entrained air is pulled down into the liquid as bubbles; these bubbles then resurface some distance away from the jet impact location. When the sprue exit is submerged, the bulk flow carries the bubbles of entrained air down and out of the sprue where the bubbles then resurface and mix with air above the liquid pool. This mixture of air is then driven into a collection bag by the pressure created by the volume of entrained air entering the space above the pool, and the air displaced by the rising liquid. Figure 3 shows a diagram of the air collection and measurement system.

Air collection is delayed until the sprue exit is submerged by liquid. Before the sprue exit is submerged, it is possible for entrained air to escape up through the sprue. Air collection is stopped before the ladle is completely empty to avoid the vortex that forms in the ladle when the liquid level is low. At the conclusion of an experiment, the total volume of air collected, V_t, is drawn from the collection bag with a small vacuum pressure (0.99 bar) through an *OMEGA* FMA1743 airflow meter [11]. To obtain V_t, the measured airflow rate is integrated over the duration that the air is extracted from the collection bag. The volume of air entrained, V_e, is then calculated from:

$$V_e = V_t - V_l \tag{1}$$

where V_l is the volume of air that is displaced by the liquid poured into the mold. The volume of air entrained is then reported as the ratio V_e/V_l.

The volume of liquid poured into the mold, V_l, is found by taking the difference in the mass measured by the load cells at the beginning and end of air collection. That mass is divided by the density of the liquid poured. For water experiments, a density of 996 kg/m^3 was used. The temperature dependent density of A356

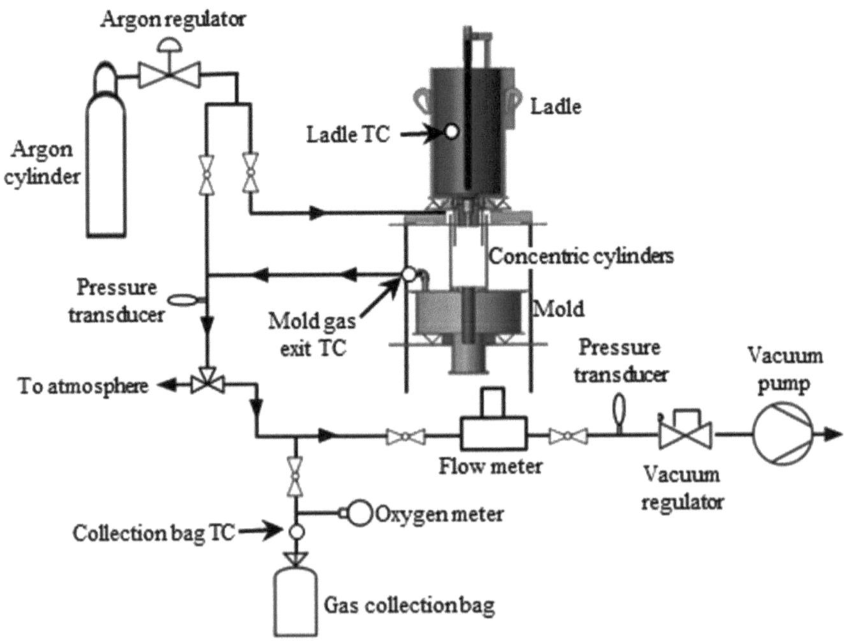

Fig. 3 Diagram of the piping network and gas measurement system used for both the water and aluminum experiments

Table 1 A356 aluminum alloy composition, in weight percent

Alloy	Si	Mg	Fe	Ti	Zn	Cu	Mn
A356	7.00	0.37	0.10	0.08	0.01	0.01	0.01

aluminum alloy was calculated using the *JMatPro®* software package [12]. The composition of the A356 alloy is given in Table 1.

Several experiments are conducted in an argon atmosphere to obtain the volume of gas entrained when no oxygen is consumed by the aluminum to form oxides. An argon atmosphere is maintained around the liquid jet with two concentric cylinders that are constantly supplied with argon, as shown in Fig. 3. Before each experiment, the mold, collection bag, and piping network are flushed with argon, and an oxygen meter is used to confirm that the percent of oxygen present in the system is below 0.5%. An oxygen meter is also used to measure the oxygen content of the collected gas. For all argon experiments, the oxygen content of the collected gas was below 0.5%. The oxygen content of the collected gas is also measured for the experiments performed in an ambient air atmosphere.

Results and Discussion

Water Experiments

The system is first tested with water as the working fluid. Several mold-filling simulations are performed using the measured flow rate from the water experiments. The predicted instantaneous air entrainment rates, Q_e/Q_l, from the simulations are integrated over the same time that air is collected during the experiments, to find the volumetric air entrainment ratio, V_e/V_l, which is then compared to the measured ratio.

Previous studies did not utilize a sprue to measure air entrainment of a plunging jet [1–5]. Therefore, to understand the effect of the sprue, simulations were performed with and without a sprue, using flow rates obtained from the water experiments. The impact location is depressed by the jet in the case with a sprue, as shown in Fig. 4a, relative to the case without a sprue in Fig. 4b. The lower impact location increases the jet velocity at impact, u_j and decreases the jet diameter at impact, d_j.

Figure 5 shows the entrainment results for the water simulations and experiments. In Fig. 5a, the predicted instantaneous entrainment rate, Q_e/Q_l, is plotted as a function of time. The predicted entrainment rates are integrated over the period shown in Fig. 5a to find the predicted volumetric air entrainment ratios that are compared in Fig. 5b. The predicted entrainment ratio without a sprue is 0.24, and 0.31 with a sprue. This shows that by depressing the impact location the sprue increases air entrainment.

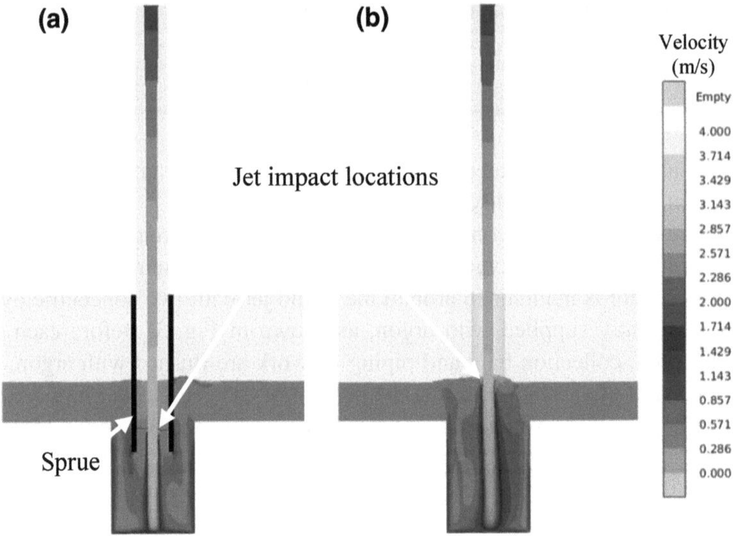

Fig. 4 Velocity contours from water-filling simulations comparing the jet impact locations **a** with a sprue and **b** without a sprue, midway through gas collection. The sprue and impact surface have been highlighted with black in (**a**)

The average air entrainment ratio measured in the water experiments shown in Fig. 5b is 0.31, which agrees with the predicted entrainment ratio from the simulation with a sprue. It can also be seen in Fig. 5b that the results from the simulation without a sprue are in good agreement with a well-established experimental correlation for a low turbulence jet plunging into an open pool [2]:

$$Q_e/Q_l = 0.003\frac{u_j^2}{gd_j} \tag{2}$$

where g is the acceleration due to gravity. The flow rate measured during the water experiments, and the nozzle diameter, are used to calculate u_j and d_j; then, Eq. 2 is used to find Q_e/Q_l, which is then integrated over the time that air is collected to obtain V_e/V_l.

Water experiments were conducted in both air and argon atmospheres to investigate if the type of gas has an effect on entrainment. Figure 5c shows the measured entrainment ratios for experiments conducted at two nozzle heights in both atmospheres. It is clear from Fig. 5c that a plunging jet entrains the same volume of argon as air. A decrease in nozzle height of 10 cm decreases the entrainment ratio by 30%. With the system thoroughly tested with water, and the accuracy of the simulations verified, experiments were conducted with A356 aluminum alloy.

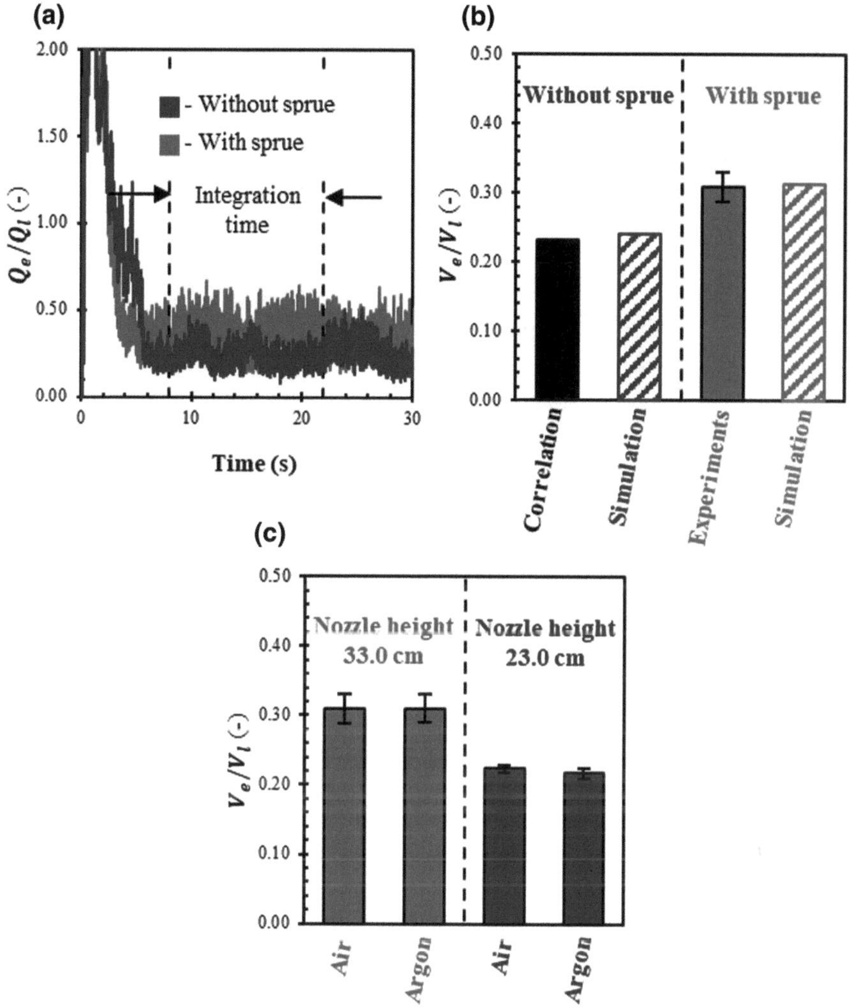

Fig. 5 Results from the water simulations and experiments. **a** Instantaneous air entrainment rates from the simulations which are integrated to obtain **b** volumetric air entrainment ratios and compared to a known correlation and experimental results. **c** Two different nozzle heights were tested in both air and argon atmospheres. Error bars represent a 95% confidence interval for the mean value of three experiments performed under nominally the same conditions

A356 Aluminum Alloy Experiments

Six experiments were carried out with liquid aluminum alloy. Table 2 shows the parameters for each experiment. Four experiments were conducted with a 33.0 cm nozzle height, two of which with an argon atmosphere and two with an ambient air atmosphere. Two experiments were conducted in an argon atmosphere with a

Table 2 List of experiments and parameters

Experiment	1	2	3	4	5	6
Atmosphere	Argon	Argon	Air	Air	Argon	Argon
Nozzle height (cm)	33.0	33.0	33.0	33.0	23.0	23.0
Average jet diameter, d_j (cm)	1.5	1.5	1.5	1.5	1.6	1.6
Average jet velocity, u_j (m/s)	3.08	3.07	3.06	3.11	2.74	2.79

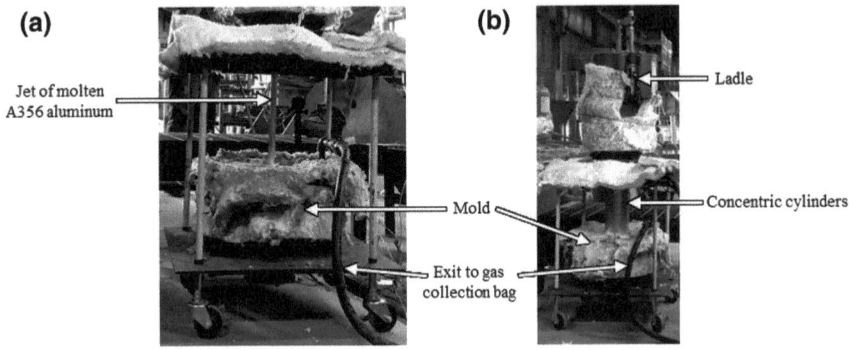

Fig. 6 Photographs of **a** the jet of molten aluminum pouring into the mold and **b** the ladle, and mold with concentric cylinders to maintain an argon atmosphere around the jet of liquid aluminum

nozzle height of 23.0 cm. The average jet velocity, u_j, and average jet diameter, d_j, provided in Table 2 were calculated from the flow rates measured with the load cells during gas collection.

A photograph of the liquid aluminum jet is shown in Fig. 6a. The jet is smooth with little to no disturbances visible on the surface. Figure 6b shows a photograph of the ladle and mold during an experiment with the jet surrounded by an argon atmosphere within the concentric cylinders. Ceramic fiber blankets kept the ladle hot between pours. The ceramic fiber blankets around the mold ensured that the aluminum stays liquid long enough to allow for removal of the lid and sprue before the aluminum solidified. This significantly reduced the time between experiments.

Figure 7 shows a summary of the volume and temperature measurements for each experiment. The volume of aluminum in the ladle shown in Fig. 7a was found by converting the weight measured by the load cells to a volume of aluminum using the temperature measured in the ladle to determine the density of the liquid aluminum. It can be seen that the flow rates were virtually the same in each experiment.

Several temperature measurements were recorded during the experiments, as shown in Fig. 7b. They include the temperature of the aluminum in the ladle, the temperature of the gas exiting the mold, and the temperature of the gas in the collection bag. The temperature of the liquid aluminum in the ladle ranged from 620–770 °C between experiments. The temperature of the gas that exited the mold,

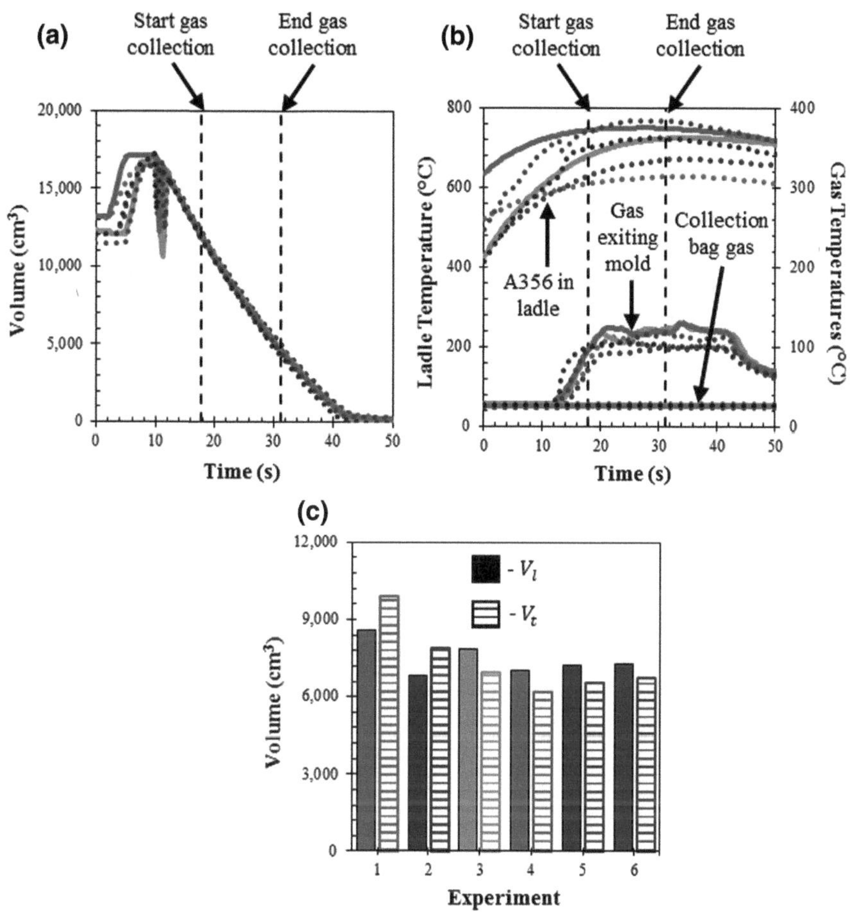

Fig. 7 Measurements taken, during the pouring of liquid aluminum, of the **a** volume of aluminum in the ladle, and **b** temperatures of the liquid aluminum in the ladle, gas exiting the mold, and within the collection bag. Dotted lines are experiments conducted in argon atmospheres and solid lines are experiments conducted in air atmospheres. Measurements taken after the pouring of liquid aluminum, of **c** the volume of aluminum poured during air collection and total volume collected gas measured at room temperature for the six experiments conducted

T_{ex}, had a range of 100–120 °C between experiments. The temperature of the gas in the collection bag was constant for all experiments and equal to the ambient temperature of the foundry, 27 °C.

The volumes of aluminum poured during gas collection are directly compared to the volumes of gas measured in the collection bag in Fig. 7c. In all experiments but the first two, the total volume of gas collected, V_t, is less than the volume of aluminum poured, V_l. Because the volume of gas entrained, V_e, is calculated by subtracting the volume of gas displaced by the aluminum, V_l, from V_t, the volumetric entrainment ratio, V_e/V_l, would be negative for many of the experiments, which is impossible. However, the entrained and displaced gases were cooled before reaching the collection bag, which reduces their volume. Therefore, it was necessary to correct V_t to account for the temperature change of the entrained and displaced gases.

In order to perform the temperature correction, the temperatures of the entrained and displaced gas are needed. Measuring the exact temperature of the gas as it is entrained or displaced is impossible. As a first approximation, it is reasonable to assume that the temperature of both the entrained and displaced gases is equal to the measured temperature of the gas exiting the mold. As can be seen in Fig. 7b, this temperature remained relatively constant from about the beginning of gas collection until the end of pouring. For each experiment, the steady value of the temperature of the gas exiting the mold, T_{ex}, was used to correct the total volume of gas in the collection bag, V_t, measured at the ambient foundry temperature, T_{amb}, according to

$$V_{t,corr} = V_t \frac{T_{ex}}{T_{amb}} \tag{3}$$

where $V_{t,corr}$ is the corrected total volume of gas. The temperature corrected volume of gas entrained, $V_{e,corr}$, is then found using Eq. 1 by subtracting V_l from $V_{t,corr}$. To obtain the temperature corrected gas entrainment ratio, $V_{e,corr}$ is divided by V_l.

Figure 8 compares the temperature corrected entrainment ratios for all experiments with simulations performed using the average aluminum flow rates measured in the experiments. Each solid bar represents the average corrected entrainment ratio from two experiments performed under nominally the same conditions. The striped bars denote the predicted entrainment ratios for the 33.0 and 23.0 cm nozzle heights.

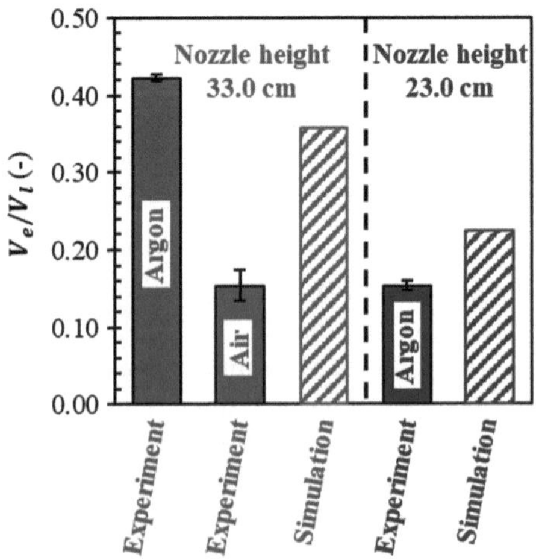

Fig. 8 Temperature corrected entrainment ratios from aluminum experiments compared to simulations with the same flow rate. Simulations were performed using thermophysical properties of A356. Error bars represent a 95% confidence interval for the mean value of two experiments performed under nominally the same conditions

The entrainment ratios for the experiments conducted with a 33.0 cm nozzle height are 0.42 in an argon atmosphere and 0.15 in air, as shown by the two leftmost bars in Fig. 8. Recall that in the water experiments, the type of gas did not affect the entrainment ratio. A possible explanation for this large difference could be the consumption of oxygen to form oxide films when pouring aluminum in ambient air.

The average oxygen level measured in the collection bag for the ambient air experiments was 19.65% ± 0.1%, which is a reduction of 1.35% from the 21.0% O_2 present in ambient air. It can be shown that this reduction corresponds to 52% (by volume) of the oxygen in the entrained air. The fact that not all of the oxygen in the entrained air is consumed may be attributed to oxide films shielding the aluminum from further oxidation. The present measurement of the oxygen reduction can be used to calculate the mass of aluminum oxide that forms due to air entrainment. A mass fraction of 16.7 ppm aluminum oxide (Al_2O_3) in the aluminum alloy poured is obtained when 52% of the oxygen in the entrained air is consumed. However, this oxygen consumption cannot account for the large difference in entrainment ratios between the argon and air experiments. If no oxygen were consumed, the entrainment ratio for the ambient air experiments would only increase by 0.02, to about 0.17. Another possible reason for the large difference between the measured entrainment ratios for argon and air may be the oxide film that is likely to be present on the surface of the plunging jet in the air atmosphere. This solid film may be reducing the surface turbulence on the jet, which would lead to less gas being entrained.

The simulations for the 33.0 and 23.0 cm nozzle heights gave entrainment ratios of 0.36 and 0.22, respectively. These predicted ratios can be compared directly to the measured ratios of 0.42 and 0.15 for the same two nozzle heights in the argon experiments. It may be concluded that the present entrainment model predicts at least approximately the entrainment ratios measured for liquid aluminum alloy in an argon atmosphere. The differences between the simulations and the argon experiments may be due to uncertainties in the temperature corrections of the measured gas volumes, but they could also be caused by the higher surface tension of aluminum compared to water. The simulation model was originally calibrated for water only [9, 10].

The entrainment ratio for the argon experiments decreases from 0.42 to 0.15 when the nozzle height is lowered from 33.0 cm to 23.0 cm. This corresponds to a 64% decrease. For the present experiments conducted with water and similar flow rates, the decrease was only 30% (Fig. 5c). Gas entrainment commences above a certain critical jet velocity, which is about 1 m/s for water [1–5]. Lowering the nozzle height reduces the jet velocity at impact. There is a possibility that because of its higher surface tension, aluminum has a higher critical jet velocity for gas entrainment than water. Additional aluminum experiments in argon, with an even further reduced nozzle height, would be needed to investigate the critical velocity for gas entrainment in liquid aluminum.

Conclusions

An experimental setup was developed to quantitatively measure gas entrainment for a plunging jet of liquid aluminum alloy. The setup was first tested with water as the working fluid and the measured volumetric gas entrainment ratios agreed well with results from corresponding computer simulations and an existing experimental entrainment correlation. It was established that for water, the entrainment ratios are the same for air and argon atmospheres. Experiments with liquid A356 aluminum alloy were also performed in both argon and air atmospheres and for two nozzle heights. It was necessary to consider and correct for the temperature of the gases during the entrainment process.

It was found that the gas entrainment ratio for the liquid aluminum experiments performed in an air atmosphere was much lower than in an argon atmosphere. This difference may be attributed to solid oxide films present on the jet surface in the air experiment, which could reduce surface turbulence and decrease air entrainment. The oxygen content in the entrained air was reduced by 52%, but this oxygen consumption cannot explain the much lower entrainment ratio compared to argon. Lowering the nozzle height decreased the entrainment ratio by a larger magnitude with liquid aluminum than with water, alluding to the possibility that liquid aluminum has a higher critical jet velocity for the commencement of air entrainment than water. Predicted entrainment ratios from computer simulations were found to approximately agree with the ratios measured in the argon experiments, but additional model calibration is needed.

While the present experiments shed considerable light on air entrainment during pouring of aluminum alloys and the formation of oxide inclusions, considerable further research is necessary. Additional experiments should be conducted for a larger range of jet velocities, and the entrainment simulation model should be improved to account for the presence of oxide films. Experiments with liquid metals other than aluminum would also be of great value.

Acknowledgements The authors wish to thank David Weiss, Dan Hoefert, and the rest of the team at Eck Industries for providing the use of their foundry and support during the aluminum experiments. F. V. Guerra acknowledges CONACYT for the scholarship provided to him during his postdoctoral stay at the University of Iowa.

This American Metalcasting Consortium (AMC) project is sponsored by the Defense Logistics Agency Troop Support, Philadelphia, PA and the Defense Logistics Agency Research & Development (R&D) Office, Ft. Belvoir, VA.

References

1. Ervine DA et al (1980) Effect of turbulence intensity on the rate of air entrainment by plunging water jets. Proc Inst Civ Eng 69(2):425–445
2. Sene KJ (1988) Air entrainment by plunging jets. Chem Eng Sci 43:2615–2623

3. Brattberg T, Chanson H (1998) Air entrapment and air bubble dispersion at two-dimensional plunging water jets. Chem Eng Sci 53(24):4113–4127
4. Van De Sande E, Smith JM (1976) Jet break-up and air entrainment by low velocity turbulent water jets. Chem Eng Sci 31(3):219–224
5. Biń AK (1993) Gas entrainment by plunging liquid jets. Chem Eng Sci 48(21):3585–3630
6. Wanstall C, Griffin J, Bates CE (1993) In: Proceedings of 47th SFSA technical and operating conference, Chicago, IL, Paper No. 1.1
7. Bates CE, Griffin J (1994) Research report no. 106. Steel founders' society of america (SFSA) Crystal Lake, IL, USA
8. MAGMAsoft, MAGMA Gmbh, Kackerstrasse 11, 52072 Aachen, Germany
9. Majidi SH, Beckermann C (2017) Modelling of air entrainment during pouring of metal castings. Int J Cast Met Res 30(5):301–315
10. Majidi SH, Griffin J, Beckermann C (2018) Simulation of air entrainment during mold filling: comparison with water modeling experiments. Metall Mater Trans B 49(5):2599–2610
11. FMA 1700 & FMA 1800 Mass Flowmeter, OMEGA, Stamford, CT, USA, 2018
12. JMatPro, Sente Software Ltd, Surrey Technology Center, Surrey GU2 7YG, United Kingdom

Connecting Oxide Bifilms' Properties from Atomistic Simulations with Virtual Casting of Aluminum

Jialin Liu, Qigui Wang and Yue Qi

Abstract Aluminum oxide bifilms, formed during melt turbulence flow, can have a significant detrimental effect on material properties after they are entrapped in the final cast products. Recently, molecular dynamics (MD) simulations were used to simulate the formation and fracture mechanisms of bifilms at the nano-scale, which are hard to obtain experimentally. The results showed that the fracture occurred at the Al/oxide interface instead of the oxide/oxide interface for both amorphous oxide and crystalline α-Al_2O_3, which represent the "young" and "old" oxides referred in aluminum casting. The fracture energy is higher for the α-Al_2O_3 bifilm. However, if OH-termination contamination occurs due to residue hydrogen gas and water trapped in the aluminum oxide bifilm interface, the OH-termination oxide bifilm fractured at the oxide/oxide interface and with a much-reduced fracture energy. This is consistent with the general picture that oxide bifilms will initiate cracks, especially fatigue cracks in cast aluminum products. For macroscopic models, crack initiation and propagation can be modeled by cohesive zone method. Therefore, we propose a simple size bridging relationship to connect the MD-predicted oxide bifilms fracture energy and fracture strength with future finite element modeling.

Keywords Oxide bifilms · Molecular dynamics · Interfaces · Aluminum alloys · Castings

J. Liu (✉) · Y. Qi
Department of Chemical Engineering and Materials Science, Michigan State University, East Lansing 48824, USA
e-mail: jlliu@msu.edu

Y. Qi
e-mail: yueqi@egr.msu.edu

Q. Wang
General Motors Corporation, Global Propulsion Systems, Pontiac, MI 48340, USA
e-mail: qigui.wang@gm.com

M. Tiryakioğlu et al. (eds.), *Shape Casting*, The Minerals, Metals & Materials Series,
https://doi.org/10.1007/978-3-030-06034-3_4

Introduction

Aluminum alloys are widely used in industries due to their lightweight, corrosion resistance, good strength, and ductility. Casting is one of the most critical aluminum manufacturing processes especially in automobile manufacturing since the late nineteenth century [1–4]. However, due to the rapid oxidation of aluminum in air, the surface liquid aluminum will be oxidized to alumina immediately after exposing to air. The thickness of alumina is only a few nanometers thick which imposes significant difficulty for experimental in situ studies. Furthermore, in shape casting, strong turbulence can be produced particularly in gravity pouring. The turbulence will trap the thin oxides and fold them into a double layer structure, "oxide bifilms" [5, 6]. The mechanical properties of the final cast product can be greatly degraded by the entrapped oxide bifilms [7, 8]. Campbell pioneered the concept and the study on "bifilms" and proposed that the reasons of decreased mechanical properties are the initiating of cracks due to the residue gas left in bifilm and facilitating of crack propagation due to bifilm's brittleness [7, 9]. But no direct experimental observation can be made to support the proposed deformation mechanisms.

The structure and chemistry of the oxide bifilms evolve with time. During casting, the freshly formed alumina is called "young oxides" while those stay in a furnace for a long time are called "old oxides." It is believed that alumina will form amorphous structure first, then gradually transform into $\gamma - Al_2O_3$ and eventually $\alpha - Al_2O_3$. After exposing to air, the surfaces of old oxides are more likely to be contaminated by impurities like water vapor, hydrogen, and CO_2 gas. Especially, hydrogen gas absorption and trapping in the old oxide bifilms have been proposed [6, 10, 11]. This will lead to the precipitation of hydroxyl groups on the surface during solidification. Both DFT calculation [12] and experimental study [13] showed that hydroxyl groups on the aluminum oxide surface are stable even at the melting temperature of aluminum.

So far, the deformation and fracture of oxide bifilm in cast aluminum has not been clearly elaborated due to the complexity of its variation in structure, chemistry details and its thinness. It is also important to note that the deformation of oxide thin films with nano-meter thickness can be dramatically different than its brittle bulk form. For example, experiments show that aluminum oxide nanoparticles smaller than 40 nm would not show brittle fracture under compression and the aluminum oxide layer exhibited "liquid-like" deformation [14–16]. The latter was first predicted by molecular dynamics simulations [17, 18], and thus, the direct experimental observations validated the accuracy of atomistic simulations.

Therefore, to understand the deformation mechanism and ultimately predict and improve the mechanical properties of casting product, we first implemented ReaxFF reactive forcefield based molecular dynamics (MD) method on three representing structures to simulate the aging evolution and deformation of bifilms. We used amorphous alumina and $\alpha - Al_2O_3$ to represent the structural difference between young oxides and old oxides, and added OH-termination on old oxides for comparison.

MD Simulation Method

To compare the change of mechanical properties during the aging process, the two extreme structures and conditions, young oxides and old oxides, were constructed in this study. As shown in Fig. 1a, to build young oxides, an 8-nm-thick Al (1 0 0) slab was oxidized directly in MD simulation in O_2 environment gas at the melting point of Al. Then, a 4-nm-thick slab was generated from the previous slab as the structure of monolayer young oxides. As shown in Fig. 1b, the old oxides' structure was built by attaching 1-nm-thick alpha alumina to Al (1 0 0) slab and performing equilibration. However, the interface between Al and alumina (Al/oxide) was poorly bonded due to the limited inter-diffusion at the melting point of Al. To enhance the Al/oxide interface and mimic the real aging process, the Al/oxide interface was heated to the melting point of alumina. The OH-contaminated surface was simulated by adding hydrogen atoms on top of surface oxygen atoms to reach –OH surface coverage of ~30%. The relaxed structure is shown in Fig. 1c. These fully relaxed monolayer slabs were subjected to the folding process to form bifilm, by duplicating the monolayer, flipping it, and placing it on top of the original monolayer slab (Fig. 1d). 1GPa pressure was applied to top and bottom at the

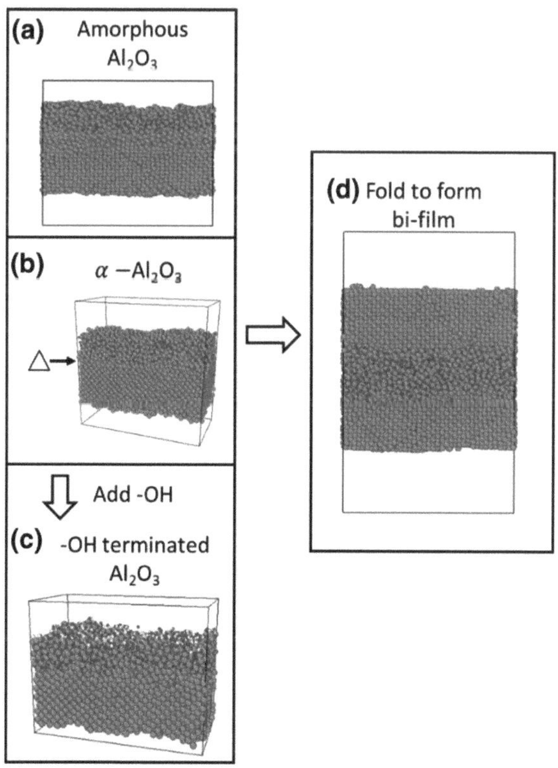

Fig. 1 Procedure to build aluminum oxide bifilm structures with different formation history

Table 1 Details and predicted mechanical properties for each bifilm structure

Interface structure	Young oxides	Old oxides	OH-terminated old oxides
Number of Al	13416	15770	15770
Number of O	3077	3758	3758
Number of H	–	–	400
Cell size/Å	78.49*39.93 × 140	81.18 × 41.64 × 140	80.86 × 41.49 × 140
Ultimate Tensile Strength/GPa	1.10	1. 55	1.20
Work of separation/J/m^2	0.43	0.53	0.30

melting point of Al while old oxides and surface contamination structure at room temperature. The details for each bifilm structure are listed in Table 1.

All calculations were performed with MD method using ReaxFF reactive forcefield [19, 20] in LAMMPS package [21, 22]. To simulate the deformation of bifilm, uniaxial tensile stress was applied to the three slab models. A constant strain rate deformation was achieved by increasing the strain by 0.5% along surface normal direction, then relaxing the structure for 1 ps. The stress at each strain state was calculated until the bifilm structure formed a crack that penetrated the whole sample.

Results and Discussion

Deformation Behavior of Bifilm

The ultimate tensile strength and work of separation calculated from the deformation simulation of three slab models are listed in Table 1. In the aging process, the ultimate tensile strength increased by 0.45 GPa from young to old oxides. The OH terminated old oxide bifilms show similar ultimate tensile strength with the young oxides.

The structures of the three bifilms before and after deformation are shown in Fig. 2. The fracture positions for (a) young and (b) old oxides are at the metal/oxides interface (the "wet" interface), while the fracture position for the OH-terminated old oxides is at the oxides/oxides interface (the "dry" interface). This contradicts with the common understanding that Al/Al_2O_3 interface is well bonded and stronger than the "dry" side. In fact, the metallic bond formed at the Al/Al_2O_3 interface is intrinsically weaker than the ionic bond formed at the Al_2O_3/Al_2O_3 interface. Wang et al. [23] also showed same conclusion using density functional theory (DFT) method. They calculated the work of separation at 0 K to be 0.79 J/m^2 which agrees well with our MD result of 0.43 and 0.53 J/m^2 for young

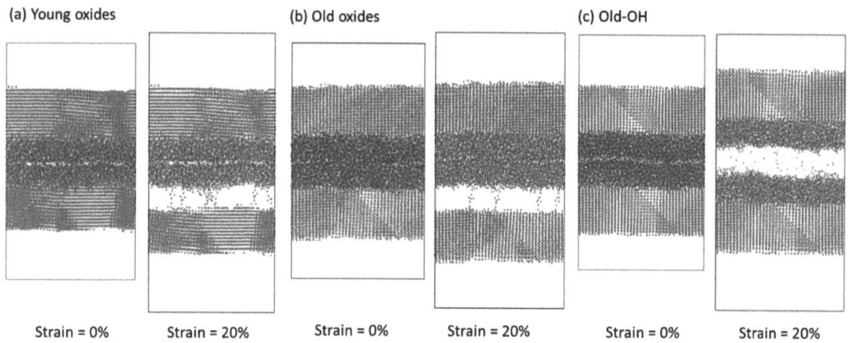

Fig. 2 Structure of each bifilm before and after deformation

and old oxides. For OH-terminated old oxides, the –OH groups on the "dry" side passivate the active site for Al–O bond between the two surfaces and prevent the reconnection of Al–O bond during the deformation. So, the work of separation drops to 0.30 J/m^2 which explains why the fracture occurs at the oxide/oxide interface. Comparing with experimental observations, Wang [8] reported that young oxides usually appeared on one side of the fracture surface while old oxides appeared on both sides after deformation. This is consistent with our prediction of the fracture location. It is demonstrated that the surface impurities such as hydrogen gas trapped in the oxide bifilm have more significant impact on the fracture mechanism compared with phase transition and folding temperature.

Microscale Property of Bifilms

Finite element model (FEM) is widely used in industry to simulate the macroscopic deformation of material. The cohesive zone model (CZM) was developed to simulate the crack formation and opening along with plasticity deformation [24]. The MD method as a smaller-scale model can pass useful material properties information to the CZM model for engineering design of better casting products. The MD-predicted ultimate tensile strengths are all significantly larger than typical values obtained in the experiment due to the lack of dislocation [25] and the limited simulation cell size. Also in MD simulation, the small cell size forbids the gradual propagation of cracks, while in FEM models the smallest propagation is determined by the mesh size. Xia et al. [26] proposed a method to predict the interface property at FEM scale by assuming the fracture energy is reserved across different length scales. After applying their approach to our system, the predicted fracture strengths of the young oxides, the old oxides, and the OH-terminated old oxides under 100-nm mesh size are 50.7 MPa, 71.1 MPa, and 63.7 MPa, respectively. If the mess size is at the scale of a millimeter, the oxide bifilm fracture strength would be on the order of 5×10^{-7} MPa, which is close to zero for all three structures. This means that fatigue crack initiation life is zero if the crack is initiated from defects like porosity and oxides [10, 27].

Summary

The evolution history of bifilm in aluminum casting is simulated using the carefully designed procedure in structure construction. MD method is implemented to model the deformation mechanism of young, old, and –OH-terminated oxides. The results showed that during the aging process, the ultimate tensile strength of nano-scale bifilm oxides is increased from 1.1 GPa to 1.55 GPa without surface contamination. With the presence of impurities such as hydrogen gas in the casting process, the ultimate tensile strength of the nano-scale bifilm is dropped to 1.2 GPa. Also, the fracture energies for the young, the old, and the –OH-terminated old oxides are 0.43, 0.53, and 0.30 J/m^2, respectively. The excess hydrogen gas will compete with Al–O bond formation on "dry" side of the bifilm and have greater impact on the fracture strength and energy than the folding history and temperature. The fracture strengths of the bifilms are scaled up to FEM length scale as parameters in CZM model used for virtual casting and life prediction. More details can be found in [28].

References

1. Campbell J (2003) Castings, 2nd edn. Elsevier Butterworth-Heinemann
2. Lovatt AM, Bassetti D, Shercliff HR, Bréchet Y (2000) Process and alloy selection for aluminium casting. Int J Cast Met Res 12:211–225. https://doi.org/10.1080/13640461.2000.11819358
3. Lee A, Lu Y, Roche A, Pan TY (2016) Influence of nano-structured silanols on the microstructure and mechanical properties of A4047 and A359 aluminum casting alloys. Int J Met 10:322–328. https://doi.org/10.1007/s40962-016-0044-4
4. Sigworth GK (2008) The modification of Al–Si casting alloys: important practical and theoretical aspects. Int J Met 2:19–40. https://doi.org/10.1007/BF03355425
5. Cao X, Campbell J (2005) Oxide inclusion defects in Al–Si–Mg cast alloys. Can Metall Q 44:435–448. https://doi.org/10.1179/cmq.2005.44.4.435
6. Campbell J (2017) Melting, remelting, and casting for clean steel. Steel Res Int 88:1600093. https://doi.org/10.1002/srin.201600093
7. Campbell J (2016) The consolidation of metals: the origin of bifilms. J Mater Sci 51:96–106. https://doi.org/10.1007/s10853-015-9399-9
8. Wang QG, Davidson CJ, Griffiths JR, Crepeau PN (2006) Oxide films, pores and the fatigue lives of cast aluminum alloys. Metall Mater Trans B 37:887–895. https://doi.org/10.1007/BF02735010
9. Campbell J (2017) Melting, remelting, and casting for clean steel. Steel Res Int 88:1600093. https://doi.org/10.1002/srin.201600093

10. Wang QG, Apelian D, Lados DA (2001) Fatigue behavior of A356-T6 aluminum cast alloys. Part I. Effect of casting defects. J Light Met 1:73–84. https://doi.org/10.1016/S1471-5317(00)00008-0
11. Mitrasinovic A, Robles Hernández FC, Djurdjevic M, Sokolowski JH (2006) On-line prediction of the melt hydrogen and casting porosity level in 319 aluminum alloy using thermal analysis. Mater Sci Eng A 428:41–46. https://doi.org/10.1016/j.msea.2006.04.084
12. Digne M, Sautet P, Raybaud P et al (2002) Hydroxyl groups on gamma-alumina surfaces: a DFT study. J Catal 211:1–5. https://doi.org/10.1016/S0021-9517(02)93741-3
13. Sharma PK, Jilavi MH, Burgard D et al (1998) Hydrothermal synthesis of nanosize alpha-Al_2O_3 from seeded aluminum hydroxide. J Am Ceram Soc 81:2732–2734. https://doi.org/10.1111/j.1151-2916.1998.tb02687.x
14. Calvié E, Réthoré J, Joly-Pottuz L et al (2014) Mechanical behavior law of ceramic nanoparticles from transmission electron microscopy in situ nano-compression tests. Mater Lett 119:107–110. https://doi.org/10.1016/j.matlet.2014.01.002
15. Calvié E, Joly-Pottuz L, Esnouf C et al (2012) Real time TEM observation of alumina ceramic nano-particles during compression. J Eur Ceram Soc 32:2067–2071. https://doi.org/10.1016/j.jeurceramsoc.2012.02.029
16. Yang Y, Kushima A, Han W et al (2018) Liquid-like, self-healing aluminum oxide during deformation at room temperature. Nano Lett 18:2492–2497. https://doi.org/10.1021/acs.nanolett.8b00068
17. Sen FG, Alpas AT, Van Duin ACT, Qi Y (2014) Oxidation-assisted ductility of aluminium nanowires. Nat Commun 5:3959. https://doi.org/10.1038/ncomms4959
18. Sen FG, Qi Y, Van Duin ACT, Alpas AT (2013) Oxidation induced softening in Al nanowires. Appl Phys Lett 102:051912. https://doi.org/10.1063/1.4790181
19. Van Duin ACT, Dasgupta S, Lorant F, Goddard WA (2001) ReaxFF: a reactive force field for hydrocarbons. J Phys Chem A 105:9396–9409. https://doi.org/10.1021/jp004368u
20. Chenoweth K, van Duin ACT, Goddard WA (2008) ReaxFF reactive force field for molecular dynamics simulations of hydrocarbon oxidation. J Phys Chem A 112:1040–1053. https://doi.org/10.1021/jp709896w
21. Aktulga HM, Fogarty JC, Pandit SA, Grama AY (2012) Parallel reactive molecular dynamics: numerical methods and algorithmic techniques. Parallel Comput 38:245–259. https://doi.org/10.1016/j.parco.2011.08.005
22. Plimpton S (1995) Fast parallel algorithms for short-range molecular dynamics. J Comput Phys 117:1–19. https://doi.org/10.1006/jcph.1995.1039
23. Wang XG, Smith JR, Evans A (2002) Fundamental influence of C on adhesion of the Al_2O_3/Al interface. Phys Rev Lett 89:286102. https://doi.org/10.1103/PhysRevLett.89.286102
24. Needleman A (1987) A continuum model for void nucleation by inclusion debonding. J Appl Mech 54:525. https://doi.org/10.1115/1.3173064
25. Liu J, Huang Z, Pan Z et al (2017) Atomistic origin of deformation twinning in biomineral aragonite. Phys Rev Lett 118:105501. https://doi.org/10.1103/PhysRevLett.118.105501
26. Xia S, Qi Y, Perry T, Kim KS (2009) Strength characterization of Al/Si interfaces: a hybrid method of nanoindentation and finite element analysis. Acta Mater 57:695–707. https://doi.org/10.1016/j.actamat.2008.10.011
27. Wang QG, Jones PE (2007) Prediction of fatigue performance in aluminum shape castings containing defects. Metall Mater Trans B 38:615–621. https://doi.org/10.1007/s11663-007-9051-4
28. Liu J, Wang Q, Qi Y (2018) Atomistic simulation of the formation and fracture of oxide bifilms in cast aluminum. Acta Mater 164:673–682. https://doi.org/10.1016/J.ACTAMAT.2018.11.008

Numerical Process Modelling and Simulation of Campbell Running Systems Designs

Chengcheng Lyu, Michail Papanikolaou and Mark Jolly

Abstract In the 1980s, John Campbell developed a new casting process from his research in the industry over a number of years. The Cosworth process was for delivering very high-quality aluminium components for the automotive industry. The process was very capital-intensive and not very flexible for smaller companies delivering lower volumes of product. However, the principles behind the process have been taken and used to develop a range of different so-called running systems to help improve the quality of castings. Some of these designs have been published in 'Castings Handbook' [1] authored by Campbell. This paper presents the results of an MSc project during which a number of the proposed designs from Campbell's Mini Casting Handbook [2] for certain features in running systems have been modelled using a validated CFD software.

Keywords Gravity sand casting · Running system design · Numerical simulation · Vortex gate · Trident gate · Bubble trap · Multiple-in-gate · Air entrainment · Gate velocity

Introduction

In order to add control to the whole casting process, a set of rules has been proposed by Campbell commonly known as the '10 Rules of Casting' [3, 4]. These rules have been applied to a whole range of alloys including aluminium, zinc, copper and cast iron. A poorly designed running system contributes to most of the defects of the final casting as most of the damage occurs in this region [1]. That is because all metal into the cavity has to go through the running system first. Moreover, the very beginning seconds of casting (the priming stage) is the most dangerous period of

---

C. Lyu · M. Papanikolaou · M. Jolly (✉)
Manufacturing Department, Cranfield University, Cranfield MK43 0AL, UK
e-mail: m.r.jolly@cranfield.ac.uk

M. Tiryakioğlu et al. (eds.), *Shape Casting*, The Minerals, Metals & Materials Series,
https://doi.org/10.1007/978-3-030-06034-3_5

generating various defects, such as air entrainment, inclusion or misruns [5]. The problem of gravity filling is extremely critical because gravity naturally accelerates the liquid metal flow and consequently causes damage if not controlled properly. Campbell [1] has summarized five characteristics of a good running system. Namely, they should:

1. be lightweight. This means the ratio of finished casting weight to total cast weight should be as high as possible;
2. control the velocity of metal into the casting which should be as low as possible, preferably, under the critical value;
3. avoid entraining any impurities into the system (such as slag, air or sand);
4. eliminate surface turbulence;
5. be easy to remove. It is better to avoid complicated curves so that breaking off the mould is simple.

While Campbell's description is more systematic, some other researchers proposed that a good running system should fill quickly (but tranquilly) and should boost directional solidification [6], or should solidify quickly and uniformly [7]. Regardless of solidification process, among these different expressions, one core principle is believed to be similar; that is, the level of surface turbulence is the fatal factor for the quality of the casting.

It is noticeable that casting rule 2 and characteristic 2 both emphasize the importance of the critical gate velocity being less than 0.5 ms^{-1} and this does not vary much for different alloys while exceeding this value triggers surface turbulence and the consequent defects. However, this value is extremely easily violated for a gravity casting system because the critical velocity 0.5 ms^{-1} for an aluminium alloy is merely equivalent to a drop of about 12.5 mm. This is almost impossible to achieve in a gravity-poured system except possible in an open shallow mould.

To address the conflict of necessary fall height and the rigorous requirement of low velocity for a closed mould techniques must be developed to enable a low velocity before metal enters the ingates. And in fact, after deep understanding of the importance of this critical value, efforts have always been made by researchers and founders to get close to it.

Earlier researchers proposed that expansion in cross section helps reduction of velocity, and this is the fundamental design principle of the so-called unpressurised system. This is theoretically right according to the continuity equation. However, this type of system has been suspected to be unable to ensure good casting quality because of more likely air entrainment and vortex flow. This has been proven by some researchers through both simulation and experiment [8]. Researchers also tried to deal with the turbulent metal flow after it is poured into the pouring basin. According to Campbell, abandoning the conical basin should be the first step to take because of its extraordinary ability to generate an air guzzling vortex [9] (which Campbell described as an 'air pump'). While air is the most evil enemy of good castings, especially for alloys prone to active reaction with oxide, such as aluminium, titanium and magnesium alloys, the reason that the conical basin is still in use is because of its cheap and practical application. It is shocking that in the investment casting sector, who claim to have the highest added value and premium casting process, the majority still use a conical pouring basin. Campbell proposed different shapes of offset pouring basin in his handbook [1], without step, with sharp step and with a radiused step. The latter is the one chosen for all the running systems in this research.

Rather than endlessly struggling to eliminate turbulence in the pouring basin, which may end up in achieving little, a technique named 'contact pouring' is strongly advocated by Campbell. This is achieved by using a small bottom pour ladle, usually capable of holding 10–20 kg metal, and is topped up from beneath the surface by a robot dipping into the metal. A stopper plugs the hole in the base of the ladle, and when the ladle is full it is moved to contact the sprue and the stopper is raised allowing the metal to enter the system. Combined with a well-designed downsprue, this can deliver turbulence- and oxide-free molten metal into the system [10]. The challenge of this technique is to align the exit of the ladle to the entrance of the sprue and ensure reliable seal to avoid leakage and a passage for air entrainment.

After the metal enters the system, to reduce its velocity, filters are explored by simulation and experiments [11], and right angle bends are also suggested to be an effective way to reduce velocity, but they may incur some other problems like entrainment of air and oxide [12]. Unfortunately, none of these methods is as ideal as counter-gravity gating systems such as the Cosworth or Crimson process.

It is common practice to assess and optimize a newly designed running system. Researchers or founders have used various techniques, classified as direct observation, modelling by water analogy or computer-aided simulation [6]. The first two trial-and-error methods require long times and high cost [13]. Since the first published work on numerical simulation in the late 1980s, many software packages have been developed for foundries and have been used increasingly [14]. Computational fluid dynamics (CFD) software is considered to be the most cost effective tool for both design and optimization of a running system [15]. The traditional 'trial-and-error' in the workshop therefore is transformed into the 'proof-of-concept' by computer [16]. A range of simulation softwares are available today for casting engineers with different features. For this paper, FLOW-3D[1] has been used to conduct simulations and FlowSight for post-processing.

Running System Designs

The aim of this paper is to assess the effectiveness of different running systems for gravity sand casting. To simplify calculation, the casting part is a very simple geometry: an aluminium cube with a length of 100 mm. A basic running system

[1]https://www.flow3d.com/.

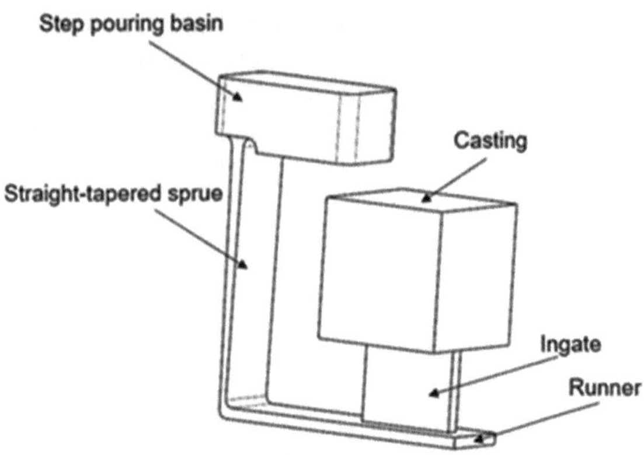

Fig. 1 Basic running system (BRS)

Table 1 Basic running system design dimensions

Parameters	Value
Initial volume flow rate Q_0	382 $cm^3 s^{-1}$
Depth of pouring basin H_b	50 mm
Width of pouring basin W_b	65 mm
Length of pouring basin L_b	120 mm
Entrance area of sprue A_2	590 mm^2 (59 mm × 10 mm)
Height of the sprue H_S	200 mm
Exit area of sprue A_3	207 mm^2 (53 mm × 4 mm)
Area of runner A_4	424 mm^2 (53 mm × 8 mm)
Length of runner L_R	200 mm
Width of ingate W_{in}	10 mm
Length of ingate L_{in}	75 mm
Height of ingate H_{in}	60 mm
Casting size	100 mm cube

(BRS) was devised using both a published nomogram [1] and the method described by Jolly [12], as illustrated in Fig. 1. Table 1 gives the dimensions and values used for BRS.

As well as BRS, nine other systems were simulated with different features which included the placement of filters, bubble trap, a vortex gate, a trident gate and four designs of multiple gates (Fig. 2). These designs were taken from pages 46, 49, 54 and 55 of Campbell's Mini Casting Handbook [2]. Simulation was proved to be an effective tool as after optimizing the mesh none of the simulations took more than an hour to run. All the simulations were run assuming an A356 alloy with an initial pouring temperature of 735 °C. Predicted filling times ranged from 4 s to 6.6 s which is reasonable as the initial system was designed to fill in 5 s.

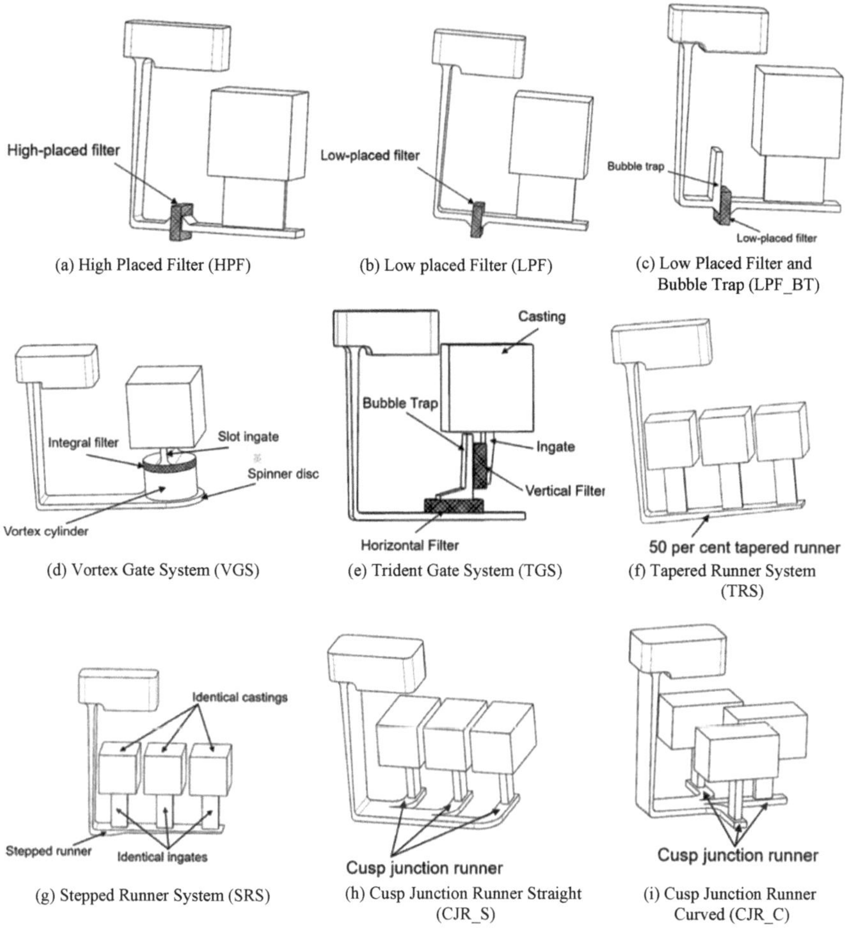

(a) High Placed Filter (HPF) (b) Low placed Filter (LPF) (c) Low Placed Filter and Bubble Trap (LPF_BT) (d) Vortex Gate System (VGS) (e) Trident Gate System (TGS) (f) Tapered Runner System (TRS) (g) Stepped Runner System (SRS) (h) Cusp Junction Runner Straight (CJR_S) (i) Cusp Junction Runner Curved (CJR_C)

Fig. 2 Isometric CAD images of the running system designs developed from features described in Campbell's Mini Casting Handbook [2] and used for comparative CFD simulations

Simulation Results and Analysis

The results of the ten systems were compared in three groups. In most cases, entrained air volume fraction (EAVF) and velocity were tracked and used as comparative measure of the effectiveness of the systems. In this paper, only a selection of the results are presented as detailed below:

1. EAVF results with BRS as a benchmark, compared with HPF, LPF as well as LPF_BT to examine the performance of filter and bubble trap (Fig. 2a–c);
2. Velocity results with BRS as a benchmark, compared with VGS and TGS to display how well the latter two work in terms of reducing velocity (Fig. 2d–e);

3. EAVF results for four multiple-in-gate systems (SRS, TRS, CJ_S and CJ_C) compared with each other to explore the best design for delivering metal into three identical castings (Fig. 2f–i).

Figures 3, 4 and 5 are selected images from a range of simulations. Figure 3 shows a comparison of the predicted entrained air volume fraction (EAVF) for the basic running system (BRS), high placed filter (HFS), low placed filter (LFS) and low filter with bubble trap (LFS_BT). These comparisons are for the time at which the running system is close to being primed; in most cases just under 1 s, and for the time when the casting has just filled (just over 4 s). Figure 4 shows a comparison of the predicted metal velocities at 1.1 s and 3.3 s between the BRS and the vortex and trident gates. Particle tracking is included and shows the distribution of particles introduced into the pouring stream above the pouring basin. Figure 5 illustrates the effect that different multi-gate running system designs have on the EAVF at times early in the filling and just before the castings are filled.

Discussion of Results

Looking at Fig. 3, it can be seen that the inclusion of a filter ensures that the running system primes quickly and in all cases faster than BRS. At 0.885, BRS is not primed, and even though the runner bar is only 8 mm in depth, there is clearly a rolling back wave developing which would lead to surface oxide and air entrainment. It is clear that HPF, LPF and LPF_BT all work well in reducing EAVF comparing with BRS. After 4.12 s, the castings are filled and the filtered castings show a much lower level of air entrainment than BRS. The bubble trap does appear to work as it has a higher level of air entrainment than the runner bar although the detailed analysis of this can only really be seen in the full simulations. Unrestricted jetting at the back of filter in HPF system is not observed as is detailed in the literature, but this may be because of the method for modelling the filter which just applies a permeability across a volume representing the filter. Jetting would only be observed in a simulation if the actual physical geometry of the filter was modelled [17].

Figure 4 demonstrates clearly the damping effect of both the vortex and trident gates on the velocity of the liquid metal compared with BRS. Both these gating systems have filters included in the design, but there is even less splashing and obvious turbulence compared with all the previous filtered systems in Fig. 3. The filling times are also considerably longer but that is due to the additional volume of the running and gating. Hence, the yield of these systems will be poorer than any of the systems in Fig. 3.

Figure 5 clearly shows that the cusp design of runner to distribute metal between similar multiple castings is far superior to either the stepped or tapered runner systems. The entrained air volume fraction in both cusp designs is lower, and the

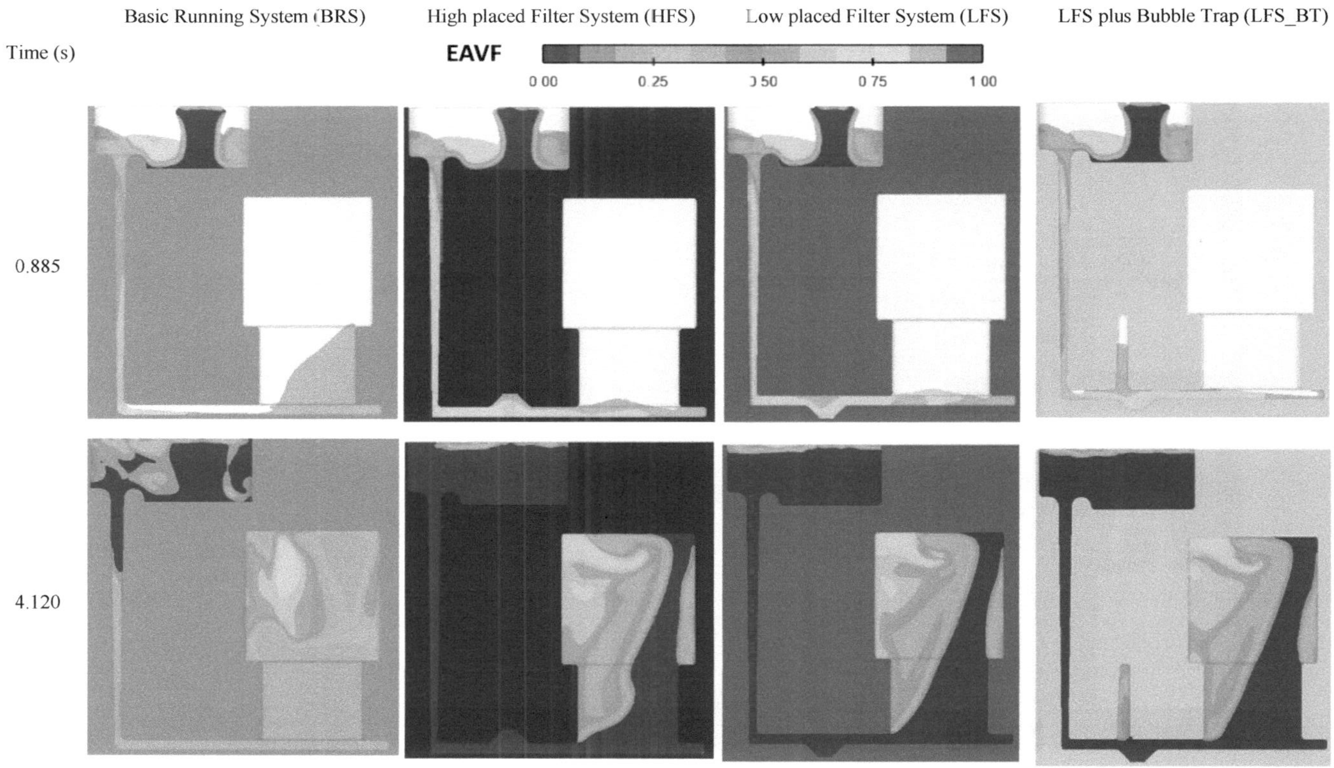

Fig. 3 Predicted entrained air volume fraction (EAVF) results from the beginning and end of filling showing the effects of different filter placement and a bubble trap. Overall, it can be seen that having a filter in the runner significantly reduced the amount of air entrainment by preventing a rolling back wave in the runner seen in BRS. The bubble trap has a limited effect in this instance but does slightly reduce the air entrainment

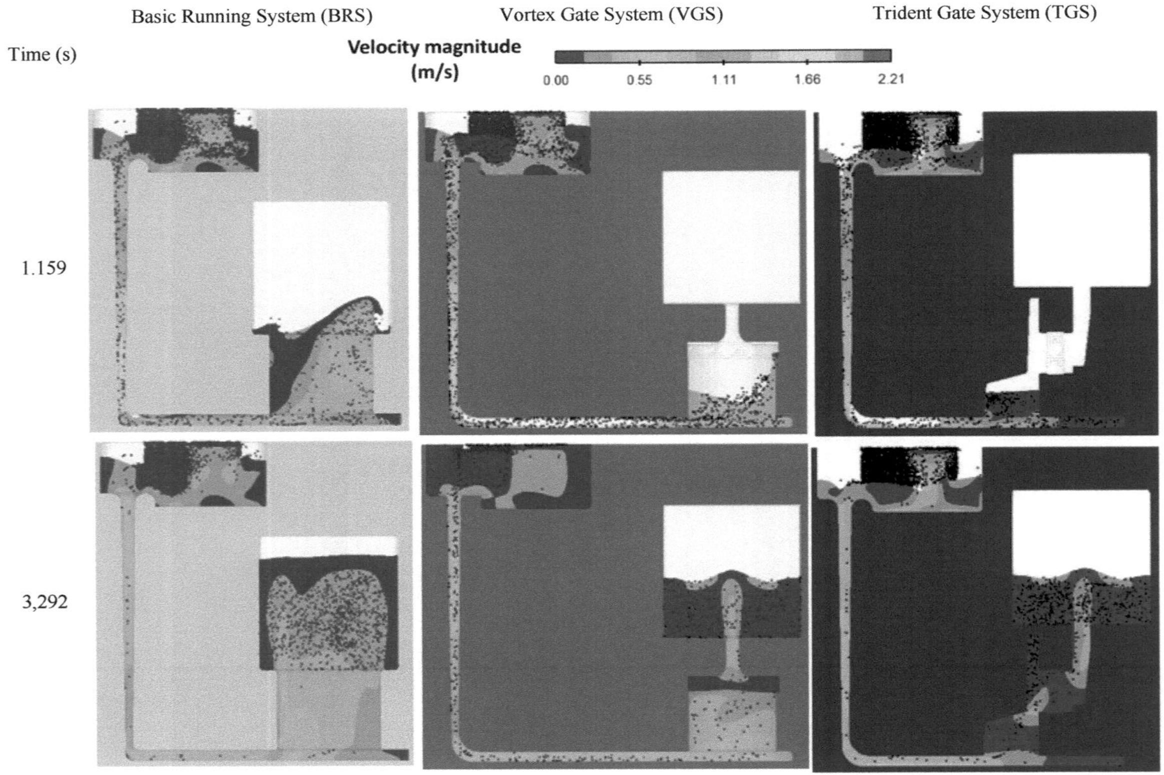

Fig. 4 Comparison of basic, vortex and trident gates showing the considerable reduction in velocity achieved by the latter two designs. However, both the vortex and trident gates showed much longer running system priming times with the vortex gate being the slowest. The predicted filling times for the systems were 4.0, 4.95 and 6.59 s for BRS, VGS and TGS, respectively

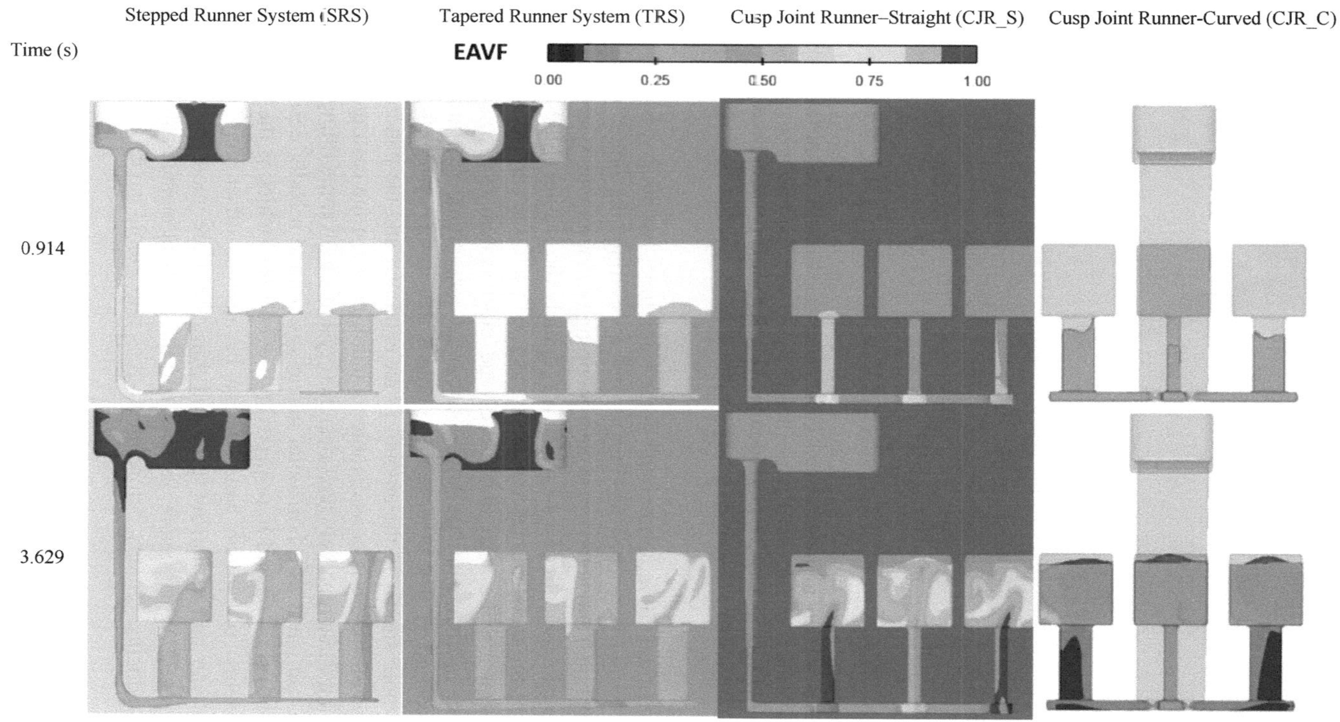

Fig. 5 Predicted entrained air volume fraction (EAVF) for the four different multiple gate systems, namely stepped, tapered straight cusp and curved cusp clearly showing that the cusp systems are superior in as they appear to entrain less air than either the tapered or stepped runners

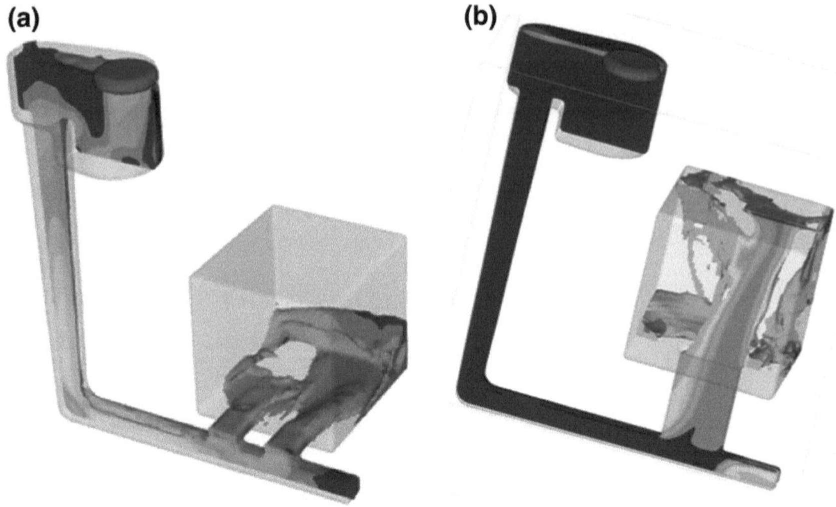

Fig. 6 **a** Horizontal ingates (wider sprue and thicker runner compared with BRS). Metal impinges with the far end of cavity from the two horizontal ingates and causes a destructive backflow and turbulence. **b** Vertical ingate (wider sprue, thicker runner and narrower ingate compared with BRS). The out-of-control metal jets from the vertical ingate and even impacts the top face of cavity and then falls back, resulting in a completely disordered metal flow and large volume of bubbles

individual castings fill at approximately the same time. Essentially, by splitting the runner bar before it reaches the ingates of any casting, the liquid metal can be equally distributed and better controlled. Stepping or tapering the system just allows the metal to take the path of least resistance, namely the casting at the far end of the runner where the metal can jet into the casting. Close to the complete filling of the casting, it can also be seen that the curved cusp system even outperforms the straight cusp system with the lowest entrained air volume fraction of the four systems investigated. The other point to note is that in this system the castings are also much more similar in the amount of air entrained and the pattern, whereas in the other systems there is considerable variation between the castings.

Comparing these relatively well-designed systems with some filling systems still in practical use (such as a horizontal ingate system and a badly designed vertical ingate system) (Fig. 6a and b, respectively), even BRS (which is the worst among the ten systems) is a much better design because its offset step pouring basin entrains little air and its narrow sprue and thin runner are good enough to control the flow of metal. What's more, its wide and thin vertical ingate is also an effective device for reducing velocity before metal goes into mould cavity.

Conclusions

After detailed comparison and analysis of ten running systems discussed in Campbell's Mini Casting Handbook [2], it is concluded that:

- Filters in HFS, LFS and LFS_BT all work well in reducing air entrainment;
- The bubble trap is an effective way to absorb air in the metal;
- VGS and TGS both perform well in reducing metal velocity, but filters in these two systems show little effect on reducing air entrainment, which needs more investigation;
- The best system to introduce metal into multiple ingates is the cusp junction, which distributes metal more evenly.
- The specific cusp design is important in ensuring that the casting being filled is reproducible.
- All of the running systems investigated performed better than poorly designed systems with parallel rather than tapered sprue and a thick runner bar.

References

1. Campbell J (2015) Complete casting handbook, 2nd edn. Butterworth-Heinemann, Oxford
2. Campbell J (2017) Mini casting handbook. Aspect Design, Malvern
3. Campbell J (2004) Castings practice: the ten rules of casting. Butterworth-Heinemann, Oxford
4. Jolly M (2005) Prof John Campbell's ten rules for making reliable castings. J Mater 57(5):19–28
5. Nafisi S, Mosleh B, Kazazie GR, Boutorabi MA Reviewing aluminum gating systems. Modern Casting, pp. 35–39 (2010)
6. Ingle PD, Narkhede BE (2018) A literature survey of methods to study and analyze the gating system design for its effect on casting quality. Materials Today 5(2):5421–5429
7. Miller RA, Lu SC, Rebello AB (1997) Simple visualization techniques for die casting part and die design. Final report, US Department of Energy, Office of Scientific and Technical Information
8. Kheirabi A, Baghani A, Bahmani A, Tamizifar M, Davami P, Shabani MO, Mazahery A (2015) Understanding the occurrence of the surface turbulence in a nonpressurized bottom gating system: numerical simulation of the melt flow pattern. In: Proceedings of the institution of mechanical engineers: part l: journal of materials: design and applications, vol. 232, no. 3, pp. 230–241, 17 December 2015
9. Campbell J (2015) Sixty years of casting research. Metall Mater Trans (A) 46(11):4848–4853
10. Campbell J (2012) Stop pouring, start casting. Int J Metalcast 6(3):7–18
11. Hsu YF, Lin HJ (2011) Foam filters used in gravity casting. Metall Mater Trans B 42 (6):1110–1117
12. Jolly MR (2003) Chapter 18: Castings. In: Milne I, Ritchie RO, Karihaloo B (eds) Comprehensive structural integrity, vol 1. Elsevier, Oxford, pp 377–466
13. Nimbulkar SL, Dalu RS (2016) Design optimization of gating and feeding system through simulation technique for sand casting of wear plate. Perspect Sci 8:39–42
14. Esparza C, Guerrero M, Rios-Mercado R (2006) Optimal design of gating systems by gradient search methods. Comput Mater Sci 36(4):457–467

15. Hu BH, Tong KK, Niu XP, Pinwill I (2000) Design and optimisation of runner and gating systems for the die casting of thin-walled magnesium telecommunication parts through numerical simulation. J Mater Process Technol 105(1–2):128–133
16. Fu MW, Yong MS (2009) Simulation-enabled casting product defect prediction in die casting process. Int J Prod Res 47(18):5203–5216
17. Jolly MR, Gebelin J-C (2003) Experimentation and numerical modeling of metal flow through filters. In: Tenth international conference on modelling of casting welding and advanced solidification processes X (MCWASPX), Sandestin, FA, USA, 2003

Synchrotron X-Ray Real-Time Studies of the Nucleation and Growth of Intermetallic Phases in Solidification

Jiawei Mi

Abstract In this paper, real-time, direct observation of the nucleation of primary Al_2Cu phases on entrained oxide films during the solidification of an Al-35%Cu alloy, and the change of growth mode from planner to dendritic type structure of Al_3Ni phases in an Al-15%Ni alloy under a pulse magnetic field are presented and discussed. Real-time synchrotron X-ray radiography provides clear and unambiguous evidence to confirm that the entrained oxide films can indeed directly nucleate primary Al_2Cu intermetallic phases. Combining radiography and tomography together, the critical conditions of changing growth mode from a planner to dendritic type for the Al_3Ni phases were also clearly revealed. The research demonstrates that synchrotron X-ray-based real-time characterisation techniques are the indispensable tools for quantitative solidification research.

Keywords Synchrotron X-rays · In situ study · Intermetallic phases · Solidification · External fields

Introduction

In 1999–2002, I completed a PhD project in the Casting Research Group at the University of Birmingham, UK, supervised by Professor John Campbell and Dr. Richard Harding. At that time, one important research direction in Campbell' group was to use real-time X-ray video radiography to study liquid metal flow and mould filling processes in casting with the aim of minimising entrainments of air bubbles or oxide films, and the formation of casting defects. Many evidences obtained from real-time X-ray radiography observation [1–4] have confirmed that, in Al-based alloys, surface turbulence of liquid melts could lead to the entrainment of oxidised surface films into the bulk liquid melts during mould filling. The entrained films

J. Mi (✉)
School of Engineering and Computer Science, University of Hull,
Hull HU6 7RX, UK
e-mail: j.mi@hull.ac.uk

M. Tiryakioğlu et al. (eds.), *Shape Casting*, The Minerals, Metals & Materials Series,
https://doi.org/10.1007/978-3-030-06034-3_6

could act as cracks in the solidified castings or preferable sites for nucleating porosity, bubbles, hot tearing, etc. Cao reported that oxide films could act as preferred substrates for the nucleation and growth of some Fe-rich phases [5]. Cao also calculated the planar disregistries between some typical oxides and α-Fe phases (primary or script type) and argued that MgO, $MgO \cdot Al_2O_3$, γ-Al_2O_3, and α-Al_2O_3 might all be good substrates for the nucleation of α-Fe phases. Although some indirect evidences from scanning electron microscopy observation were obtained to support such argument, direct observation of the nucleation of α-Fe phases from oxide films was not possible using the laboratory-based X-ray radiography technique at that time.

In the late 1990s, Professor Ragnvald Mathiesen and Lar Arnberg [6] of the Norwegian University of Science and Technology began the pioneering work of using coherent and monochromatic synchrotron X-rays available at the beamlines ID22 and ID18 of European Synchrotron Radiation Facility (ESRF) to study in situ the cellular and dendritic solidification of metal alloys [6]. The spatial and temporal resolutions achieved at that time were $\sim$2.5 μm and $\sim$0.7 s, respectively, in a field of view of up to 1 mm^2. Since then, worldwide research on using synchrotron X-rays to study solidification processes has grown rapidly. The author has just published a comprehensive review on the advances and developments in this field [7]. These in situ and real-time studies provided much greater insights into how dendritic structures and intermetallic phases evolve and adapt according to the coupled physics of heat/solute transfer, fluid flow, and external fields [7]. In the past 8 years or so, the research in the author's group at University of Hull, UK has been concentrated on studying the dynamic evolutions of dendrites and intermetallic phases in highly dynamic solidification environments, i.e. under ultrasonic waves [8–11] and pulse electromagnetic fields [12, 13]. The in situ experiments were mainly carried out in the third-generation synchrotron X-ray facilities, for example the ESRF in Grenoble, France, the Advanced Photon Source (APS) at Argonne National Laboratory, USA, the Diamond Light Source (DLS), UK, the Swiss Light Source (SLS) at the Paul Scherrer Institute, Switzerland, and Synchrotron SOLEIL, France. In this paper, two case studies are presented to demonstrate the new insights revealed by using the powerful synchrotron X-ray radiography and tomography techniques.

Direct Observation of the Nucleation of Intermetallic Phases on Oxide Films

The experiments were carried out at the Pression Structure Imagerie par Contraste à Haute Énergie (PSICHÉ) beamline of Synchrotron SOLEIL, France [14]. The experimental setup was described in detail in [14], and therefore do not repeat here. Hypereutectic Al-35% Cu alloy was used and melted in a custom-made quartz crucible inside a two-zone furnace. The temperature of the bottom half was set at

600 °C to ensure that the alloy was fully molten (the liquidus temperature of Al-35% Cu alloy is 554 °C). The temperature of the top half was set at 630 °C to establish a positive thermal gradient (~1 K/mm) in the melt to force the primary Al_2Cu intermetallic to grow from bottom to top. After both halves of the furnace reached their target temperatures, they were held at these temperatures for ~5 min to ensure the alloy was fully molten and also to allow sufficient aluminium oxide film growth on the melt surface. Then, solidification process started by setting the top and bottom furnace halves to a cooling rate of ~2 K/min. The growth of primary Al_2Cu intermetallic phases was recorded using a Vision Research Miro 310 M high-speed camera attached to the X-ray imaging optics (5×) at the beamline. The field of view was 5120 × 2624 μm^{2}, and the nominal pixel resolution was 4 μm/pixel. When the primary Al_2Cu phases growth front appeared at the bottom of the field of view, a niobium sonotrode (1 mm in diameter) driven by a Hielscher UP100H ultrasound generator was inserted into the melt from the top to apply ultrasound and hence to force the surface oxide film into the bulk liquid. The ultrasonic transducer power used was 100 W with a frequency of 30 kHz, and the ultrasound duration was around 5 s. After that, the ultrasound was switched off and the sonotrode was lifted out of the melt, and the subsequent solidification of primary Al_2Cu intermetallic phases was recorded at 1 fps.

An image sequence extracted from a radiography video recording the typical primary Al_2Cu intermetallic growth on the fragments of oxide film is shown in Fig. 1. The corresponding video can be found in the online version of [14].

The first frame of the recording is defined as t = 0 s. As shown in Fig. 1a, the dark phase is the primary Al_2Cu intermetallic and the grey matrix is the alloy melt. The light grey particles marked by numbers are fragments of aluminium oxide films. In total, 10 fragments of aluminium oxide film were successfully introduced to the melt by ultrasound in this experiment and they remained rather stationary in the field of view which facilitated observation of their effect on intermetallic

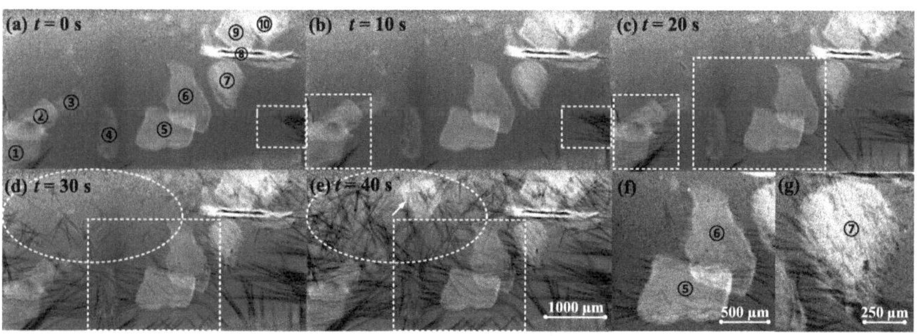

Fig. 1 Images of the primary Al_2Cu phases growing on oxide particles from (**a**) to (**e**); the field of view is 5120 × 2624 μm^2 and images were recorded at 1 fps. The t = 0 s frame was when the sonotrode was switched off and lifted up from the melt; **f** an enlarged image of the phases growing on oxide particles No. 5 and No. 6 in (**a**) and (**d**); **g** an enlarged image of the phases growing on oxide particle No. 7 in (**a**) and (**d**) [14]

growth. The identification of those fragments as aluminium oxide film was based on their X-ray attenuation compared to that of the melt, and the analysis is detailed in [14].

As solidification proceeded, slight growth of the primary Al_2Cu intermetallic was observed on No. 1 and No. 2 fragments of oxide film at the bottom left corner in Fig. 1b. Following that, distinct growth of the primary Al_2Cu intermetallic on the fragments of oxide film (No. 1, 2, 4, 5, 6, 7, 8, 9 and 10) was manifested in Fig. 1c and d. An enlarged image of the intermetallic growth on the fragments of oxide film No. 5 and No. 6 is shown in Fig. 1f. As the growth on these fragments of oxide film continued, the primary Al_2Cu intermetallic phases finally impinged with each other as shown in Fig. 1e, leaving no space for further growth. The average intermetallic length was measured as 320 $\pm$ 20 μm. Note that the intermetallic growth on the fragments of oxide film is essentially in three dimensions as can be observed on oxide particle No. 7 as shown in Fig. 1(g). Also, a showering of equiaxed primary Al_2Cu intermetallic followed by further equiaxed growth was observed at the top left corner of the field of view as highlighted by a dashed ellipse in Fig. 1d and e. This is a result of prior fragmentation of growing Al_2Cu intermetallic particles by ultrasonic processing.

Direct Observation of the Growth of Intermetallic Dendrites in a Pulse Magnetic Field

The experiments were conducted at the DLS. A novel pulse magnetic solidification apparatus (Fig. 2a) was used in the in situ studies [12, 13] using synchrotron X-ray radiography and tomography techniques. The apparatus consists of three main parts. The first part has two small-scale electric resistance furnaces with each equipped with an independent PID temperature controller. The second part is a helix coil made by enamelled copper wires of 2 mm in diameter for generating electromagnetic fields. The third part is a pulse electric current generator with a control system as described in details in [13]. The two furnaces together with a quartz tube (for containing the experimental alloys) were put inside the copper coil. Pulse electric current generated by the pulse generator was passed through the coil to create a pulse magnetic field (PMF) inside the alloy samples. Typical pulse magnetic flux densities generated by different voltages discharged from the generator are shown in Fig. 2b, which were measured by positioning the axial probe of a Gauss metre at the centre of the coils and recoded by an oscilloscope connected to a desktop computer [13]. This solidification apparatus is designed for processing bulk ally samples with a diameter of up to 30 mm in order to test samples with solidification conditions similar to those found in industry.

An Al-15%Ni was chosen as the experimental alloys to study the evolution of primary Al_3Ni intermetallic phases under different pulse magnetic flux intensities. The metal alloy samples were held inside a capillary quartz tube (inner diameter of

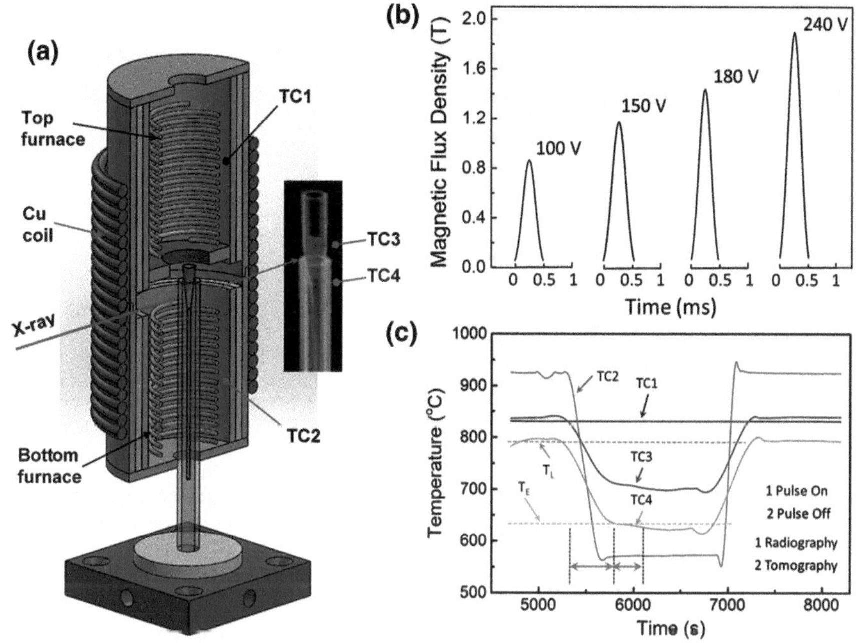

Fig. 2 **a** A CAD drawing, showing the pulse magnetic field solidification apparatus with an enlarged inset on the right, showing the special quartz tube sample holder and the liquid metal alloy. **b** The measured magnetic flux density with discharging voltage of 100, 150, 180 and 240 V, respectively. **c** The measured temperature profiles at four points (TC1 top furnace, TC2 bottom furnace, TC3 and TC4 close to the metal sample), and the period when pulse was applied and when radiography and tomography were taken during the solidification experiments [13]

1 mm and 0.1 mm wall thickness) and positioned along the central axis of the two furnaces as highlighted in Fig. 2a. The temperatures of the two furnaces (at the locations marked by TC1 and TC2) were controlled independently to achieve the desired temperature during melting and cooling. Two K-type thermocouples of 2 mm apart were positioned very close to the quartz tube (at the locations marked by TC3 and TC4 in Fig. 2a insert) to measure the temperatures at the location where X-ray images were taken. All temperatures were recorded using a data logger (TC-08; Pico Technology, UK) connected to a desktop computer during the experiments. The peak magnetic flux density applied into the sample was 1.5 T (an initial discharge density of 180 V as shown in Fig. 2b). A polychromatic X-ray beam was used in the experiments. It was generated by an undulator of 5 mm gap and was reflected from the platinum stripe of a grazing incidence focusing mirror and high-pass filtered with 1.3-mm pyrolytic graphite, 3.2-mm aluminium and 20-μm steel. Images were recorded by a CMOS (2560 × 2160 pixels) pco.edge 5.5 (PCO AG, Germany) detector which was coupled to a 500-μm-thick CdWO4 scintillator and a light microscope. An objective lens of 4× was used, achieving an effective spatial resolution of 0.81 μm/pixel in a field of view of 2.1 mm × 1.8

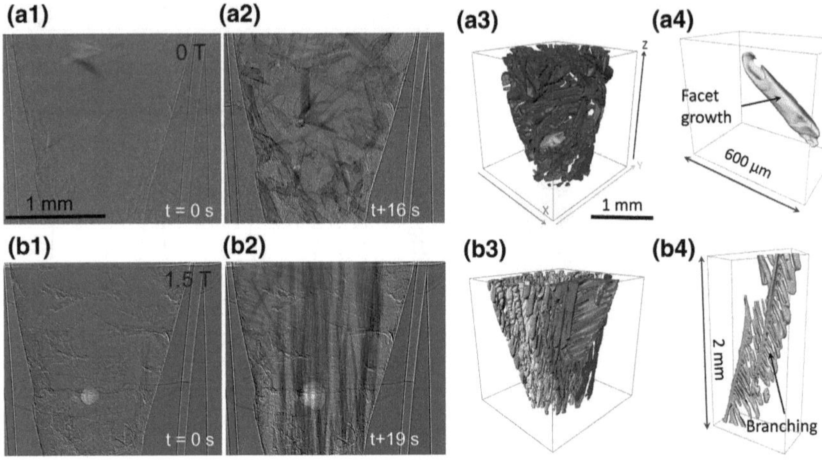

Fig. 3 Synchrotron X-ray radiography (a1, a2, b1 and b2) and tomography (a3, a4, b3 and b4), showing the Al_3Ni intermetallic phases of an Al-15%Ni alloy during solidification and their corresponding 3D morphology without PMF (top row, 0 T), and with PMF (bottom row, 1.5T). a4 and b4 show the transition from facet growth to dendritic growth due to PMF

mm. The exposure time was set at 0.1 s in order to obtain sufficient signal–noise ratio. For tomography experiments, a total of 2001 equally spaced projection images were acquired over 180-degree continuous rotations for each sample. All datasets were denoised, segmented and rendered using an open-source image processing software, Fiji and Avizo 9.

Figure 3 shows the solidified Al_3Ni phases revealed by radiography. Without PMF, Al_3Ni phases are almost randomly orientated (Fig. 3a2). With PMF of 1.5 T, almost all Al_3Ni phases are vertically aligned along the magnetic flux direction (Fig. 3b2).

In addition to the rich information obtained by radiography during the growth of Al_3Ni phases, tomography reveals in more details the change of 3D morphology due to the application of PMF. Without PMF, Al_3Ni phases exhibit planar surface (Fig. 3a3 and a4), indicating a facet growth mode. With PMF of 1.5 T, numerous dendritic phases formed (Fig. 3b3 and b4) with rich side branches, demonstrating that PMF has a strong effect on disrupting the planar liquid–solid interface, leading to the formation of dendrite structures. Interested readers are referred to the detailed explanation in [13].

Summary

The growth of primary Al_2Cu intermetallic on fragments of oxide film in a hypereutectic Al-35% Cu alloy was clearly observed and revealed for the first time. This direct observation provides conclusive evidence for the longstanding argument that oxide particles in a real metallic alloy melt can act as nucleation substrates for primary phases during solidification. Pulse magnetic fields are shown to be effective in orientating and aligning the Al_3Ni phases towards the magnetic flux direction and change the growth of the Al_3Ni phases from a planar mode to dendritic mode. Synchrotron X-ray techniques allow the highly dynamic phenomena in metal solidification processes to be studied with spatial and temporal resolution that are not possible before, opening a new era for solidification science research.

Acknowledgements The author would like to acknowledge the financial support from the UK Engineering and Physical Science Research Council (EPSRC) for the Ultra-Cast project [Grants Nos. EP/L019965/1, EP/L019884/1, EP/L019825/1]. The awards of synchrotron X-ray beam time by Synchrotron SOLEIL, France [Proposal No. 20160076], and Diamond Light Source (MT13488-1, MT13488-2), UK, are also gratefully acknowledged.

References

1. Fox S, Campbell J (2000) The visualisation of oxide film defects during solidification of aluminium alloys. Scripta Mater 43:881–886
2. Mi J, Harding RA, Campbell J (2002) The tilt casting process. Int J Cast Metal Res 14 (2002):325–334
3. Mi J, Harding RA, Wickins M, Campbell J (2003) Entrained oxide films in TiAl castings. Intermetallics 11:377–385
4. Mi J, Harding RA, Campbell J (2004) Effects of the entrained surface film on the reliability of castings. Metall Mater Trans A 35A:2893–2902
5. Cao X, Campbell J (2003) The nucleation of Fe-rich phases on oxide films in Al-11.5Si-0.4 Mg cast alloys. Metall Mater Trans A 34A:1409–1420
6. Mathiesen RH, Arnberg L, Mo F, Weitkamp T, Snigirev A (1999) Time resolved X-ray imaging of dendritic growth in binary alloys. Phys Rev Lett 83:5062
7. Eskin DG, Mi J (eds) Solidification processing of metallic alloys under external fields. Springer Series in Materials Science, vol. 273. Springer, Cham. https://doi.org/10.1007/978-3-319-94842-3_2
8. Mi J, Tan D, Lee TL (2015) In situ synchrotron X-ray study of ultrasound cavitation and its effect on solidification microstructures. Metall Mater Trans B 46B:1615–1619
9. Tan D, Lee TL, Khong JC, Connolley T, Fezzaa K, Mi J (2015) High speed synchrotron X-ray imaging studies of the ultrasound shockwave and enhanced flow during metal solidification processes. Metall Mater Trans A 46:2851–2861
10. Wang F, Eskin D, Mi J, Wang C, Connolley T (2017) A synchrotron X-radiography study of the fragmentation and refinement of primary intermetallic particles in an Al-35 Cu alloy induced by ultrasonic melt processing. Acta Mater 141:142–153
11. Wang B, Tan D, Lee TL, Khong JC, Mi J (2018) Ultrafast synchrotron X-ray imaging studies of microstructure fragmentation in solidification under ultrasound. Acta Mater 144:505–515
12. Manuwong T, Zhang W, Kazinczi PL, Bodey AJ, Rau C, Mi J (2015) Solidification of metal alloys under electromagnetic pulses and characterization of 3D microstructures using synchrotron X-ray tomography. Metall Mater Trans A 46:2908–2915
13. Du W (2018) In-situ synchrotron X-ray imaging and tomography studies of the microstructural evolution in metal solidification under pulse electromagnetic fields. PhD Thesis, University of Hull, Hull
14. Wang F, Eskin D, Connolley T, Wang C, Koe B, King A, Reinhard C, Mi J (2018) In-situ synchrotron X-ray radiography observation of primary Al_2Cu intermetallic growth on fragments of aluminium oxide film. Mater Lett 213:303–305

Determination of Liquid Metal Quality with Deep Etching Method

Furkan Tezer, Özen Gürsoy, Eray Erzi, Mert Zorağa and Derya Dışpınar

Abstract In this study, the deep etch method was applied to Al–Si casting alloys in order to visualize the presence of the bifilms. Reduced pressure test was used to quantify the bifilms in the melt. The cross section of the RPT samples was subjected to deep etching with H_2SO_4, HNO_3, NaOH, and NaCl solutions at various temperatures and concentrations. The aim was targeted to etch the matrix and reveal bifilms on these surfaces. Additionally, the oxide structure on the surface of cast specimens was altered by treatment at 500 °C for 8 h. One half was ground, and these samples were subjected to deep etching again in the same conditions. It was found that bifilms were dissolved after 5 min of deep etching in HNO_3, H_2SO_4, NaCl, and NaOH solutions.

Keywords Deep etch · Bifilm · Liquid metal quality · Porosity

Introduction

Al–Si alloys are commonly used in many fields of industry, primarily automotive, due to their good casting characteristics, heat treatment suitability, high mechanical properties, low density, and high corrosion resistance [1]. The deep etch method can be used for a variety of purposes. The main aim of this method is to reveal the structure by removing the metal from the surface. Various acids and bases can be applied in different compositions for metal removal from the surface. Nafisi [2] applied %10 HF to examine the silicon modification by adding Ti, B, and Sr. Louvis [3] claimed that oxide layers are formed when 40 wt% NaOH is used for the A6061 alloy. Alexander [4] applied this method to determine the degree of conversion of the Al_6(Fe, Mn) phase to a cubic α-Al-(Fe, Mn)-Si. Makhlouf [5] tried this method to describe the mechanisms and crystallography of the Al–Si eutectic reaction. Ghomashchi [6]

F. Tezer · Ö. Gürsoy · E. Erzi · M. Zorağa · D. Dışpınar (✉)
Engineering Faculty, Metallurgical and Materials Engineering Department, Istanbul University Cerrahpasa, 34320 Avcilar, Istanbul, Turkey
e-mail: deryad@istanbul.edu.tr

M. Tiryakioğlu et al. (eds.), *Shape Casting*, The Minerals, Metals & Materials Series,
https://doi.org/10.1007/978-3-030-06034-3_7

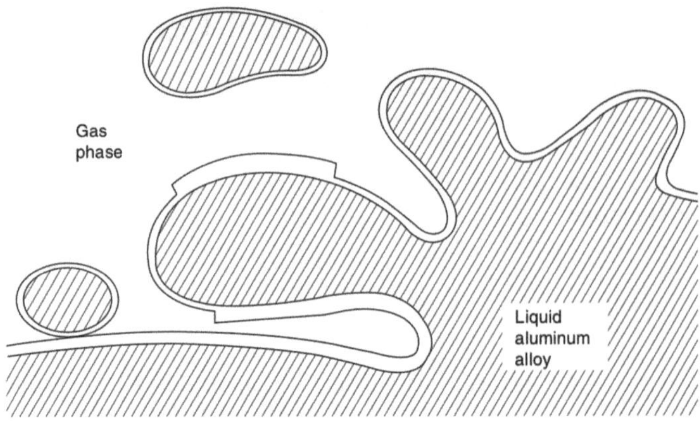

Fig. 1 Formation mechanism of bifilms [8]

applied deep etch using %10 NaOH and studied AlTiSi intermetallics. In another study, Nafisi [7] investigated the silicon modification by adding Sr and using %10 HF. Nevertheless, it has not been seen before in the literature that the deep etch method is used to examine bifilms. If the oxide formed on the surface is dragged into the liquid aluminum by turbulence as shown in Fig. 1, the dry and wet faces of the oxide collide with each other and fold over and catch the air inside.

This defect acts as a crack in the liquid, and a ceramic–ceramic unbounded interface occurs and becomes harmful for the casting piece. This defect is called a double oxide film: bifilm. Stationary conditions during solidification permit the folded bifilms to reshape and open in the form of a straight crack.

In this way, bifilms become more harmful. Bifilms could not be detected for a long time because they were very thin. This problem was solved by RPT test [9]. When a 100 g of sample is solidified under 100 mbar vacuum, bifilms are opened up. Campbell [10] describes this mechanism as folding/unfolding and furling/unfurling. The cross section of the RPT sample is subjected to image analysis to measure the size, number, and dimension of bifilms which is called bifilm index [11]. Bifilms can be remained in liquid aluminum for a long time. This also affects the mechanical properties negatively. Thus, the hydrogen gas content and bifilms must be considered to increase the metal quality. Bifilm index can be calculated as follows:

$$\text{Bifilm index } = \sum (porelength) = \text{L}_\text{b} \tag{1}$$

Dispinar [12] defines the bifilm index range as follows:
0–25 mm: best quality
25–50 mm: good quality
50–100 mm: average quality
100–150 mm: bad quality
>150 mm: don’t cast!

Experimental Method

Firstly, the tensile bar sample obtained from the AlSi10 Mg alloy was divided into seven pieces. And these seven samples were ground with sandpaper. Four samples were deeply etched in 2-M HNO_3 solution at room temperature for durations of 1, 6, 24, and 48 h. The other three samples were deeply etched in a 2-M HNO_3 solution at 50 °C for 1, 4, and 6 h. After that, the specimens were cleaned with acetone for 5 min to examine in SEM.

In the second experiment, 1-M HNO_3, 1-M H_2SO_4, and 1-M NaOH solutions were selected. Al12Si alloy was used, and two samples were placed at 50 °C in each solution. These samples were deeply etched for 1 and 2 h.

In the third experiment, 1-M HNO_3, 1-M H_2SO_4, and 1-M NaCl were selected. Nine samples from AlSi12 alloy were used. Three samples were placed in each solution at 50 °C. These samples were deeply etched for 5, 15, and 30 min. Also, six samples of aluminum alloy dross containing 6% Cu and 0.5% Sn were deeply etched for 15 and 30 min with two samples in each solution.

In the fourth experiment, four samples obtained from the tensile bar were used. Half of the surfaces of these samples were ground with 600 grit paper. For the other half, the old oxide film layer was formed on the surface by keeping it at 500 °C for 8 h in electric furnace. In this way, it was aimed to understand how the two surfaces reacted with the acid. The samples were deeply etched for 5 and 10 min in 1-M H_2SO_4 and 1-M HNO_3 solutions at 50 °C.

Results and Discussion

Figures 2 and 3 show the HNO_3 deeply etched samples at 25 and 50 °C. As can be seen, aluminum matrix was completely dissolved and only the Si phase became apparent.

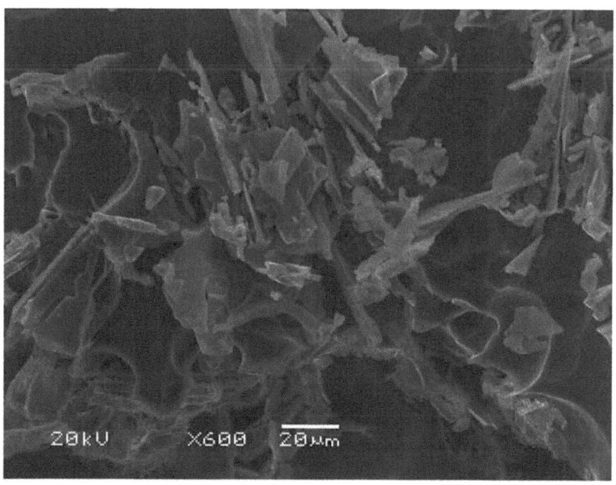

Fig. 2 SEM image of 2-M HNO_3 at 25 °C, 48 h

Fig. 3 SEM image of 2-M HNO_3 at 50 °C, 4 h

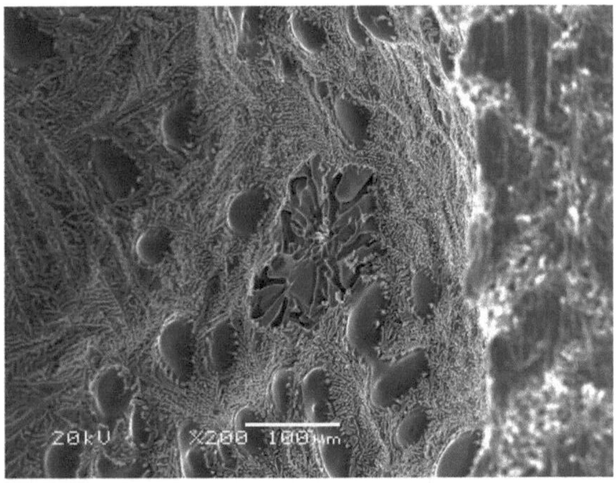

Fig. 4 SEM image of 1-M H_2SO_4 at 50 °C, 2 h

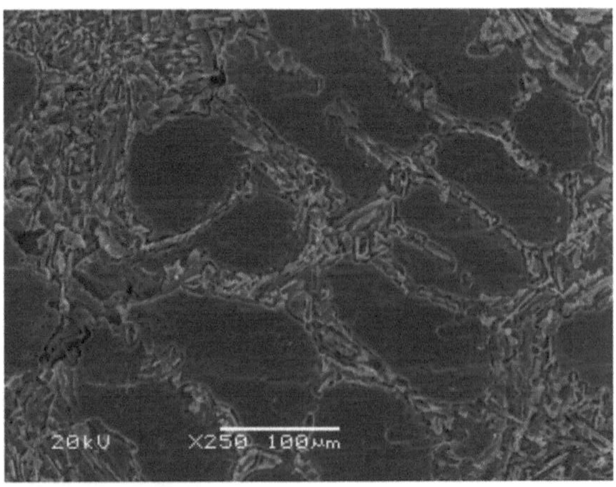

Fig. 5 SEM image of 1-M HNO_3 at 50 °C, 1 h

Part of the aluminum and oxide layer has been dissolving from the surface; consequently, the eutectic Si particles are clearly visible in Fig. 3.

Chinese script Fe phase has been dissolved in 1-M sulfuric acid as shown in Fig. 4, and the eutectic Si particles became apparent. O'Reilly [13, 14] had studied the characterization of Fe intermetallics.

As shown in Fig. 5, the aluminum in the dendrites has been dissolved and the eutectic Si particles are revealed after 1-h etching.

The modification of silicon is shown in Fig. 6. Again, the acid used in the deep etch method had dissolved the aluminum matrix but silicon phases remained undissolved.

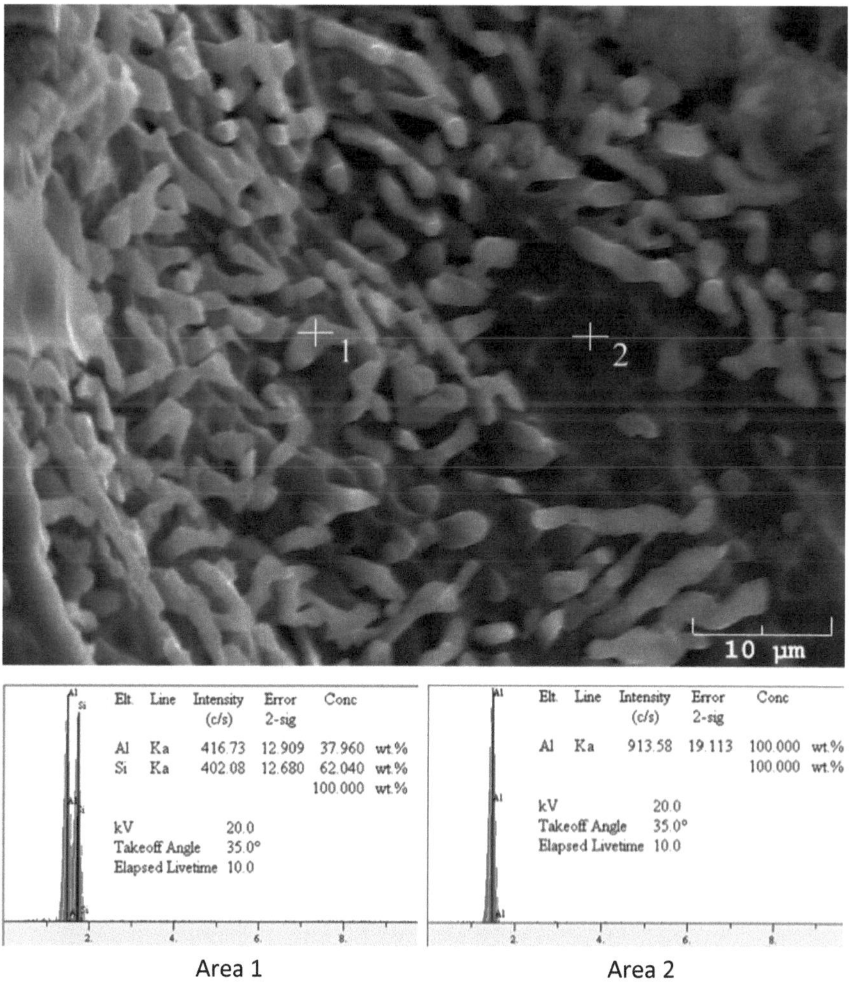

Fig. 6 Second test EDS image in 1-M H_2SO_4, 50 °C, 2 h × 2000

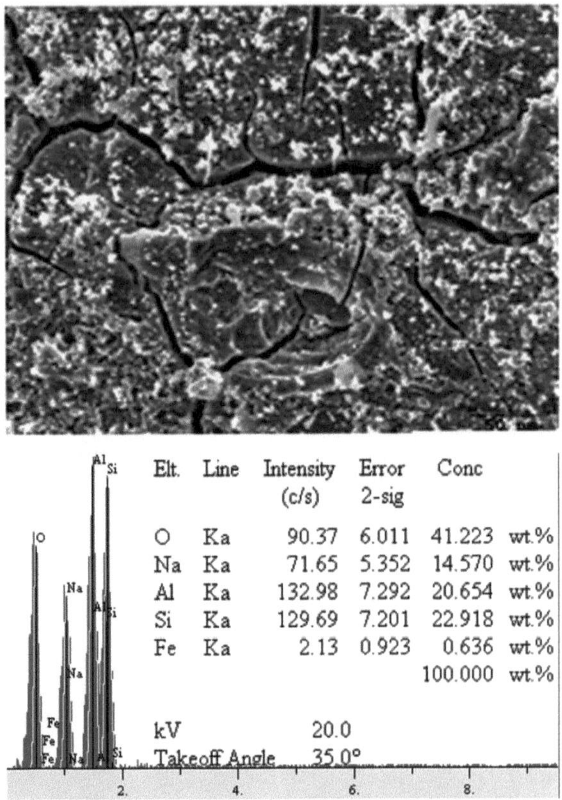

Fig. 7 EDS image of 1-M NaOH at 50 °C, 2 h × 350

When base solution was used (NaOH), it can be seen from Fig. 7 that the sodium aluminate compound has accumulated on the aluminum surface. Bifilms could not be detected for this reason.

The silicone modification was observed in samples deeply etched with nitric acid for 30 min in Fig. 8. Different responses within the porosity and on the surface are quite interesting. Inside the pores, there was large fraction of eutectic phase, and on the surface outside the pores, the eutectic ratio was lower.

When deep etch is applied with 1-M NaCl, different phases were formed owing to corrosive effect on dendrites in Fig. 9.

It was found that the oxide layer was cracked into pieces as shown in Fig. 10 after 30 min of holding in HNO_3. This situation resembled the effect of fluxes on oxides.

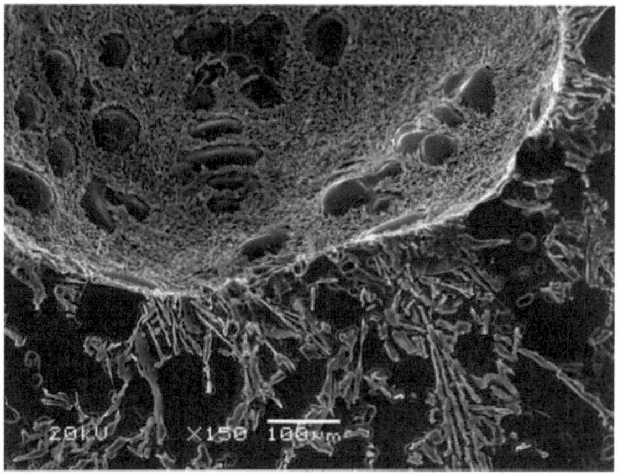

Fig. 8 SEM image of 1-M HNO_3 at 50 °C, 30 min

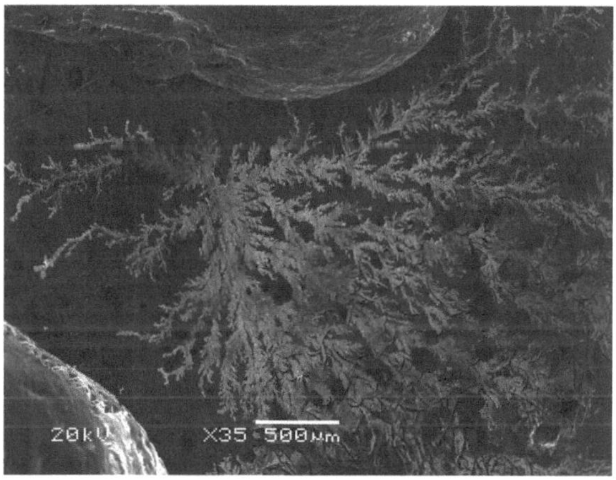

Fig. 9 SEM image of 1-M NaCl at 50 °C, 30 min

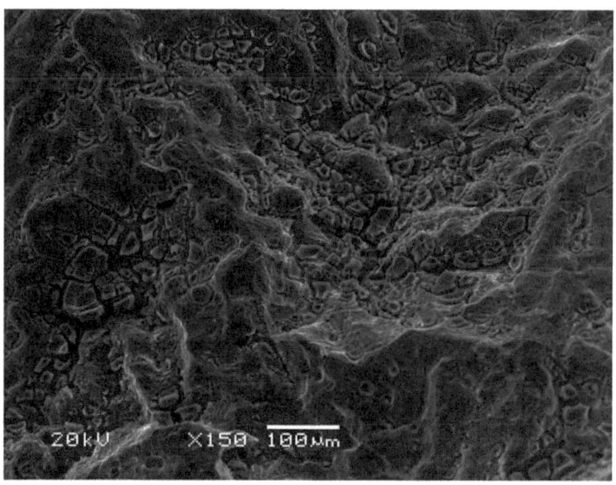

Fig. 10 Dross sample of 1-M HNO_3 at 50 °C, 30 min

Fig. 11 SEM image of 1-M H_2SO_4 at 50 °C, 10 min

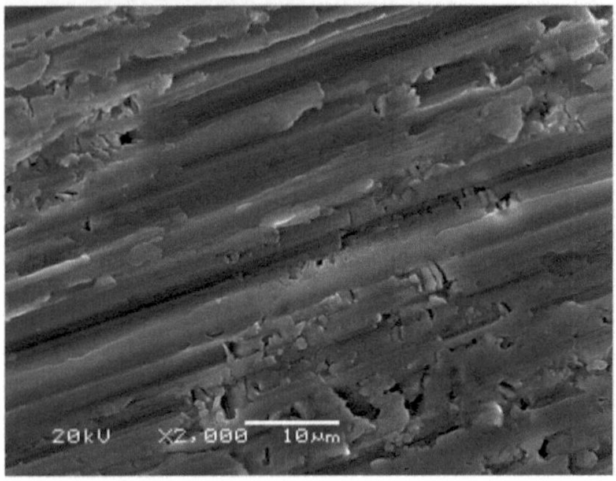

(A) H_2SO_4 , 50°C, 10 min, ground zone

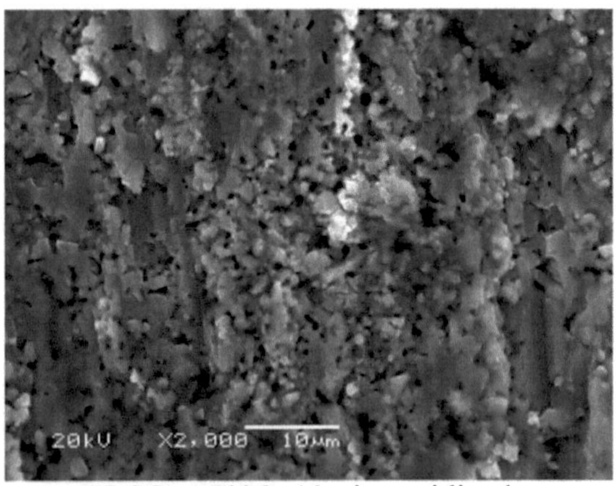

(B) H_2SO_4 , 50°C, 10min, oxidized zone

Fig. 12 SEM image of 1-M HNO_3 at 50 °C, 10 min

(A) HNO_3, 50°C, 10 min, ground zone

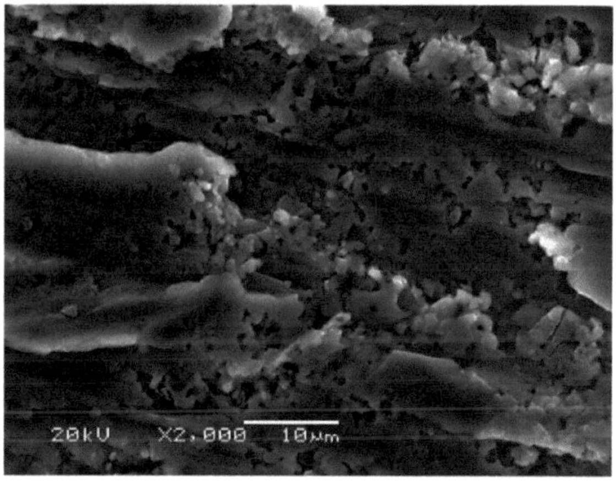

(B) HNO_3, 50°C, 10 min, oxidized zone

The sample surfaces were deliberately oxidized by holding the samples at 500 ° C for 8 h. Then, one half was ground in order to check the etching effect on different oxide structures. In Fig. 11, it was detected that the ground zone is less dissolved than the oxidized zone.

In Fig. 12, it is detected that the ground zone is less dissolved than the oxidized zone which is similar to Fig. 11.

Fig. 13 SEM image of 1-M H_2SO_4 at 50 °C, 5 min

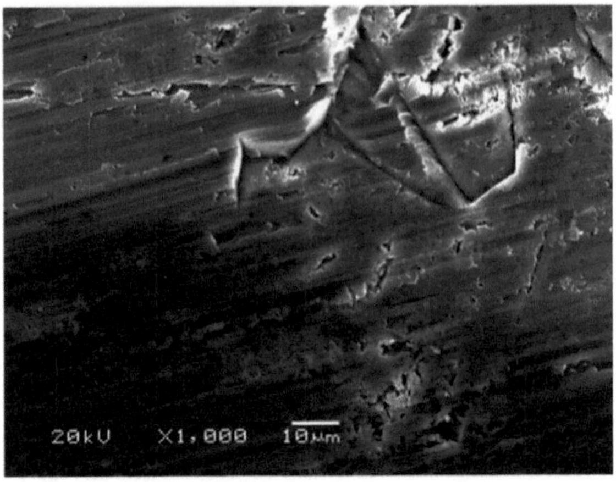

(A) H_2SO_4 , 50°C, 5 min, ground zone

(B) H_2SO_4 , 50°C, 5 min, oxidized zone

In Figs. 13 and 14, it can be seen that similar event was observed as with Figs. 11 and 12. The ground zone is more determined than oxidized zone in terms of 1-M HNO_3 and 1 M H_2SO_4 resistance for 5 min.

Fig. 14 SEM image of 1-M HNO_3 at 50 °C 5 min

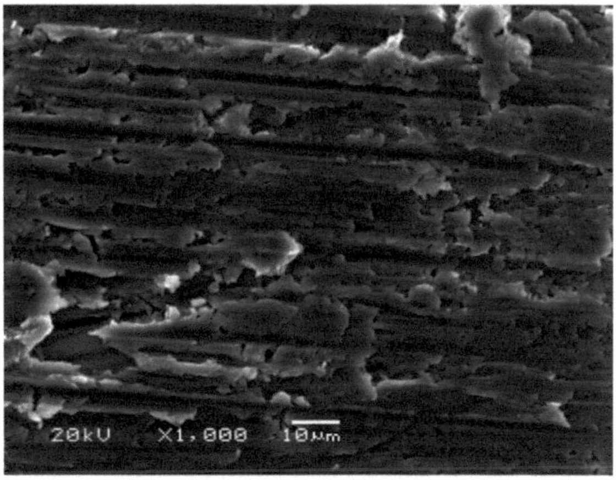

(A) HNO_3, 50°C, 5 min, ground zone

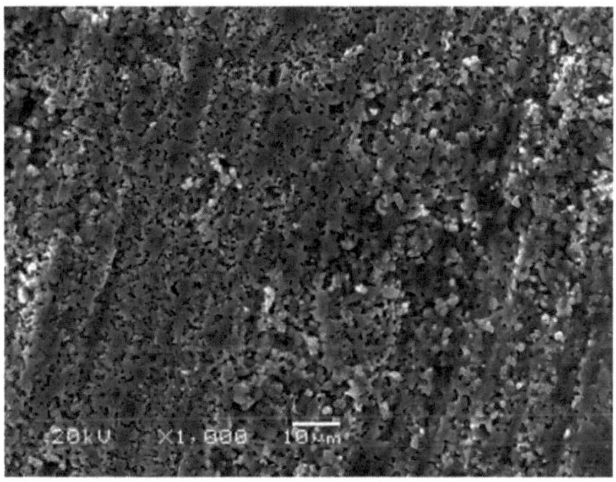

(B) HNO_3, 50°C, 5 min, oxidized zone

Conclusion

According to SEM images, bifilms were dissolved at 50 °C and room temperature in acid solutions owing to high deep etch times and concentration regardless of the duration of the tests.

When the concentration of the solution and the time of the deep etch was decreased, no bifilms were seen. The cause of this is the deep etch time (1 h) was observed to be too long. Also, when examined at the SEM images, the inner part of the porosity is controlled by the eutectic silicon. On the surface, it can be observed clearly that dendrites are dominant.

It was observed that different phases have been occurring owing to the corrosive effect on the surface of the dendrites in the deeply etched sample with 1-M NaCl solution in 30 min. The oxide layer was completely dissolved from the surface. On the other hand, in the grinded surface, less oxide was dissolved. According to the parameters used in this study, it has been observed that oxides and bifilms have been dissolving continuously when various concentrations of HNO_3, H_2SO_4, NaCl, and NaOH were used. Therefore, less deep etch durations are suggested for revealing bifilms such as less than 5 min.

It is believed that this method can be applied for the characterization of silicon modification.

References

1. Gürsoy Ö et al (2015) The time dependent effect of adding Al-5Ti-1 B grain refiner on A356 aluminium casting alloy. Paper presented at the 7th Aluminium Symposium, İstanbul, Turkey
2. Nafisi S, Ghomashchi R (2006) Combined grain refining and modification of conventional and rheo-cast A356 Al–Si alloy. Mater Charact 57(4–5):371–385
3. Louvis E, Fox P, Sutcliffe CJ (2011) Selective laser melting of aluminium components. J Mater Process Technol 211(2):275–284
4. Alexander DTL, Greer AL (2002) Solid-state intermetallic phase tranformations in 3XXX aluminium alloys. Acta Materialia 50(10):2571–2583
5. Makhlouf MM, Guthy HV (2001) The aluminum–silicon eutectic reaction: mechanisms and crystallography. J Light Metals 1(4):199–218
6. Ghomashchi R (2012) The evolution of Al–Ti–Si intermetallic phases in Ti-added A356 Al–Si alloy. J Alloy Compd 537:255–260
7. Nafisi S, Ghomashchi R (2006) Effects of modification during conventional and semi-solid metal processing of A356 Al–Si alloy. Mater Sci Eng A 415:273–285
8. Dispinar D, Campbell J (2004) Critical assessment of reduced pressure test. Part 1: porosity phenomena. Int J Cast Metals Res 17(5):280–286
9. Tan E et al (2011) Reproducibility of reduced pressure test results in testing of liquid aluminum gas levels. Paper presented at the 6th international advanced technologies symposium (IATS'11), Elazığ, Turkey, 16–18 May 2011
10. Dispinar D, Akhtar S, Nordmarka A, Di Sabatino M, Arnberg L (2010) Degassing, hydrogen and porosity phenomena in A356. Mater Sci Eng A 527:3719–3725
11. Dispinar D, Campbell J (2004) Critical assessment of reduced pressure test. Part 2: quantification. Int J Cast Metals Res 17(5):287–294
12. Dispinar D, Kvithyld A, Nordmark A (2012) Quality comparison between molten metal from remelted sheets: mill finish and coated. In: 141st TMS annual meeting, pp 1031–1035, USA, 11–15 March 2012
13. Kumar S, Grant PS, O'Reilly KAQ (2016) Evolution of Fe bearing intermetallics during DC casting and homogenization of an Al–Mg–Si Al alloy. Metall Mater Trans A 47(6):3000–3014
14. Feng S, Liotti E, Lui A, Kumar S, Mahadevegowda A, O'Reilly KAQ, Grant PS (2018) An in-situ method to estimate the tip temperature and phase selection of secondary Fe-rich intermetallics using synchrotron X-ray radiography. Scripta Mater 149:44–48

Effect of Fe-Rich Intermetallics on Tensile Behavior of Al–Cu 206 Cast Alloys at Solid and Near-Solid States

K. Liu, X. Cao, A. Bolouri and X.-G. Chen

Abstract Iron is one of the most common impurity elements in Al–Cu 206 cast alloys as it often causes the precipitation of Fe-rich intermetallic phases during solidification due to its extremely low solid solubility in aluminum. The characteristics of the Fe-rich intermetallics, such as type, morphology, size, and distribution, have significant influences on the tensile behaviors of the Al alloys. In the present work, two Al–Cu 206 cast alloys containing different types of Fe-rich intermetallics (dominated by either platelet β-Fe or Chinese script α-Fe) were cast and their tensile tests were performed at both solid (room temperature) and near-solid (2.8 vol. % liquid) states. It is found that the tensile properties in both solid and near-solid states are improved when the Fe-rich intermetallics change from platelet to Chinese script morphologies. During the solid-state tensile deformation, the failure occurs mainly along the platelet β-Fe intermetallics/Al matrix interface or within the Chinese script α-Fe particles. In the near-solid state, the alloy containing mainly Chinese script α-Fe is found to have more free flow paths for liquid feeding, leading to improved tensile properties. By contrast, the platelet β-Fe can cause the blockage of the liquid flow paths, leading to the degraded tensile properties and worsened susceptibility to hot-tearing.

Keywords Al–Cu alloy · Fe-rich intermetallics · Tensile properties · Solid and near-solid state · Hot-tearing

K. Liu (✉) · X. Cao · X.-G. Chen
Department of Applied Science, University of Quebec at Chicoutimi, Chicoutimi, QC G7H 2B1, Canada
e-mail: kun.liu@uqac.ca

X. Cao
Aerospace Manufacturing Technology Center, National Research Council Canada, Montreal, QC H3T 2B2, Canada

A. Bolouri
Department of Engineering Design and Mathematics, University of the West of England, Bristol BS16 1QY, UK

M. Tiryakioğlu et al. (eds.), *Shape Casting*, The Minerals, Metals & Materials Series,
https://doi.org/10.1007/978-3-030-06034-3_8

Introduction

Al–Cu 206 cast alloys have been widely used in modern aerospace and automobile industries due to their excellent mechanical properties including elevated temperature tensile strength. In these alloys, however, iron is considered to be one of the most common impurities and thus an extremely low upper limit for Fe is usually allowed, e.g., 0.15 wt% for 206.0 or even 0.07 wt% for A206.0 alloys [1] (all alloy compositions are in weight percent unless indicated otherwise). This has significantly increased the manufacturing costs and severely limited their wide applications. To this end, it is important to understand the role of Fe to increase the allowable upper limit of Fe in these alloys.

Due to the low solubility of Fe in solid aluminum, most Fe will precipitate as the Fe-rich intermetallics during the solidification. Recently, various Fe-rich intermetallics have been reported in Al–Cu 206 cast alloys at high Fe contents [1–4]. These Fe-rich phases include platelet β-Al_7Cu_2Fe and Al_3Fe as well as Chinese script α-$Al_{15}(FeMn)_3(SiCu)_2$, Al_mFe, and Al_6Fe over the Fe levels of 0.3–0.5% [1, 2, 4]. It is well accepted that the platelet Fe-rich intermetallics are detrimental to the mechanical properties of Al cast alloys due to the fact that they can act as the stress risers and crack initiators [5]. Therefore, they are usually required to be transferred into the less-harmful Chinese script Fe-rich intermetallics through controlling the chemical composition and solidification condition [1, 5–7]. In addition, Al–Cu 206 cast alloys are susceptible to the hot-tearing defect during solidification, which is closely related to the tensile stress arising from the thermal gradient and contraction during solidification [8, 9]. It is reported that the Fe-rich intermetallics can significantly influence the hot-tearing susceptibility [10]. For example, Chinese script α-Fe can improve the hot-tearing resistance, but the reason is not clear. Recently, the near-solid tensile testing method has been well developed to simulate the stress–strain conditions during solidification of aluminum alloys [11, 12], providing a practical approach to investigate the influence of Fe-rich intermetallics on the hot-tearing susceptibility.

In the present work, two Al–Cu 206 cast alloys with a high Fe content (0.3%) and different Mn and Si concentrations were specially designed to obtain dominantly either platelet β-Fe or Chinese script α-Fe intermetallic phases. The tensile tests were then performed at solid state (room temperature) and near-solid state (with a liquid volume of 2.8 vol. %). The microstructures and the fracture surfaces of the failed tensile samples were fully analyzed to disclose the role of Fe-rich intermetallics during tensile deformation at both the solid and near-solid states.

Experimental

As indicated in Table 1, two Al–Cu 206 cast alloys with different dominant Fe-rich intermetallic phases were prepared in the present work. Alloy 311 was designed with dominant platelet β-Fe (Al_7Cu_2Fe), while Alloy 333 had mainly Chinese script α-Fe ($Al_{15}(FeMn)_3(SiCu)_2$). The casting parameters and process can be found in the literature [1].

For the room temperature (RT) solid-state tensile testing specimens, the traditional T7 temper (solution treatment: 505 °C for 2 h + 520 °C for 8 h and artificial aging: 185 °C for 5 h) was applied. After the T7 treatment, these samples were machined to the shape and dimensions as indicated in Fig. 1a. The tensile testing was then performed at room temperature at a strain rate of 10^{-3} s^{-1} using an INSTRON 8801 testing system. During the tensile testing, an extensometer with a gauge length of 25 mm was attached to the tensile sample to record the displacement. The tensile properties, namely ultimate tensile strength (UTS), yield strength (YS) at 0.2% offset strain, and elongation (El), were obtained from the averaged value of 3 tensile tests.

The near-solid tensile testing was performed on a Gleeble 3800 thermomechanical testing unit using the samples machined from as-cast condition with the rod shape and dimensions shown in Fig. 1b. The testing temperatures for Alloys 311 and 333 with 2.8 vol. % liquid are 551 °C and 541 °C, respectively (Table 1). The temperature was monitored by using three K-type thermocouples which were spot-welded on the surface at the middle and at two sides with a distance of ±8 mm from the middle (Fig. 1b). A ceramic tube was adopted to homogenize the temperature in the hot middle zone of the test sample. The tensile testing was performed at a strain rate of 10^{-3} s^{-1}, and a minimum of three samples were tested. Details about the near-solid tensile testing can be found in [11].

The microstructures in the as-cast and tempered conditions as well as the fracture surfaces of the failed tensile specimens were analyzed using a scanning electron microscope (SEM, JSM-6480LV). The volume percent of the Fe-rich intermetallics was measured using an optical microscope (OM, Nikon Eclipse ME 600) equipped with an image analyzer.

Table 1 Chemical composition of the experimental alloys

Alloys#	Elements (wt%)						Temperatures at 2.8 vol.% liquid[a]
	Cu	Mg	Fe	Mn	Si	Al	
311	4.66	0.27	0.30	0.12	0.19	Bal.	551 °C
333	4.62	0.28	0.29	0.29	0.33	Bal.	541 °C

[a]The temperature at a liquid volume of 2.8% was obtained from the differential scanning calorimeter (DSC) curves. More experimental details can be found in the literature [11]

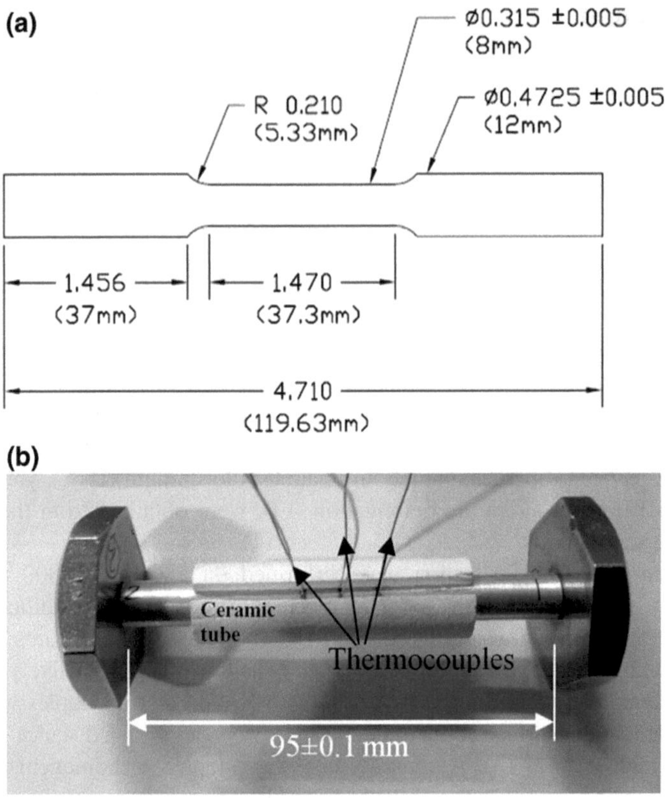

Fig. 1 Shape and dimensions of the tensile samples tested at RT (solid state) (**a**) and near-solid state (**b**) (units in inches and millimeters)

Results and Discussion

Fe-Rich Intermetallics in as-Cast and T7 Heat-Treated Conditions

Figure 2 shows the microstructures of the two experimental alloys in the as-cast and T7 heat-treated conditions. In addition to Al_2Cu phase, the two as-cast alloys contain different dominant Fe-rich intermetallics, i.e., platelet-like β-Fe (Al_7Cu_2Fe) for Alloy 311 (Fig. 2a) and Chinese script α-Fe ($Al_{15}(FeMn)_3(SiCu)_2$) for Alloy 333 (Fig. 2b), which can be attributed to the higher Mn and Si contents in Alloy 333 than those in Alloy 311 [1, 3]. The area percentage of the dominant Fe-rich intermetallics measured from the test bar surface is 1.1 vol. % for β-Fe in Alloy 311 and 1.4 vol. % for α-Fe in Alloy 333, respectively. Besides, it is also found that most platelet β-Fe is interlocked with Al_2Cu phase, while Chinese script α-Fe is distributed in the matrix with little mutual mixture between α-Fe and Al_2Cu. This

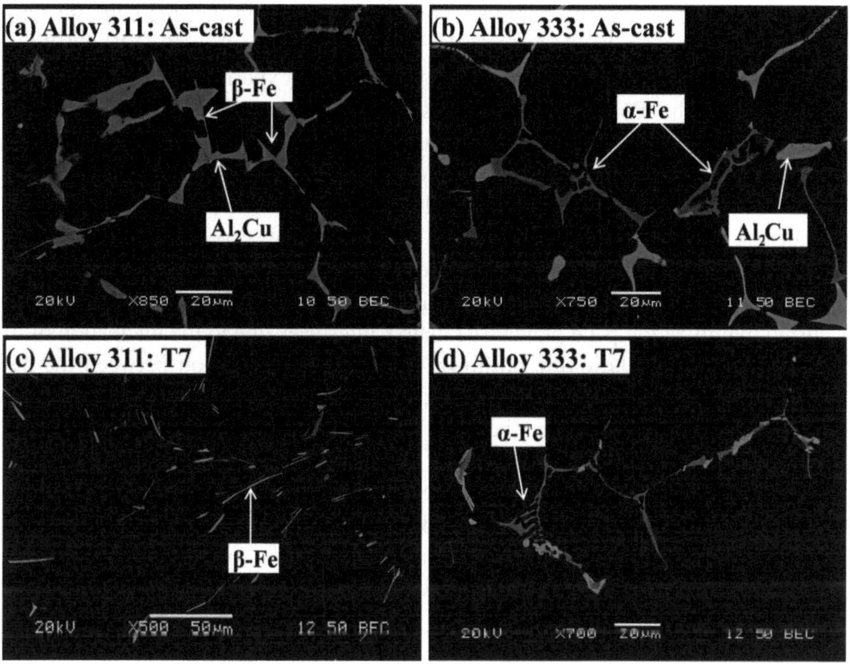

Fig. 2 Microstructure of the two experimental alloys in as-cast (**a**, **b**) and T7 (**c**, **d**) conditions

can be attributed to their different precipitation temperatures and reactions. It is reported [1] that the precipitation temperature of β-Fe (~570 °C) is slightly higher than that of Al_2Cu (~540 °C), explaining the mutual interlock between the two phases. In contrast, the precipitation temperature of α-Fe (~625 °C) is much higher than those of β-Fe (~570 °C) and Al_2Cu (~540 °C). Therefore, during solidification, the α-Fe is formed prior to the β-Fe and Al_2Cu phases and can randomly distribute within the matrix including the interdendritic regions.

In the T7 heat-treated condition, almost all the Al_2Cu phases were dissolved into the matrix, but little change occurred for both Fe-rich intermetallics. Similarly to the as-cast condition, only platelet β-Fe and Chinese script α-Fe phases are present in tempered Alloy 311 (Fig. 2c) and Alloy 333 (Fig. 2d), respectively.

Solid-State Tensile Behavior

Figure 3 shows the typical engineering stress–strain curves of the two experimental alloys obtained at RT (solid state). The UTS, YS, and El for Alloy 333 are 470 MPa, 426 MPa, and 3.6% compared with 425 MPa, 403 MPa, and 1.2% for Alloy 311, respectively. Therefore, all tensile properties (UTS, YS, El) are improved for Alloy 333 compared to Alloy 311, indicating that the Chinese script α-Fe has a less harmful influence on the tensile properties than the platelet β-Fe.

Figure 4 shows the fracture surfaces of the two experimental alloys and the distribution of the Fe-rich intermetallics on the mating fracture surfaces. More brittle fracture surface is observed for Alloy 311 (Fig. 4a), while mixed (both brittle and ductile) fracture is characterized for Alloy 333 (Fig. 4b). Figure 4c–f shows the distribution of Fe-rich intermetallics on the mating fracture surfaces. The two β-Fe platelets (A in Fig. 4c and B' in Fig. 4d) are clearly observed on one side of the fracture surfaces, but only the matrix is exposed on the corresponding side of the mating surfaces (A' in Fig. 4d and B in Fig. 4c), indicating that the tensile cracking occurred mainly at the β-Fe platelet/Al matrix interfaces due to the nucleation of the β-Fe on one side of the oxide bifilm or debonding/decohension between the particles and matrix. Therefore, the tensile cracking can easily initiate and propagate along the longitudinal interfaces between the β-Fe plates and Al matrix, leading to the brittle fracture as shown in Fig. 4a.

By contrast, the distribution of Chinese script α-Fe on the fracture surface in Alloy 333 behaves differently from the β-Fe. As shown in Fig. 4e and f, the α-Fe particles are observed on the corresponding locations of both mating fracture surfaces, indicating that the tensile cracking appears mainly within the particles

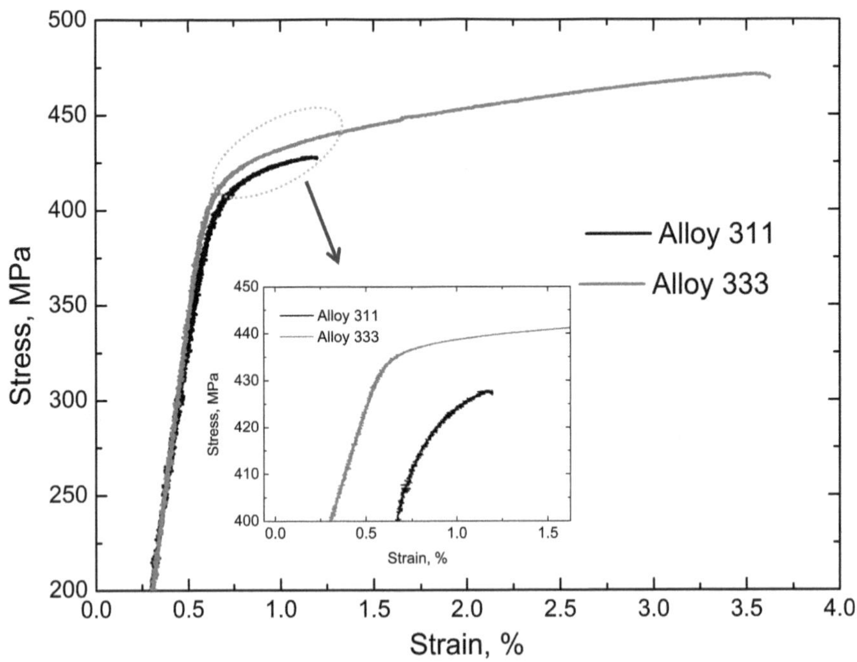

Fig. 3 Typical engineering stress–strain tensile curves for Alloys 311 and 333 tested at RT (solid state)

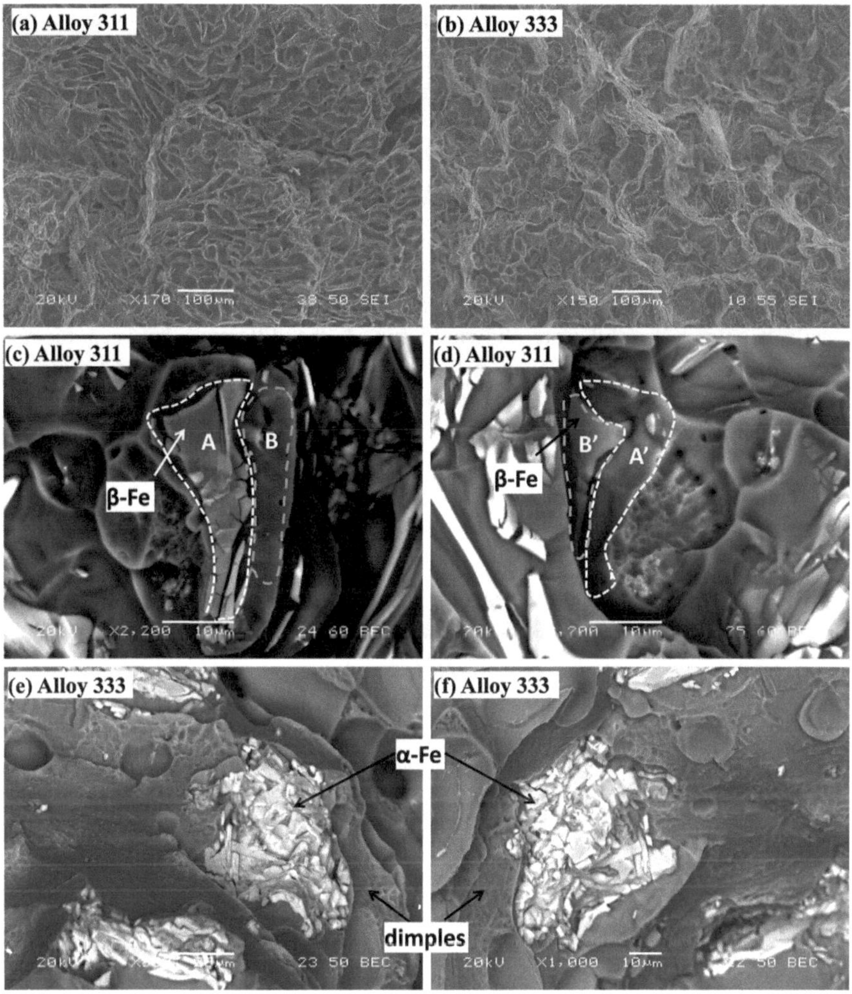

Fig. 4 Fracture surfaces of the experimental alloys (**a**, **b**) and distribution of Fe-rich intermetallics on the mating fractures (**c**–**f**)

themselves. Meanwhile, some small dimples are observed in the neighborhood of the α-Fe particles (Fig. 4e and f), indicating the ductile nature of the Al matrix fracture. As mentioned in the literature [1], the α-Fe precipitates are at relatively high temperature during solidification and thus can grow from their parent nucleus into well-developed branches/dendrites in different directions. These dendrites are mutually interwoven with the Al matrix, making the crack difficult to propagate only along the particle/matrix interfaces. Fracture within the α-Fe particles leads to the premature failure, but microvoids cannot easily develop around these large intermetallic particles. By contrast, the Al matrix can be deformed under a tensile stress, leading to the formation of some fine dimples (Fig. 4e and f). Therefore, both brittle (for intermetallics) and ductile (for matrix) fractures appear in Alloy 333 (Fig. 4b).

Near-Solid Tensile Behavior at 2.8 Vol. % Liquid

The typical stress–displacement curves for Alloys 311 and 333 tested at the near-solid state with 2.8 vol. % liquid are shown in Fig. 5. Compared with the tensile behavior at solid state shown in Fig. 2, the difference between the two experimental alloys appears even at the very beginning of the tensile testing. As shown in Fig. 5a, the tensile stress increases more slowly with the displacement for Alloy 333 than Alloy 311 as indicated by the lower slope of the stress–displacement

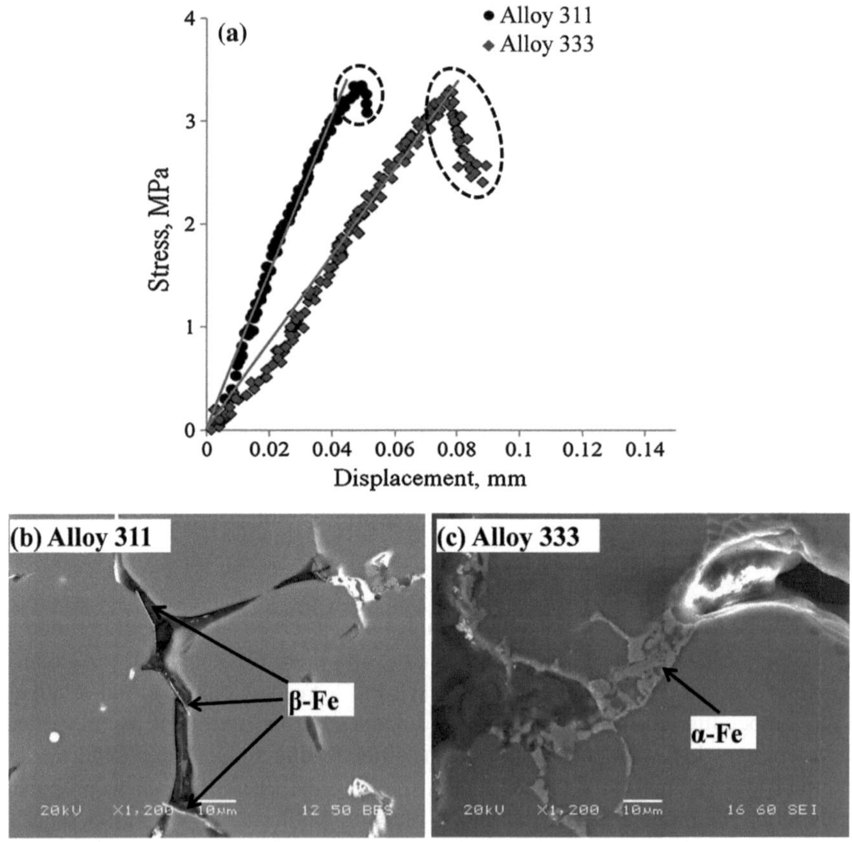

Fig. 5 Near-solid stress–displacement curves (**a**) and Fe-rich intermetallics along the tensile cracks close to the "near-fracture" locations for Alloy 311 (**b**) and Alloy 333 (**c**)

curve for Alloy 333. In other words, Alloy 333 can tolerate larger displacement than Alloy 311 when the same stress was applied. Since the initiation and propagation of hot cracks during casting generally occur before the maximum stress [13], it is reasonable to assume that expansion of damage in the mush zone takes place more slowly for Alloy 333 than that for Alloy 311, suggesting the lower hot-tearing susceptibility for Alloy 333. This observation is in good accordance with the fact that Chinese script α-Fe can improve the hot-tearing resistance compared to the platelet β-Fe in 206 cast alloys [10].

It is also interesting to note that more displacement to fracture appeared for Alloy 333 than that for Alloy 311 after the peak stress (marked as black circles in Fig. 5a). Over this stage, after the peak stress but prior to the full failure, additional interrupted tensile testing was performed and metallographic specimens adjacent to the "near-fracture" location were taken to disclose the distributions of intermetallic particles along tensile cracks (Fig. 5b and c for Alloys 311 and 333, respectively). It was found that the cracking paths are dominantly along the β-Fe platelets (Fig. 5b), leading to an easier premature failure (i.e., shorter displacement) for Alloy 311. By contrast, the propagation of the crack can be blocked by the Chinese script α-Fe particles (Fig. 5c), leading to delayed failure and longer displacement for Alloy 333. Therefore, Alloy 333 experienced longer displacement before the final fracture after the peak stress, as indicated in Fig. 5a. However, it is found that the peak tensile stresses are similar for both experimental alloys (3.3 MPa for Alloy 311 and 3.4 MPa for Alloy 333), which can be attributed to the same large amount of liquid (2.8 vol. %) during the tensile deformation. In this case, the liquid can completely wet the solid grains, separating them with liquid films, and thus the tensile stress is mainly controlled by the liquid phase [14].

Furthermore, the total displacement to the failure is doubled for Alloy 333 (0.09 mm), as compared with Alloy 311 (0.045 mm), which most likely resulted from the different dominant Fe-rich intermetallics. It was reported that the feeding of liquid, especially for the last liquid phase (<5 vol. %) within the mush structure, determines the tensile strength of the semisolid castings [14, 15]. If the liquid can easily flow within the mush structure and feed the regions which are being tensile deformed, the structure can accommodate more deformation until the fracture failure [12]. Therefore, the significant difference in displacement for the two experimental alloys is also expected to be related to the variations in the flow of liquid during deformation caused by the different intermetallic compounds present in experimental alloys. As shown in Fig. 6, both the morphology and the distribution of the Fe-rich intermetallics are found to have a significant influence on the flow of the liquid during tensile deformation. In the two experimental 206 alloys, the remained liquid phase is mainly low melting point Al–Al_2Cu eutectic in the interdendritic regions [1]. As expected, the re-solidified Al_2Cu after the near-solid tensile testing can be well observed on the fracture surfaces (Fig. 6). For Alloy 311, the re-solidified Al_2Cu is present mostly in the area near or even between the β-Fe platelets (red circle in Fig. 6a). This observation has been further supported by the phenomenon that platelet β-Fe tends to be always interlocked with Al_2Cu due to their similar precipitation temperatures (Fig. 2a). Therefore, the β-Fe multi-platelets present in a

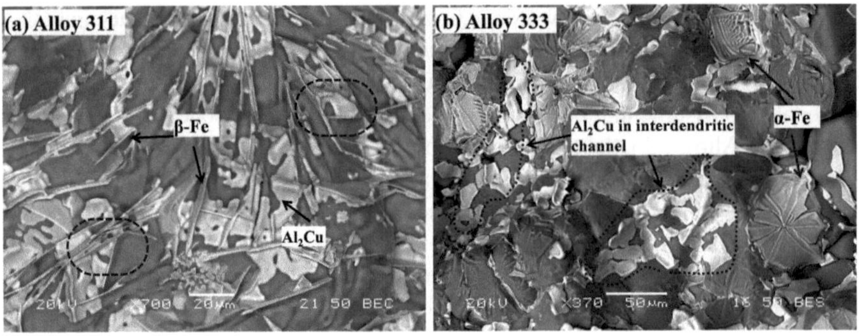

Fig. 6 Fracture surfaces after the completion of the near-solid tensile testing of the experimental alloys

bundle within the interdendritic channels can block the flow of the Al–Al_2Cu eutectic liquid and make it difficult for the remained liquid to feed the required regions during tensile deformation. In this case, the permeability of the mush structure is greatly reduced and the void/crack can easily form and propagate, explaining the faster rise and steeper slope for the stress–displacement curve and lower displacement for Alloy 311 (Fig. 5a). By contrast, the re-solidified Al_2Cu is not necessary to be coexisted with the α-Fe in the interdendritic regions for Alloy 333 (red circles in Fig. 6b), indicating the more free flow of the Al–Al_2Cu eutectic liquid during the tensile deformation. Due to the fact that the α-Fe is not necessarily located within the interdendritic channels as it precipitates at a much higher temperature than Al_2Cu [1] and its compact morphology, the α-Fe has less blocking effect on the flow of the remained Al–Al_2Cu liquid compared to the platelet-like β-Fe. Therefore, the liquid can flow more freely, feed within the mush zone structure before it is fully blocked, and thus accommodate a higher deformation before the final fracture, explaining the larger displacement and lower hot-tearing susceptibility for Alloy 333. Regarding the casting of Al–Cu 206 alloys, the increase of the liquid feeding ability during the last stage of the solidification can have beneficial effect on the crack healing process, reducing additionally the hot-tearing susceptibility.

Conclusions

- During the solid-state tensile testing at room temperature, tensile failure mainly occurs at the β-Fe/Al matrix interfaces but within the α-Fe intermetallic particles. Therefore, the cracking is easier to initiate and propagate along the longitudinal axial direction of the β-Fe platelets but more difficult within the α-Fe particles, leading to the improved strength and ductility for the alloy with dominant Chinese script α-Fe compared to that with platelet β-Fe.

- During the near-solid tensile deformation, the tensile stress increases with displacement much more slowly for the alloy dominantly with α-Fe than that with β-Fe, indicating the lower susceptibility to hot-tearing for the former.
- The liquid flow during the tensile deformation has a significant influence on the near-solid tensile properties, which is greatly affected by the morphology and distribution of the Fe-rich intermetallics. The liquid can be severely blocked by the interlocked β-Fe platelets, but it can flow more freely and feed the deformed region in the alloy with dominant α-Fe, leading to its higher displacement to fracture.

Acknowledgements The authors would like to acknowledge the financial support from the Natural Sciences and Engineering Research Council of Canada (NSERC) and Rio Tinto Aluminum, through the NSERC Industry Research Chair in Metallurgy of Aluminum Transformation at the University of Quebec at Chicoutimi, and support of National Research Council of Canada.

References

1. Liu K, Cao X, Chen XG (2011) Solidification of iron-rich intermetallic phases in Al-4.5Cu-0.3Fe cast alloy. Metall Mater Trans A 42(7):2004–2016
2. Liu K, Cao X, Chen XG (2012) Precipitation of iron-rich intermetallic phases in Al-4.6Cu-0.5Fe-0.5Mn cast alloy. J Mater Sci 47(10):4290–4298
3. Kamguo Kamga H, Larouche D, Bournane M, Rahem A (2010) Solidification of aluminum-copper B206 alloys with ion and silicon additions. Metall Mater Trans A 41 (11):2844–2855
4. Lin B, Xu R, Li H, Zhang W (2018) Formation of Fe-rich intermetallics in Al–5.0Cu–0.5 Fe alloys with different Mn additions. Mater Sci Technol 34(12):1447–1459
5. Seifeddine S, Johansson S, Svensson IL (2008) The influence of cooling rate and manganese content on the β-Al5FeSi phase formation and mechanical properties of Al–Si-based alloys. Mater Sci Eng A 490(1):385–390
6. Liu K, Cao X, Chen XG (2012) Effect of Mn, Si, and cooling rate on the formation of iron-rich intermetallics in 206 Al-Cu cast alloys. Metall Mater Trans B 43(5):1231–1240
7. Liu K, Cao X, Chen XG (2013) Formation and phase selection of iron-rich intermetallics in Al-4.6Cu-0.5Fe cast alloys. Metall Mater Trans A 44(2):682–695
8. Eskin DG, Katgerman L (2004) Mechanical properties in the semi-solid state and hot tearing of aluminium alloys. Prog Mater Sci 49(5):629–711
9. Farup I, Drezet JM, Rappaz M (2001) In situ observation of hot tearing formation in succinonitrile-acetone. Acta Mater 49(7):1261–1269
10. Kamguo Kamga H, Larouche D, Bournane M, Rahem A (2010) Hot tearing of aluminum–copper B206 alloys with iron and silicon additions. Mater Sci Eng A 527(27):7413–7423
11. Bolouri A, Liu K, Chen X-G (2016) Effects of iron-rich intermetallics and grain structure on semisolid tensile properties of Al-Cu 206 cast alloys near solidus temperature. Metall Mater Trans A 47(12):6466–6480
12. Bolouri A, Chen X-G (2017) Tensile deformation behavior of Al-Cu 206 cast alloys near the solidus temperature. Mater Sci Forum 877:90–96

13. Colley LJ, Wells MA, Maijer DM (2004) Tensile properties of as-cast aluminum alloy AA5182 close to the solidus temperature. Mater Sci Eng A 386(1):140–148
14. Fabrègue D, Deschamps A, Suery M, Drezet JM (2006) Non-isothermal tensile tests during solidification of Al–Mg–Si–Cu alloys: Mechanical properties in relation to the phenomenon of hot tearing. Acta Mater 54(19):5209–5220
15. Phillion AB, Hamilton RW, Fuloria D, Leung ACL, Rockett P, Connolley T, Lee PD (2011) In situ X-ray observation of semi-solid deformation and failure in Al–Cu alloys. Acta Mater 59(4):1436–1444

Part II
Casting Defects and Their Characterization

Determining Casting Defects in Thixomolding Mg Casting Part by Computed Tomography

Jiehua Li, Bernd Oberdorfer and Peter Schumacher

Abstract A quantitative analysis on the casting defects is of great necessity to further improve the mechanical properties of high-performance alloys, in particular dynamic fatigue properties. Computed tomography (CT) is frequently utilized to determine the casting defects of casting parts and thereby evaluate their casting qualities. In this contribution, one Mg casting part (AZ91D, Mg–9Al–1Zn) produced by thixomolding was investigated by computed tomography. No significant casting defect was observed, strongly indicating that thixomolding could produce high-quality Mg casting parts. Furthermore, newly developed reconstruction software developed in Austria (VrVia GmbH) was used to obtain more details from the same casting part. This investigation demonstrates that computed tomography is an efficient method to determine casting defects in near-net-shape casting parts.

Keywords Mg alloy · Thixomolding · Computed tomography · Casting defects

Introduction

High-performance Mg alloys (i.e. AZ91) have been widely used in automotive industry due to their lightweight. To date, high-pressure die casting is one of the most important casting technologies to produce high-quality near-net-shape Mg casting parts. However, various casting defects (i.e. porosity and inclusion) very often occur during high-pressure die casting due to its higher filling process and thereby gas entrapment. Although necessary vacuum system together with conventional high-pressure die casting can be used to reduce the formation of casting defects and improve the mechanical properties, the application of vacuum system also increases the costs in terms of facility itself and necessary maintenance.

J. Li (✉) · P. Schumacher
Institute of Casting Research, Montanuniversität Leoben, 8700 Leoben, Austria
e-mail: jiehua.li@unileoben.ac.at

B. Oberdorfer · P. Schumacher
Austrian Foundry Research Institute, 8700 Leoben, Austria

M. Tiryakioğlu et al. (eds.), *Shape Casting*, The Minerals, Metals & Materials Series,
https://doi.org/10.1007/978-3-030-06034-3_9

Compared with high-pressure die casting, thixomolding is another very promising to produce high-quality near-net-shape Mg casting parts due to its lower process temperature and smooth filling process. Furthermore, during thixomolding, α-Mg dendrites can be broken and distributed evenly within the α-Mg matrix, which is favourable for improving the performance of castings. The formation of casting defects (i.e. pore and/or shrinkage cavities) can be significantly decreased. However, appropriate technology parameters have to be provided; otherwise, the inherent merits of thixomolding could not be utilized perfectly. In order to optimize the technology parameters, it is of great necessity to determine casting defects of thixomolding Mg casting part as a feedback. In order to determine casting defects of thixomolding Mg casting part, computed tomography (CT) has very often been used [1–8]. The main aim of the present investigation is to determine casting defects and thereby evaluate the quality of one near-net-shape thixomolding Mg casting parts using CT.

Experimental

One near-net-shape thixomolding Mg casting part (AZ91D) was investigated by CT, with the aim to determine casting defects and thereby evaluate the quality. The sample was scanned on a Phoenix X-ray v|tome|x c equipped with a 240 kV cone-beam X-ray tube operated at 70 kV. For the volume reconstruction, a modified Feldkamp algorithm for the filtered back projection as implemented by the system supplier was used [1]. Porosities were thus determined.

Results

Figure 1a shows 3D image of porosities in the near-net-shape thixomolding Mg casting part (AZ91D). Clearly, no significant casting defect was observed, strongly indicating that thixomolding could produce high-quality Mg casting parts. Figure 1b and c shows some small porosities but with a low number density.

Figure 2 shows the optical microscopy image (Fig. 2a) and SEM image (Fig. 2b) of the near-net-shape thixomolding Mg casting part (AZ91D), respectively. Clearly, after thixomolding, α-Mg dendrites were broken and distributed evenly within the α-Mg matrix, which is different from the presence of α-Mg dendrites in the samples produced by high-pressure die casting.

Figure 3 shows the hardness of the near-net-shape thixomolding Mg casting part (AZ91D). For comparison, the hardness values of the samples produced by high-pressure die casting and alloying with RE are also shown. Clearly, thixomolding AZ91D shows a much higher hardness (86 compared with 57.6 and 58.3), which can be due to the microstructure change (Fig. 2) and less porosities (Fig. 1). Again, it further confirms the advantage of thixomolding.

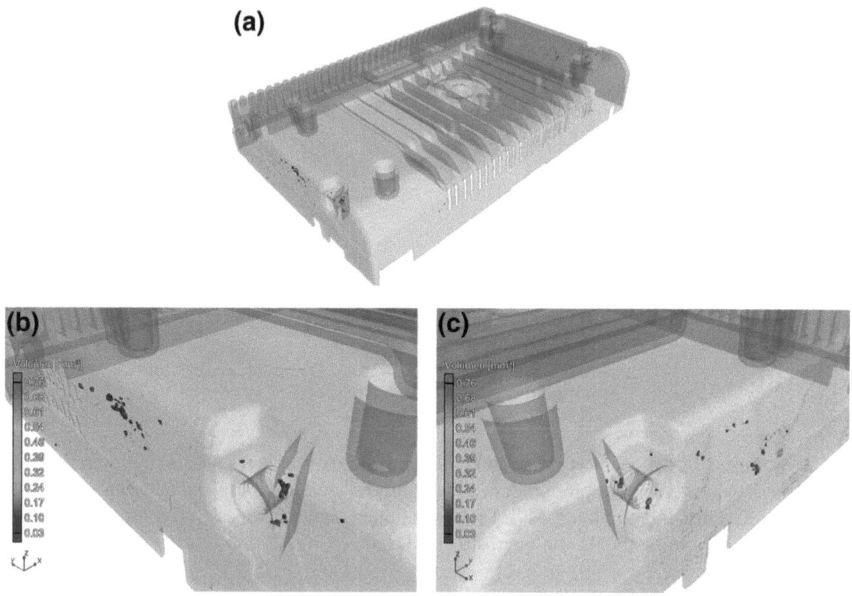

Fig. 1 **a** Three dimensional image of porosities in the near-net-shape thixomolding Mg casting part (AZ91D), and **b**, **c** show some small porosities

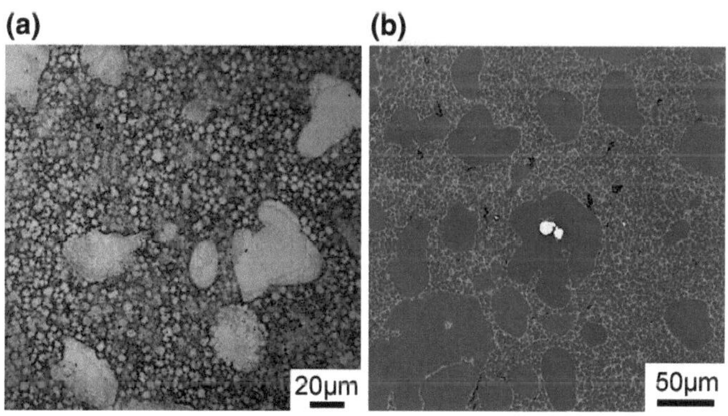

Fig. 2 Optical microscopy image (**a**) and SEM image (**b**) of the near-net-shape thixomolding Mg casting part (AZ91D)

Finally, it should also be noted that some possible artefacts may be present during reconstruction using commercial reconstruction software. For example, for the casting parts with different thicknesses, a misleading about the determination of porosities has been found [2, 3]. In order to solve this problem and other possible reconstruction problems, new reconstructed software has been developed in Austria (VrVia GmbH) under the support of FFG funding. Figure 4a shows 3D image of

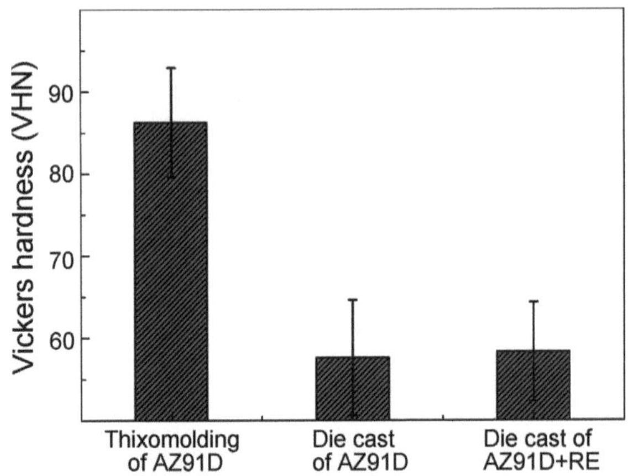

Fig. 3 Hardness of the near-net-shape thixomolding Mg casting part (AZ91D). For comparison, the hardness values of the samples produced by high-pressure die casting and alloying with RE are also shown. Clearly, thixomolding AZ91D shows a much higher hardness (86 compared with 57.6 and 58.3), which can be due to the microstructure change (Fig. 2) and less porosities (Fig. 1)

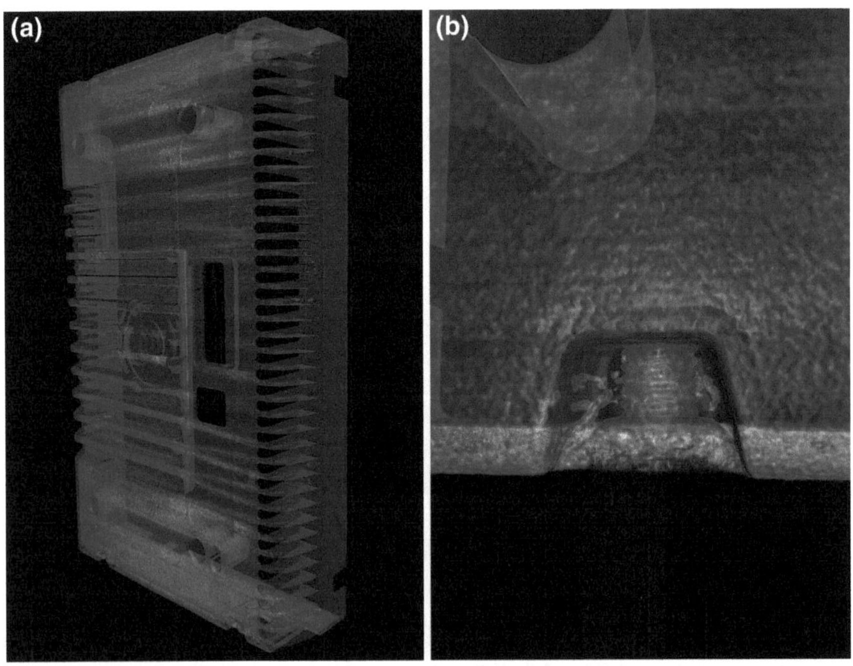

Fig. 4 **a** Three-dimensional image of porosities in the near-net-shape thixomolding Mg casting part (AZ91D) and **b** shows some small porosities reconstructed using newly developed software. A transparent mode was used to show the whole casting part (**a**) and a local region with the presence of porosities (**b**)

porosities in the near-net-shape thixomolding Mg casting part (AZ91D) reconstructed using newly developed software. Figure 4b shows the porosity in one local region. A transparent mode was used to show the whole casting part (Fig. 4a) and a local region with the presence of porosities (Fig. 4b). Clearly, newly developed software shows a great advantage compared with the commercial software.

Conclusion

One Mg casting part (AZ91D, Mg–9Al–1Zn) produced by thixomolding was investigated by CT. No significant casting defect was observed, strongly indicating that thixomolding could produce high-quality Mg casting parts. Furthermore, newly developed reconstruction software was used to obtain more details from the same casting part. This investigation demonstrates that computed tomography is an efficient method to determine casting defects in near-net-shape casting parts.

References

1. Rose J, Pabel T, Geier GF, Schumacher P (2010) Method for the estimation of porosity analysis of CT-data. Giesserei-rundschau 57:244–246
2. Li JH, Oberdorfer B, Habe D, Schumacher P (2018) Determining casting defects in near-net shape casting aluminum parts by computed tomography. Front Mech Eng 13(1):48–52
3. Li JH, Ludwig TH, Oberdorfer B, Schumacher P (2018) Solidification behaviour of Al Si based alloys with controlled additions of Eu and P. Int J Cast Met Res 31:319–331
4. Mayo SC, Stevenson AW, Wilkins SW (2012) In-line phase-contrast X-ray imaging and tomography for materials science. Materials 5:937–965
5. Withers PJ (2015) Fracture mechanics by three-dimensional crack-tip synchrotron X-ray microscopy. Phil Trans R Soc A 373:20130157
6. Samavedam S, Sakri SB, Hanumantha RD, Sundarrajan S (2014) Estimation of porosity and shrinkage in a cast eutectic Al-Si alloy. Can Metall Q 53:55–64
7. Nicoletto G, Konečná R, Fintova S (2012) Characterization of microshrinkage casting defects of Al-Si alloys by X-ray computed tomography and metallography. Int J Fatigue 41:39–46
8. Cabeza S, Mishurova T, Garcés G, Sevostianov I, Requena G, Bruno G (2017) Stress-induced damage evolution in cast $AlSi_{12}CuMgNi$ alloy with one- and two-ceramic reinforcements. J Mater Sci 52:10198–10216

The Effect of the Addition of Transition Metals on Double Oxide Film Defects in an Al–Si–Mg Alloy

W. D. Griffiths, A. J. Caden and Q. Chen

Abstract Double oxide film defects (bifilms) are associated with a reduction in mechanical properties in Al and Mg alloys, but also with a variability in mechanical properties that is difficult to compensate for when designing castings. This paper reports the effect of the addition of 0.45 wt% Mo, or of 0.45 wt% W, to an Al-7Si-0.3 Mg aerospace and automotive alloy. The additions resulted in an increase in the Weibull modulus of the UTS (i.e. a reduction in the scatter of the measured mechanical property data), by a factor of about two (about from m = 16 to m = 32). The observation of AlN on the fracture surfaces of some tensile test bars suggests the additions of Mo or W were associated with the acceleration of the reaction between the oxygen and nitrogen in the internal atmosphere of the double oxide defects and the surrounding liquid metal. This may alter the morphology of the defects, reducing their size and reducing their impact of mechanical properties. The greatest improvement in reproducibility of tensile properties (UTS) was obtained when an addition of 0.45 wt% W was coupled with a well-designed running system, resulting in a Weibull modulus of m = 96.

Keywords Al alloys · Double oxide films · Molybdenum · Tungsten

Introduction

The exploitation of light alloys (Al, Mg and Ti) can be assisted by the adoption of the casting process for the production of components at a reduced cost, compared to other production methods. However, the process inevitably involves the transfer of liquid metal into a container, wherein it solidifies to adopt the required shape. This is accompanied by a characteristic type of defect known as a double oxide film

W. D. Griffiths (✉)
College of Engineering and Physical Sciences,
University of Birmingham, Edgbaston, Birmingham B15 2TT, UK
e-mail: W.D.Griffiths@bham.ac.uk

A. J. Caden · Q. Chen
School of Metallurgy and Materials, College of Engineering
and Physical Sciences, University of Birmingham, Edgbaston,
Birmingham B15 2TT, UK

M. Tiryakioğlu et al. (eds.), *Shape Casting*, The Minerals, Metals & Materials Series,
https://doi.org/10.1007/978-3-030-06034-3_10

defect, or "bifilm". This arises due to the oxidation of the surface of the liquid metal. Al, Mg and Ti alloys are all highly reactive, implying that if their oxide skin parts, or is removed it will almost instantaneously reform. During casting, the liquid metal with its permanent oxide skin flows into the mould in a surface turbulent fashion (in contrast to bulk turbulent flow). The net effect is the formation of a splash in which the liquid metal folds over and recombines with the bulk metal, carrying with it its oxide skin constantly forming, splitting and reforming, described by Campbell [1], and shown in Fig. 1.

The defect occurs as a doubled-over oxide skin, in which the recombining splash of liquid meal carries into the liquid metal a defect having a characteristic two-dimensional morphology, being several mm^2 in area, but only a few microns in thickness. However, the deleterious effect of the defect on mechanical properties is not so much caused by the incorporation of the doubled-over oxide film, since the oxide film is typically sub-micron in thickness, but by the incorporation into the casting of a quantity of air (or the localised mould gases), which might vary in size from spherical bubble millimetres in diameter to gas-filled crevices some tens of microns in thickness. This results in not only a reduction in mechanical properties, but also a variation in the amount of that reduction, due to variation in the size of the defect, variation in its morphology (ranging from spherical to two-dimensional) and variation to its orientation relative to the stresses in a section experienced in service. This all results in a variation in the properties to be expected in a casting, which is often compensated for by increasing the section's thickness of the casting (i.e. applying a "Safety Factor"). However, this inevitably reduces the weight-saving gained by using light alloys with their good strength-to-weight ratios.

It was suggested by Green and Campbell [2] that once a double oxide film defect formed, its interior gases would start to react with the surrounding bulk metal. In the case of Al, the interior gases might be expected to be air in the case of die casting, but may contain hydrocarbons or hydrogen in the case of a resin- or clay-bonded sand mould. In the case of Mg alloys, cast under an SF_6/CO_2 atmosphere or under other protective gases, the subsequent life cycle is difficult to deduce [3]. Ti alloys have not yet been investigated for their propensity to form double oxide film defects.

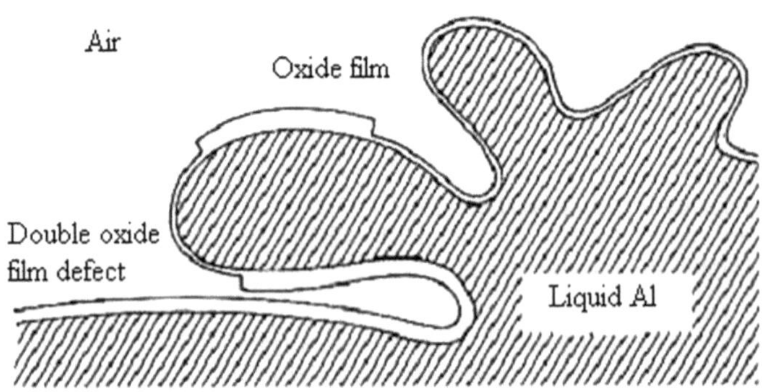

Fig. 1 Sketch of the formation of a double oxide film defect (bifilm) in an Al alloy

Green and Campbell [2] proposed that, in the case of the internal atmosphere in a double oxide film defect containing air, oxygen would be the first gas to react with the surrounding alloy, forming alumina (Al_2O_3) or spinel ($MgAl_2O_4$), followed by reaction with nitrogen, to form AlN. Once all of the oxygen and nitrogen had been consumed, the remainder would consist of insoluble and unreactive Ar, residual from the original 1vol.% of Ar in air. Once the oxygen and nitrogen had been consumed, the volume of the atmosphere inside the defect would be greatly reduced, presumably resulting in a reduced impact of the double oxide films on mechanical properties. Much research has been carried out on running system design, to prevent the formation of double oxide film defects. An alternative approach may be exploited if the defects can be reduced in size and impact by the addition of elements that alter the nature of the oxide or nitride that forms during the life cycle of the defect.

For example, consideration of the Gibbs free energy of formation of oxides suggests that the following elements (i) can form oxides preferentially to spinel ($MgAl_2O_4$); Y, Sc, Ca, Be, Ce, La. Mg, Sr and Hf, (ii) can form oxides preferentially to alumina (Al_2O_3), Li and Zr, and (iii) can form nitrides preferentially to AlN; Hf, Zr, Ti and Sc [4]. Any one of these additions, or combination of additions, may alter the nature of the oxide or nitride skin of a double oxide film defect and may accelerate the rate of reaction between the internal atmosphere and the bulk metal, reducing the impact on mechanical properties.

However, in the course of investigation of these ideas, an increase in the Weibull modulus was observed with the addition of the elements Mo and W, despite them not being on the list of elements that are oxide or nitride formers preferentially to Al. This paper presents the Weibull moduli resulting from the additions of various transition elements made in an attempt to accelerate the consumption of the internal atmosphere of the defect, reducing its scatter of properties.

Experimental Procedure

Three alloys were cast in resin-bonded sand moulds, with a running system design as shown in Fig. 2, in which it was expected that the filling of the mould would be accompanied by the entrainment of large amounts of double oxide film defects. Figure 3 shows a photograph of the pattern used for making moulds having a well-designed running system, the main characteristics of which were (i) a tapered downsprue, (ii) a 20 ppi filter based at the bottom of the downsprue, (iii) a curved runner bar after exiting the filter, and a thin runner bar, to ensure the liquid metal filled the running system completely and (iv) a reduced ingate velocity to prevent surface turbulence during filling of the test-bars.

The alloys cast were 2L99 (Al-7Si-0.3 Mg), 2L99 alloy with additions of about 0.39%Mo and about 0.44 wt% W. These additions were obtained from master alloys of composition 2L99-10 wt% W and 2L99-10 wt% Mo master alloy, both made up from pure Mo and W and 2L99 alloy, held at 1300 °C for 40 min.

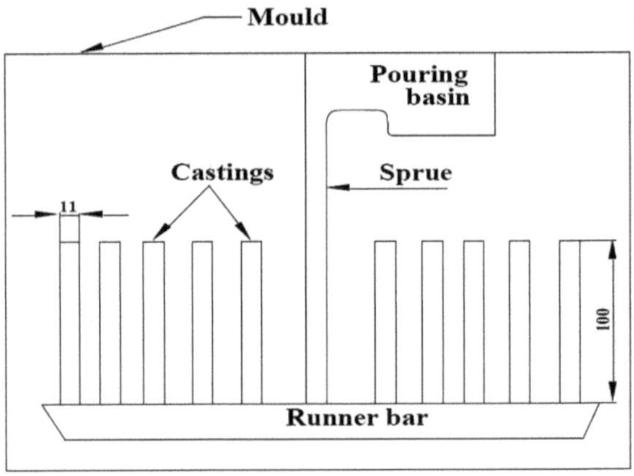

Fig. 2 Sketch of the mould used to obtain tensile test bars containing double oxide film defects, with a badly designed running system

Fig. 3 Photograph of the pattern used to obtain tensile test bars, with a well-designed running system, minimising the entrainment of double oxide film defects

About 20 kg of 2L99 alloy was cast from an induction furnace at 725 °C ± 5 °C. Lance degassing with Ar reduced the H content. In addition, the moulds had been held in the laboratory for more than one week, in order to reduce the pickup of H from the resin binder during mould filling. For the poorly designed running system, after filling the moulds were rolled over 180° to assist feeding during solidification.

A total of 77 tensile test bars were made from the 2L99 alloy with no additions, and 30 tensile test bars were made for each element addition. The tensile test bars were subsequently heat-treated to the T6 condition, namely solutionising at 540 °C for 12 h, followed by quenching in hot water at 65 °C and ageing at 155 °C for

3.5 h. The test bars obtained were of length 100 mm, having a gauge length of 37 mm, with the diameter in the gauge length of 6.75 mm.

The scatter in the UTS and %Elongation results obtained from each of the alloys were compared by calculating their Weibull moduli to show the variability of properties obtained with and without the additions of Mo and W. The fracture surfaces of some test bars were also examined, using scanning electron microscopy (SEM) and energy-dispersive spectroscopy (EDS), using a JOEL 7000 SEM equipped with Oxford INCA.

Chemical analysis by XRF showed that variations in the composition of the alloy that might have affected the scatter of results and hence the Weibull modulus, such as Mg, Fe and H contents, had all been controlled to reduce the variability in the results.

Results

Figure 4 shows the Weibull modulus plots for UTS for the 2L99 alloy, compared to 2L99 alloy with additions of about 0.45 wt% Mo and W, respectively. The Weibull modulus of the 2L99 alloy was about 16, about half that found in the case of the alloy with additions of Mo and W, which were about 33 and 31, respectively. Figure 5 shows the Weibull modulus plots for %Elongation for the alloys which were about 2.6%, in comparison with the alloys with additions of Mo and W, which were about 3.5 and 2.9, showing some improvement in ductility (about 35% and 12%, respectively).

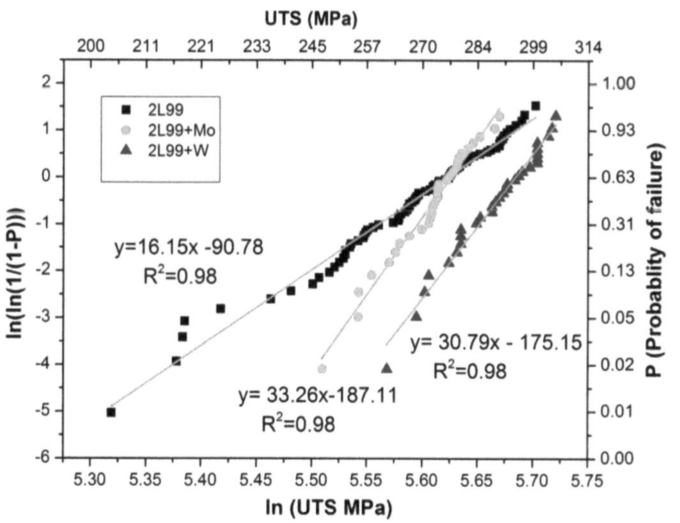

Fig. 4 Weibull modulus plot for UTS, for 2L99 alloy, 2L99+Mo alloy and 2L99+W alloy

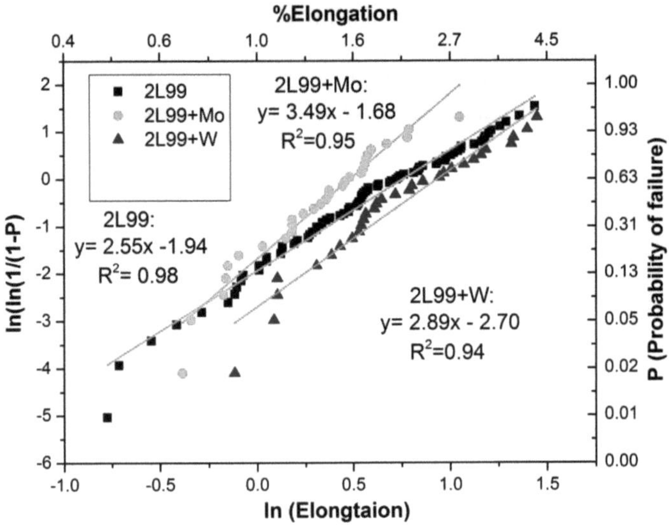

Fig. 5 Weibull modulus plot for %Elongation, for 2L99 alloy, 2L99+Mo alloy and 2L99+W alloy

Table 1 shows a summary of the Weibull moduli obtained with the different additions made including additional elements tested (Hf, Ti and Mn). The most significant result was the high Weibull modulus in the case of the well-designed running system used to cast the 2L99 alloy with W, associated with a Weibull modulus of the UTS of 96, a Weibull modulus of the %Elongation of around 10. This should be compared to the same alloy cast with a badly designed running system, with an addition of W, which produced Weibull modulus values of 31 for the UTS and 2.9 for the %Elongation.

Three other additions were tested; Hf because it was noted that it could form oxides and nitrides preferentially to Al, Ti because it was noted that it could form nitrides preferentially to Al (but not oxides), and Mn, which should not form oxides or nitrides preferentially to Al, but did lie adjacent to Mo and W in the periodic table (which themselves should not form oxides or nitrides preferentially to Al). Of these three additions, no improvement in Weibull modulus, for UTS or % Elongation, was observed.

Figure 6 shows typical SEM images of the fracture surfaces, with a typical double oxide film defect, in this case of the 2L99 alloy containing Mo. Figure 6 also shows a typical fracture surface, but in this case the area associated with fracture of the test bars is an extensive area of AlN, shown in more detail, where it is apparent that it consists of a porous region of an angular phase. The EDX spectra show significant amounts of both N and O, suggesting that, as the oxygen in the internal atmosphere should have reacted and been consumed first, the AlN subsequently formed in association with the oxide in some way.

Table 1 A summary of the Weibull moduli obtained with the different element additions

Alloy	Running system design	Weibull modulus UTS	Weibull modulus %Elongation
2L99	Badly designed running system	16	2.6
2L99	Well-designed running system	24	
2L99+Mo	Badly designed running system	33	3.5
2L99+Mo	Well-designed running system	–	–
2L99+W	Badly designed running system	31	2.9
2L99+W	Well-designed running system	96	9.2
2L99+Hf	Badly designed running system	7	6
2L99+Ti	Badly designed running system	11	2.1
2L99+Mn	Badly designed running system	12	2.8

Earlier, possible elemental additions had been proposed that might accelerate the reaction between the internal atmosphere of a double oxide film defect and its surrounding melt. Some of these were trialled in the same way, with the addition of about 0.45 wt% being added to a 2L99 melt via a master alloy. Currently, the optimum addition rate for the elements is not known, the value of 0.45 wt% being arrived at serendipitously, but was also used in the case of these other elemental additions. Further work is required to determine the optimum addition rate for any of the additions made.

Figure 7 shows an example of a fracture surface containing a feature about 3 mm in diameter, showing several EDX analyses. Location 2 was associated with 26 wt% Hf and 35 wt% O (4 at.% Hf and 52 at.% O, respectively), while location 3 was associated with 20 wt% Hf and 40 wt% O (3 at.% and 55 at.%, respectively), suggestive of a Hf–O compound. Spectra from locations 1 to 4, located either side of the feature, were free of hafnium, but contained 17 wt. O and 11 wt% O, respectively, suggesting an associated oxide film. Location 5 had 53 wt% Hf but about 2 wt%. O, but 17 wt% Si and 27 wt% Al, suggesting a Hf-bearing intermetallic.

Discussion

The addition of about 0.45 wt% Mo or W was associated with an increase in the Weibull modulus of the UTS and %Elongation, i.e. a reduction in the scatter of the results, which was found in both cases. This was associated with an increase of about 10% in the mean UTS, although the improvement in %Elongation was much less. The addition of 0.45 wt% Mo or W was arrived at by chance, by the observation of the change in the appearance of the surface oxide skin of the liquid metal after an addition had been made. The optimum addition rate has yet to be established.

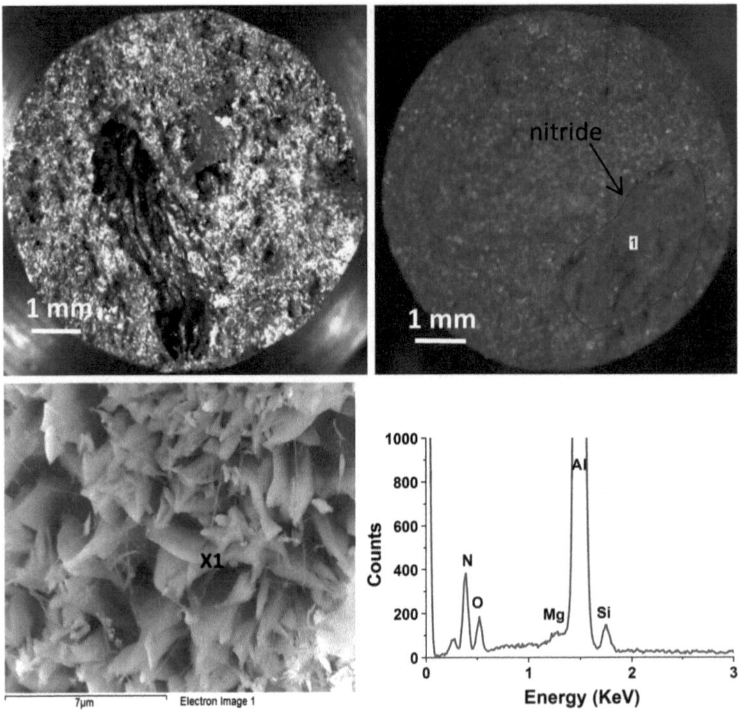

Fig. 6 Top-left; typical example of a double oxide film defect in a 2L99+Mo alloy. Top-right; fracture surface of a 2L99+Mo alloy showing an area of AlN. Bottom-left; higher magnification of AlN. Bottom-right; EDX of location X1 in left-hand image, showing the presence of oxygen and nitrogen

Fig. 7 SEM image of a feature on the fracture surface of an 2L99+Hf alloy, containing Hf

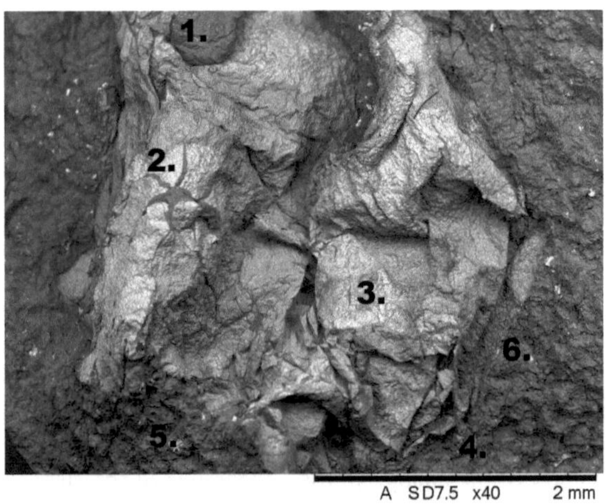

The mechanism by which the scatter in mechanical properties is reduced, and the mean mechanical properties increased, has also yet to be established although several mechanisms suggest themselves. In previous work, by Cao and Campbell [5], exploring the effect of the addition of Fe, suggested that Fe-rich compounds could nucleate on oxide films and then being denser could settle out, removing the oxide films and hence improving mechanical properties. In the experiments reported here, Mo- and W-bearing intermetallics were observed to form in the 2L99-10Mo and 2L99-10 W master alloys, and while Mo and W in solution might lead to further precipitation of intermetallics on oxide films, the introduction of the master alloys in order to raise the Mo and W contents of the castings, having been held at temperatures of about 1300 °C for 40 min, might be thought to introduce oxides as much as remove them.

The original strategy, to add alloying elements to accelerate the consumption of the interior atmosphere, first oxygen, then nitrogen, is supported by the observation of AlN over areas of the fracture surfaces. Only a few examples were observed in the case of the 2L99+Mo alloy, and none observed on the test-bar fracture surfaces of the 2L99+W alloy. Nonetheless, the presence of any AlN, hitherto unobserved, suggests the improvements in Weibull modulus observed in both cases were associated with a mechanism operating to consume the interior oxygen. This may be an alteration of the morphology of the oxide layer, perhaps the introduction of a porous phase such as AlN, although the exact mechanisms are not understood.

However, it should be borne in mind that alteration of the oxide film morphology may not lead to a desired effect on the oxide film. As Fig. 7 shows the addition of Hf, capable of forming both oxides and nitride preferentially to Al led to the formation of a Hf-bearing compound but without any corresponding improvement in mechanical properties, or an increase in Weibull modulus of the UTS or of the mean UTS. However, the amount of Hf added was around 0.5 wt%, to reproduce the amount of the Mo and W addition, and it is possible that more may be needed (or less) in order to achieve an improvement.

Similarly, the addition of Ti was not associated with any improvement in mechanical properties, despite TiN being preferentially formed compared to AlN. Again, some alteration in morphology (a porous phase) was observed (not shown in this paper), but the amount of porosity observed was so slight that any change in mechanical properties was perhaps not to be expected.

The greatest improvement in Weibull modulus was associated with the addition of about 0.45 wt% W to castings made with a well-designed running system. This was associated with a Weibull modulus for the UTS of 96 and 9.2 for the Weibull modulus of the %Elongation. Without the addition of Mo or W, the well-designed running system was associated with a Weibull modulus for the UTS of 24, compared to 16 with a poorly designed running system (an increase of 50%). The addition of W led to an improvement, in the mould with a poorly designed running system, of a factor of about 2. In the presence of a 0.45 wt% addition of W the well-designed running system the Weibull modulus increased, compared to the poorly designed running system, by a factor of about 3 times. In the case of the well-designed running system, the addition of 0.45 wt% W was associated with an increase in Weibull modulus by a factor of about four times. Finally, a well-designed running system, with an addition of 0.45 wt% W, resulted in an increase in the Weibull modulus by a factor of about six times.

To summarise, both additions of W and improvements in running system design enhance mechanical properties, but in these experiments the improvement stemming from better running system was enhanced in the presence of W.

Conclusions

1. The addition of about 0.45 wt% Mo increases the Weibull modulus of the UTS by about a factor of about two. The addition of about 0.45 wt% W also increases the Weibull modulus of the UTS by about a factor of about two. The reason for this is not yet understood. The effect on the %Elongation was to bring about a small increase, which may not have been statistically significant.
2. The greatest increase in Weibull modulus, in these experiments, was obtained using an addition of W, coupled with a well-designed running system. This achieved a Weibull modulus of 96 and a Weibull modulus for the %Elongation of about 9.
3. Addition of other elements, such as about 0.5 wt% Hf, Ti and Mn, produced little or no improvement in Weibull modulus. Hf had been noted as an element capable of forming an oxide or a nitride preferentially to Al; no improvement in Weibull modulus was observed.

Acknowledgements The authors would like to acknowledge Ms. Hannah-Beth Hall of the University of Birmingham for the provision of data concerning the effect of the addition of W, and Mr. Liam Duncan for the provision of data concerning the effect of the addition of Hf.

References

1. Campbell J (2011) Complete casting handbook, 2nd edn, Chap. 2, ‘Entrainment’. Butterworth-Heinemann, London, pp 19–103
2. Green NR, Campbell J (1994) Influence of Oxide Film Filling Defects on the Strength of Al-7Si-Mg Alloy Castings. AFS Trans 102:341–348
3. Li T (2017) PhD thesis, University of Birmingham, UK
4. Raeiszadeh R (2006) PhD thesis, University of Birmingham, UK
5. Cao X, Campbell J (2004) Effect of precipitation and sedimentation of primary α-Fe phase on liquid metal quality of cast Al-11.1Si-0.4 Mg alloy. Int J Cast Met Res 17(1):496–504

On Estimating Largest Defects in Castings

Murat Tiryakioğlu and Irisi Nini

Abstract The procedure developed by Beretta and Murakami has two issues that were not addressed previously: (i) the selection of the Weibull plotting position formula for linear regression fits to estimate Gumbel distribution parameters and (ii) the untested hypothesis that the estimates for the upper Gumbel percentiles are distributed normally. Monte Carlo simulations were run to determine the plotting position formula that provided the least bias and the distribution of Gumbel percentiles. It was found that among the nine formulas used in this study, the one developed by Hazen had the least bias, whereas the one by Weibull had the highest bias. Moreover, 0.999 percentiles of the Gumbel distribution were found to follow the three-parameter lognormal distribution. Empirical relationships between the estimated parameters of the three-parameter lognormal distribution and sample size are provided in the paper.

Keywords Gumbel · Beretta–Murakami method · Fatigue of cast metals · Three-parameter lognormal

Introduction

Fatigue is the most common failure type among engineered components in service and accounts for 90% of the failures reported [1, 2]. Cracks leading to final fatigue fracture form in the regions of highest stress. In components designed properly, the regions of highest stress correspond to locations of largest defects, such as pores and inclusions [3]. It has been demonstrated [4–7] that there is a direct relationship between the size of the largest defects and fatigue life of aluminum components. Wang et al. [4] showed that there is a strong relationship between fatigue strength,

---

M. Tiryakioğlu (✉)
University of North Florida, Jacksonville, FL 32224, USA
e-mail: m.tiryakioglu@unf.edu

I. Nini
School of Engineering, University of North Florida, Jacksonville, FL 32224, USA

M. Tiryakioğlu et al. (eds.), *Shape Casting*, The Minerals, Metals & Materials Series,
https://doi.org/10.1007/978-3-030-06034-3_11

σ_f, at 10^7 cycles and the size of the largest pore in various Al–Si–Mg–(Cu) alloy castings.

Murakami [3, 8, 9] developed a model to predict fatigue strength, σ_f, based on the size of inclusions and Vickers hardness, H_v, of the steel:

$$\sigma_f = 1.41\frac{120 + H_v}{\left(\sqrt{A_i}\right)^{1/6}} \tag{1}$$

where A_i is the cross-sectional area of the largest defect in the metal. Kobayashi and Matsui [10] showed that the Murakami method can be used to predict fatigue strength of A356 aluminum alloy castings. Hence, it has been demonstrated that it is possible to estimate the fatigue behavior of cast alloys if the size of the largest defect is known. Beretta and Murakami [11] introduced a method to estimate the size of the largest defect. The procedure involves the following steps:

1. Take a number (n) of polished sections on the part and determine the area of the largest defect in each polished section.
2. Fit a Gumbel distribution to the square root of the data. The cumulative distribution function, P, of the Gumbel distribution is written as [12]:

$$\mathrm{P(x)} = \exp\left(-\exp\left(-\frac{(x-\lambda)}{\delta}\right)\right) \tag{2}$$

   where λ and δ are location and scale parameters, respectively. These two parameters need to be estimated.
3. Determine the percentile of the Gumbel distribution for the defect based on its volume with respect to the volume of the component.
4. Estimate the upper confidence limit (95%) of the percentile.
5. Use the upper confidence limit of the percentile as $\sqrt{A_i}$ to estimate fatigue strength.

There are two issues with the procedure outlined above: (i) linear regression estimates and (ii) the distribution of the percentiles of the Gumbel distribution and their confidence limits. This paper is motivated by these two issues.

Background

Estimating Gumbel Parameters via Linear Regression

Taking double logarithm of Eq. (2) yields

$$-\ln(-\ln(P)) = \frac{x}{\delta} - \frac{\lambda}{\delta} \tag{3}$$

which has a linear form with a slope of 1/δ and an intercept of –λ/δ. To use Eq. (3) to estimate λ and δ, a cumulative probability, P, has to be assigned to each data point. There are several plotting position formulas in the literature. Beretta and Murakami used the one introduced by Weibull [13]:

$$P = \frac{i}{n + 1} \tag{4}$$

where *i* is the rank of each data point in ascending order. The other plotting position formulas in the literature that have been used in various applications were summarized by Cunane [14]. All formulas can be written in the form [15]:

$$P = \frac{i - a}{n + b} \tag{5}$$

The selection of the plotting formula affects the bias in the estimates of the distribution parameters, as shown for the Weibull [15–17] and Gumbel distributions [18].

Percentiles of the Gumbel Distribution

Beretta and Murakami calculated percentiles of the Gumbel distribution by using the estimated parameters. The percentiles that they calculated are based on:

$$P = 1 - \frac{V_d}{V_c} \tag{6}$$

where V_d and V_c are the volume of the largest defect and the casting volume, respectively. Beretta and Murakami reported percentiles for $V_d/V_c = 10^{-3}$, 10^{-6}, and 10^{-9}.

Because percentiles are estimated by using estimates of distribution parameters, they have their own distribution. Therefore, confidence intervals need to be attached to percentiles, based on their distribution. Beretta and Murakami assumed that the percentiles of the Gumbel distribution follow the normal distribution. No evidence for the normality of the percentile estimates was provided. They suggested that two-sided 95% confidence intervals be developed and the upper confidence limit be used as $\sqrt{A_i}$. A formula based on Cramer–Rao theorem was provided to estimate the standard error (deviation).

Monte Carlo Simulations

This study was conducted in two stages: in Stage 1, the bias in estimated parameters based on the nine plotting position formulas from the literature for ten sample sizes ranging from n = 10 to 100. Monte Carlo simulations were used to generate n data from a Gumbel distribution with parameters $\delta = 1$ and λ at five levels; 0.1, 1, 10, 100, and 1000. For one observation, n random numbers between 0 and 1 were generated to obtain a set of x values; all assigned a rank, which was then used to calculate probabilities using Eq. (5). Subsequently, Eq. (3) was used to estimate the Gumbel parameters via linear regression. The sample size was changed systematically between 10 and 100. For each sample size and λ value, the experiment was repeated 10,000 times. Estimated Gumbel parameters were normalized by dividing them by true values and their averages were calculated for each n and plotting position formula. The plotting position formulas used in this study were then evaluated for least bias for both parameters. The best plotting position formula was selected.

In Stage 2, Monte Carlo simulations were run to determine the 0.999 percentiles ($V_d/V_c = 10^{-3}$), $P_{0.999}$, for ten sample sizes varying between 10 and 100. The δ/λ ratio was first systematically changed between 0.1 and 1, based on the findings of Beretta and Murakami, to determine whether the δ/λ ratio and/or the values of δ and λ affected the estimated percentiles. The data for $P_{0.999}$ were analyzed to determine the best-fitting distribution.

Results and Discussion

Stage 1: Linear Regression

It was determined that the distribution and average of the normalized estimated parameters were not affected by the initial value of λ nor the δ/λ ratio. Therefore, all normalized data were combined to evaluate the bias generated by each plotting position formula. It was determined that all plotting position formulas provided essentially unbiased estimates of λ. The bias in estimates of δ, however, was found to be strongly correlated to the term b in the denominator in Eq. (5), as presented in Fig. 1. Among the nine plotting position formulas tested in this study, the least bias was obtained when b = 0 as in the formula that by Hazen [19];

$$P = \frac{i - 0.5}{n} \tag{7}$$

The formula proposed by Weibull and used by Beretta and Murakami (Eq. 4) had the highest bias (b = 1). Therefore, it is recommended that Eq. (4) be replaced by Eq. (7) in attempts to estimate sizes of the largest defects.

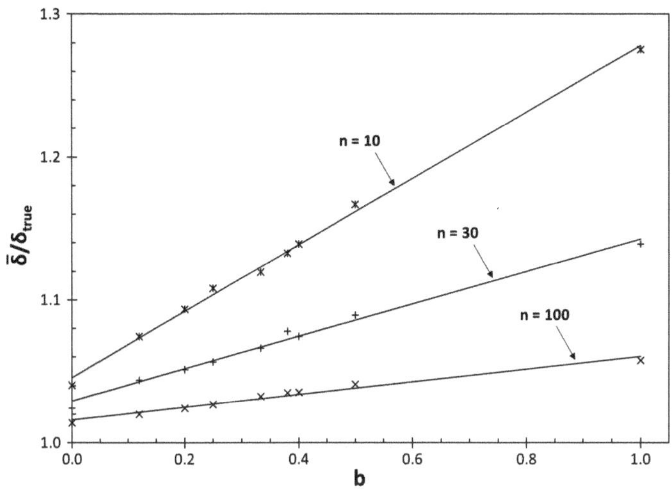

Fig. 1 Change in bias in estimates of δ with sample size and plotting position formula

Stage 2: Estimating Percentiles

It was determined that neither δ/λ ratio nor the values of δ (or λ) affected the estimated percentiles when they were normalized:

$$P_{N(0.999)} = \frac{P_{0.999} - \lambda}{\delta} \tag{8}$$

Therefore, data obtained at different values of δ/λ ratio were combined for each sample size. Consequently, 50,000 runs were evaluated for each sample size.

To determine the best-fitting distribution, data for three sample sizes, n = 10, 30, and 100, were used. The Anderson–Darling goodness-of-fit statistic (A^2) [20] was used to evaluate the fits provided by eight different distributions. Note that a lower A^2 value would indicate a better fit. The results are summarized in Table 1. For

Table 1 Goodness of fit, as described by the Anderson–Darling statistic, of each distribution to the data for n = 10, 30, and 100

Distribution		A^2	
	n = 10	n = 30	n = 100
Normal	440.3	272.3	116.2
Lognormal	17.0	10.9	11.5
Three-parameter Lognormal	7.2	5.6	2.6
Gamma	37.6	55.7	33.6
Three-parameter Gamma	23.1	13.5	4.6
Gumbel (for maxima)	26.0	82.4	199.9
Weibull	585.7	860.1	862.2
Three-parameter Weibull	397.9	366.4	212.7

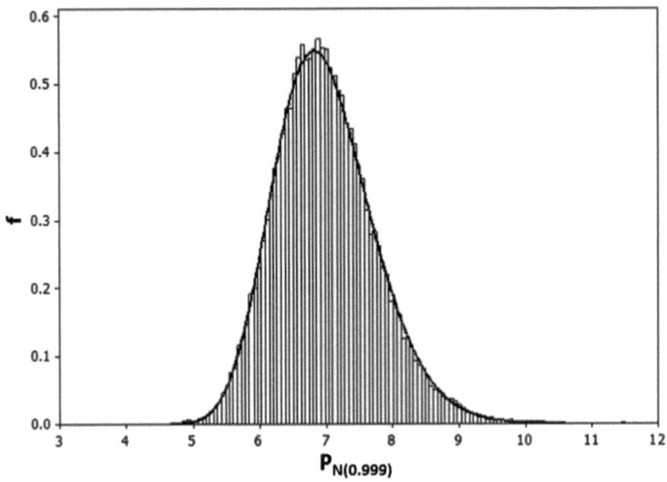

Fig. 2 Histogram for $P_{N(0.999)}$ for n = 100 and the three-parameter lognormal fit to them

three sample sizes, the lowest A^2 value was obtained with the three-parameter lognormal distribution. The fit of this distribution to data for n = 100 is shown in Fig. 2 The probability density function, f, of this distribution is written as:

$$f(x) = \frac{1}{(x - \tau)\omega\sqrt{2\pi}} \exp\left[\frac{-(\ln(x - \tau) - \theta)^2}{2\omega^2}\right] \tag{9}$$

where θ is the location parameter, ω is the scale parameter, and τ is the threshold. It is also noteworthy that the normal distribution did not provide good fits at all. Therefore, assuming that $P_{N(0.999)}$ is distributed normally is likely to introduce large errors into fatigue strength estimates as given in Eq. (1).

The change in the location parameter with the inverse square root the sample size is presented in Fig. 3. The best-fit line has the following formula:

$$\theta = \frac{2.148}{\sqrt{n}} + 1.397 \tag{10}$$

The coefficient of determination, R^2, of the fit is 0.995.

The change in the scale parameter (ω) and the threshold (τ) with sample size is presented in Figs. 4 and 5, respectively. The best-fit curve in Fig. 4 has the following formula:

$$\omega = 0.143 - 0.058 \cdot \ln(n) \tag{11}$$

with R^2 = 0.998. The best-fit line in the figure is written as:

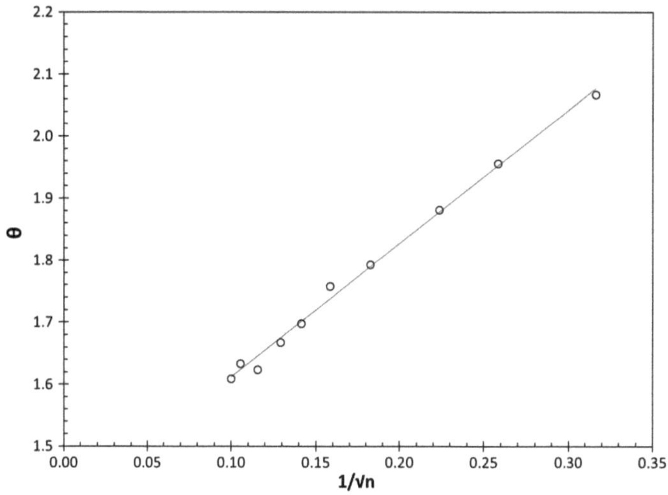

Fig. 3 Change in the location parameter with sample size

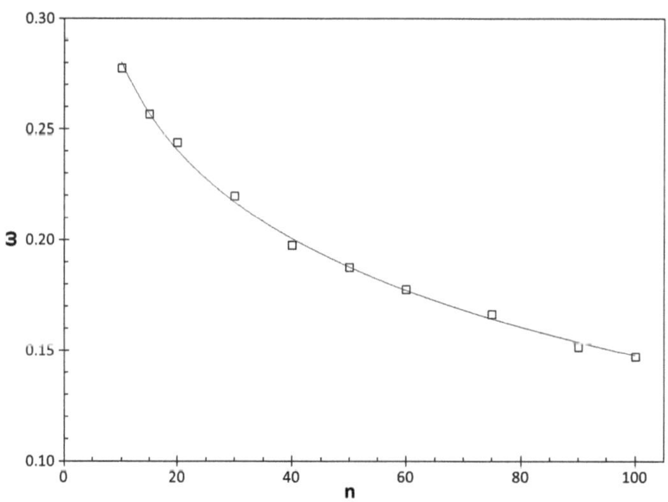

Fig. 4 Change in the scale parameter with sample size

$$\tau = 3.358 - \frac{13.564}{\sqrt{n}} \tag{12}$$

with $R^2 = 0.995$. Equation 10 through 12 can be used to estimate the three parameters of the lognormal distribution for $P_{N(0.999)}$, which can subsequently be used to estimate the upper (one-tailed) 95% confidence limit. These two

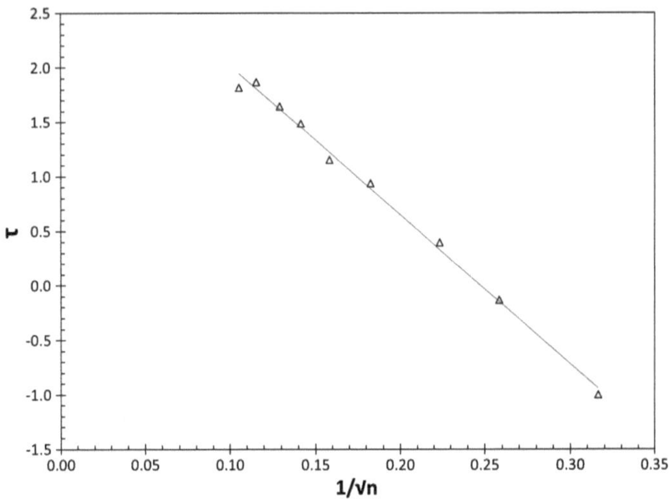

Fig. 5 Change in the threshold with the inverse square root of sample size

improvements are expected to provide more accurate estimates in efforts to predict fatigue behavior of metals.

Conclusions

- The value or the ratio of Gumbel parameters does not have any effect on the bias in normalized estimate means.
- All plotting position formulas yielded essentially unbiased estimates of λ.
- The plotting position formula that yields the least bias in the estimates of δ is the one proposed by Hazen. The one by Weibull was found to have the largest bias among the nine tested in this study.
- The 0.999 percentile estimates of the Gumbel distribution follow the three-parameter lognormal distribution. The normal distribution used by Beretta and Murakami was found to provide very poor fits to the estimates of percentiles.
- The parameters of the three-parameter lognormal distributions were found to be functions of the sample size. Empirical equations provided in this study can be used to calculate the one-tailed upper confidence limits for the 0.999 percentile.

References

1. Dieter GE (1976) Mechanical metallurgy, 2nd edn. McGraw-Hill, New York

2. Reed-Hill RE (1964) Physical metallurgy principles. Van Nostrand, Princeton
3. Murakami Y (2002) Metal fatigue: effects of small defects and nonmetallic inclusions. Elsevier
4. Wang QG, Crepeau PN, Davidson CJ, Griffiths JR (2006) Metal Mater Trans B 37:887–895
5. Tiryakioğlu M (2009) Mater Sci Eng A 520:114–120
6. Tiryakioğlu M (2009) Metal Mater Trans A 40:1623–1630
7. Zhang B, Chen W, Poirier DR (2000) Fatigue Fract Eng Mater Struct 23:417–423
8. Murakami Y, Endo M (1994) Int J Fatigue 16:163–182
9. Murakami Y, Beretta S (1999) Extremes 2:123–147
10. Kobayashi M, Matsui T (1996) Trans JSME Ser A 62 341–346
11. Beretta S, Murakami Y (1998) Fatigue Fract Eng Mater Struct 21:1049–1065
12. Gumbel EJ (1957) Statistics of extremes. Columbia University Press, New York, USA
13. Weibull W (1939) A statistical theory of the strength of materials. Generalstabens litografiska anstalts förlag
14. Cunnane C (1978) J Hydrol 37:205–222
15. Tiryakioğlu M, Hudak D (2007) J Mater Sci 42:10173–10179
16. Tiryakioğlu M (2006) J Mater Sci 41:5011–5013
17. Tiryakioğlu M, Hudak D (2008) J Mater Sci 43:1914–1919
18. Hazen A (1914) Trans Am Soc Civ Eng 77:1539–1669
19. De M (2000) Stoch Env Res Risk Assess 14:1–7
20. Anderson TW, Darling DA (1954) J Am Stat Assoc 49:765–769

Ti Grain Refinement Myth and Cleanliness of A356 Melt

Özen Gürsoy, Eray Erzi and Derya Dışpınar

Abstract Grain refinement of cast alloys provides several advantages such as establishment of finer dendritic structure, lowered porosity due to increased feeding, and thereby higher mechanical properties are reached. For Al–Si alloys, Ti is the preferred; however, Ti has the tendency to sink to the bottom of the crucible due to density difference. In this work, this phenomenon was used to evaluate the correlation of Ti nucleation on bifilms and thereby sedimenting the defects to clean the melt. Primary and secondary A356 alloys were used. After grain refinement addition, samples were collected every 10 min to investigate the change in melt quality. Sand and die moulds were used. Tensile samples were produced for mechanical characterisation. Step mould was used to check the microstructural changes. Reduced pressure test was used to quantify melt cleanliness. It was found that melt quality was increased by duration. Mechanical properties were increased regardless of the microstructure.

Keywords Grain refinement · Ti · Melt quality · Cleanliness · Bifilm

Introduction

A356 aluminium alloy is most commonly used and researched casting alloy in the aerospace and automotive industries. In practice, A356 has cast basically two different moulds: die and sand. If it is necessary to compare, it has been claimed that much higher mechanical properties are obtained by using die mould in casting. It is known that a similar effect is observed by using Al–Ti–B grain refiner master alloy [1–3]. Microstructural differences have been shown as the main reason of this.

Grain refinement with Al–Ti–B master alloy is the most commonly used method in aluminium casting processes. With the addition of grain refiner, aluminium

Ö. Gürsoy (✉) · E. Erzi · D. Dışpınar
Metallurgical and Material Engineering Department, Istanbul University-Cerrahpasa, Istanbul, Turkey
e-mail: ozen.gursoy@gmail.com

M. Tiryakioğlu et al. (eds.), *Shape Casting*, The Minerals, Metals & Materials Series,
https://doi.org/10.1007/978-3-030-06034-3_12

matrix is dissolved and released TiB_2 and $TiAl_3$ particles heterogeneously nucleate α-Al during solidification [1]. Thus, higher mechanical properties are obtained by means of fine equiaxed grains. However, due to the density differences between liquid aluminium and grain refiners, the particles show settling behaviour when they are added to liquid aluminium. Therefore, the efficiency of Al–Ti–B master alloys has been a long-going discussion by many researchers [2, 3].

Bifilms whose density is almost equal to the molten metal is suspended in liquid aluminium [4, 5]. Time dependently, settling of TiB_2 and $TiAl_3$ particles sweep down the bifilms to bottom of the crucible after grain refiner addition. This phenomenon directly influences the melt quality, microstructure, and mechanical properties.

Experimental Study

In this study, the effect of Ti-based inoculant on bifilm existence was investigated. 356 aluminium alloys were supplied from automotive industry's scraps. The chemical composition of the 356 alloys as can be seen in Table 1. The sand moulds were prepared by mixing silica sand, resin, and hardener. The charge was preheated at 400 °C for 1 h and then the casting temperature was set at 725 °C. Two different casting conditions, with (ratio of wt. 0.1%) and without Al-5Ti-1B addition, were applied. The pouring was begun 10 min after grain refinement master alloy addition. In these conditions, tensile bars and step samples are poured to analyse mechanical properties and microstructure, respectively, into either die or sand moulds. At the same time, bifilm index is determined for each condition by reduce pressure test (RPT) machine to determine the melt quality changing. No sample was heat treated.

Results and Discussion

According to spectral analyses (Fig. 1), there is a slight decreasing of Ti ratio in the melt for the as-received alloy. However, in the Al-5Ti-1B grain refiner added melt, noticeable increasing of Ti content towards the bottom zone of the molten metal was observed.

BI analyses show clear time-depending trend which is decreasing in both conditions. Interestingly, although the molten alloy had the same initial BI, Al-5Ti-1B master alloy added melt had lower quality. Master alloy addition decreases the melt

Table 1 Chemical composition of the 356 alloys (wt.%)

Si	Fe	Mg	Ti	B	Sr
7.31	0.13	0.26	0.11	0.001	0.008

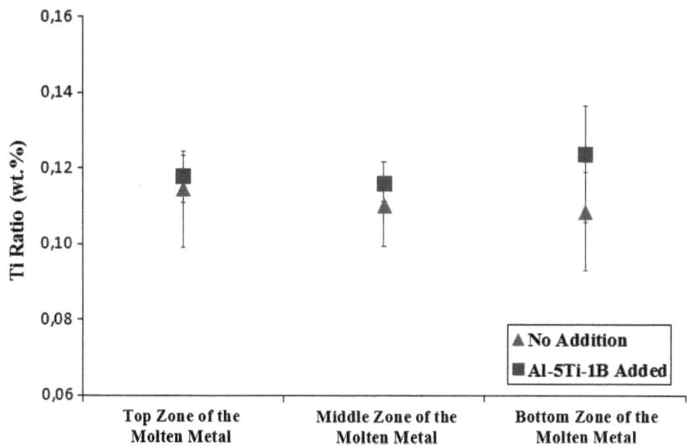

Fig. 1 Change of Ti ratio in different zones of the crucible

quality because of oxide film entrapment during addition. Also, in both melts, BIs increase in the bottom zone of the molten metal (Fig. 2). Drawn trend lines according to BI results of both melts in Fig. 2 show melt quality change depend on time with grain refinement process. These trend lines were added just to show the trend not to show the equation. This is a good evidence of bifilm accumulation in the bottom of the crucible accompanied by the Ti settlement.

Conventional results cannot be observed in SDAS analyses except for few data. Despite the grain refiner addition, there is no successful refinement. Figure 3 also shows the importance of cooling rate on grain size during solidification. SDAS data of die-cast samples are higher than sand cast samples'.

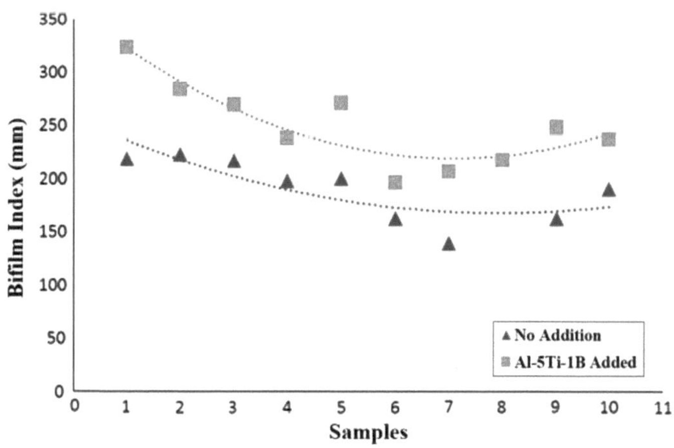

Fig. 2 BI changes of molten metals

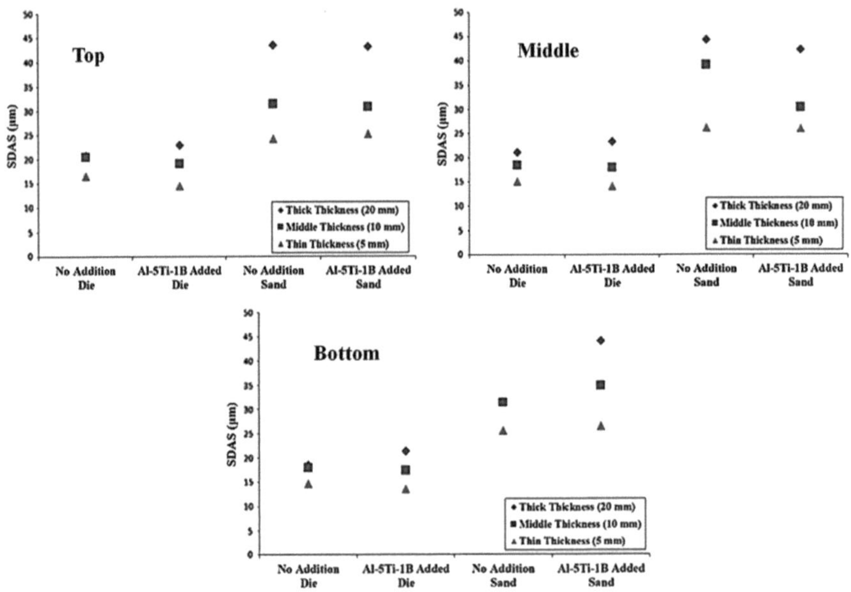

Fig. 3 Microstructural analyses of all conditions (top: samples collected from the top of the crucible; middle: middle of the crucible as melt was consumed; bottom: bottom of the crucible)

Despite the finer grains, the worst and inconsistent UTS and elongation values are observed in die casting (Figs. 3, 4, 5).

According to Weibull analyses, grain refiner addition cannot rely on a better result for UTS values, and it has an adverse effect on elongation values in die cast samples. The best tensile test results are obtained distinctly in sand casting.

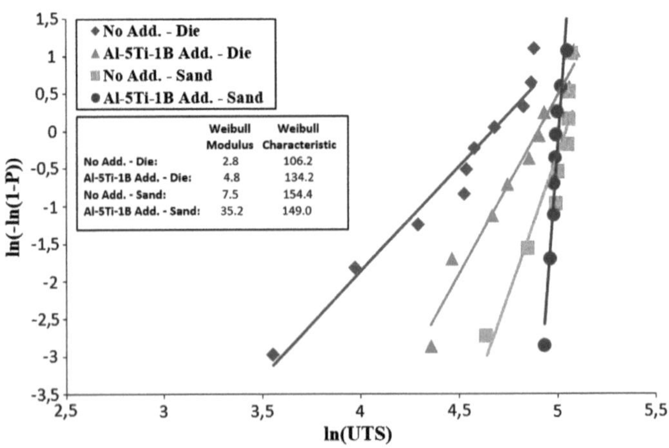

Fig. 4 Weibull analysis of UTS results of tensile tests

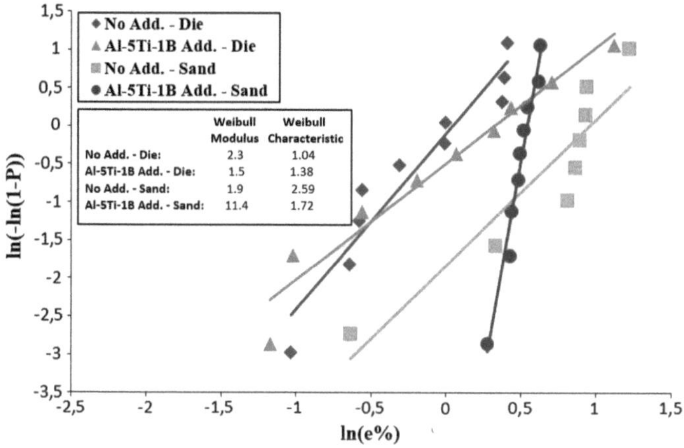

Fig. 5 Weibull analysis of elongation results of tensile tests

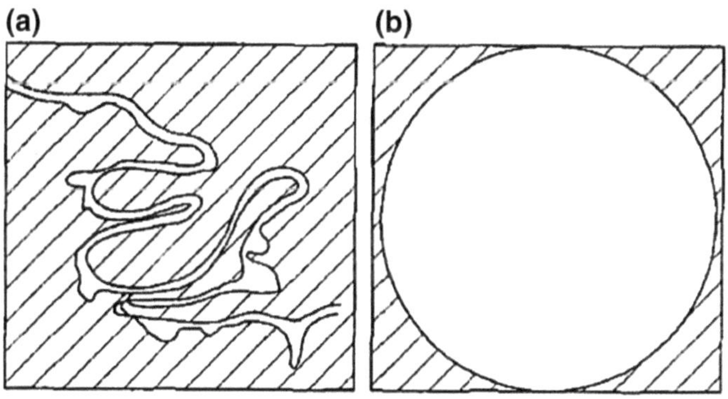

Fig. 6 Bifilm shapes: **a** convoluted bifilm and **b** orbicular bifilm (pore) [6]

Results of two different conditions (no add (i.e. as-received) and Al-5Ti-1B add.) are close to each other. Al-5Ti-1B master alloy addition in sand casting has increased stability of UTS and elongation values.

Bifilm shape affects the tensile properties in solidified structure. The form bifilm can take depending on the mould can be seen in Fig. 6. The most effective parameter of bifilm shape is cooling rate. High cooling rate prevents unfurling of bifilm. Low cooling rate may result in the unravelling of bifilms during solidification. So that, the notch effect of the failures is decreased in tensile testing compare to the compacted-scrambled bifilms.

Conclusion

- Addition of master alloy increases the bifilm index and decreases the melt quality.
- In these conditions, no clear relation between grain refiner addition and microstructure was detected.
- Duration (actually gravity) and particles from grain refiner, such as TiB_2 and $TiAl_3$, settle the bifilms.
- The shape factor of bifilm effects considerably on mechanical properties.

References

1. Mohanty PS, Gruzleski JE (1995) Mechanism of grain refinement in aluminium. Acta Metall Mater 43(5):2001–2012
2. Schaffer PL, Dahle AK (2005) Settling behaviour of different grain refiners in aluminium. Mater Sci Eng A 413:373–378
3. Limmaneevichitr C, Eidhed W (2003) Fading mechanism of grain refinement of aluminum–silicon alloy with Al–Ti–B grain refiners. Mater Sci Eng A 349(1–2):197–206
4. Dispinar D, Campbell J (2011) Porosity, hydrogen and bifilm content in Al alloy castings. Mater Sci Eng A 528(10–11):3860–3865
5. Dispinar D, Akhtar S, Nordmark A, Di Sabatino M, Arnberg L (2010) Degassing, hydrogen and porosity phenomena in A356. Mater Sci Eng A 527(16–17):3719–3725
6. Campbell J (2003) Castings. Elsevier

On the Effects of Defects and Imperfections on Tensile Toughness of a Secondary Aluminium Alloy

Jakob Olofsson, Anton Bjurenstedt and Salem Seifeddine

Abstract In order to design and produce high-quality castings with reliable performance, the effect of the melt handling and melt quality during different processing stages needs to be understood and controlled, and numerical methods to provide correct input data to structural analyses of castings enabled. This paper aims to investigate tensile properties, in particular tensile toughness, of a secondary high-pressure die casting (HPDC) aluminium alloy with different levels of defects and imperfections. The melt, which was transported in liquid state from the smelter to the foundry, has been sampled after different holding times by casting into Y-blocks. Tensile testing was performed, and the levels of defects and imperfections were characterized using measurements of porosity, bifilm index, density index, sludge factor and the amount of iron-rich intermetallics. Two different quality indices have been evaluated, and a method to apply the results in simulations of damage in a casting, containing defects, subjected to load is demonstrated.

Keywords Toughness · Defects · Quality index · Shape casting · Component casting

Introduction

Imperfections and defects in aluminium-silicon based alloys are a function of chemical composition, melt treatment, handling and pouring practices, process parameters and design. In secondary aluminium-silicon alloys, iron and manganese may form compact intermetallic compounds but also sludge if the prevailing conditions allow. Oxides and oxide films in various forms can also be pre-existing

J. Olofsson (✉) · S. Seifeddine
School of Engineering, Jönköping University, P.O. Box 1026, 551 11 Jönköping, Sweden
e-mail: jakob.olofsson@ju.se

A. Bjurenstedt
Swerea SWECAST, P.O. Box 2033, 550 02 Jönköping, Sweden

M. Tiryakioğlu et al. (eds.), *Shape Casting*, The Minerals, Metals & Materials Series,
https://doi.org/10.1007/978-3-030-06034-3_13

in the ingot and therefore already in the melt, and/or formed during the melting and pouring stages. Besides acting a defect in itself, oxides and oxide films may also act as the origin for other defects as porosity. The process parameters are also crucial; if not well optimized and controlled, air entrapment associated with the HPDC process decreases the soundness and quality of castings. All the aforementioned features will to certain extents govern the mechanical properties of the final casting produced in aluminium-silicon cast alloys.

The performance and failure of cast aluminium alloys are commonly discussed based on ultimate tensile strength and elongation to failure, which have been related to the amount and morphology of porosity [1], characteristics of eutectic silicon precipitates [2, 3], iron-rich intermetallics [4] and oxide films [5]. For critical applications, it has however been pointed out that properties as fracture toughness and tensile toughness is of higher relevance [6]. The tensile toughness, also known as strain energy density, here referred to as toughness and denoted W (MJ/m^3), is defined as the area under the tensile curve, i.e.

$$W = \int_{0}^{\varepsilon_f} \sigma d\varepsilon \quad (1)$$

Toughness is a measure of the energy absorbed by the specimen per unit volume of the material [7]. Toughness is closely related to the definition of the J-integral used in crack propagation analyses in fracture mechanics [8] and shows a strong relationship to both elongation to fracture and fracture toughness [6]. Toughness can thus be considered as a parameter which measures the elastic–plastic fracture response and is firmly based on fracture mechanics [7].

The mechanical performance can also be discussed in terms of structural integrity and quality indices. Several different quality indices have been proposed with different applicability [9]. The original quality index was developed by Drouzy [10]. This quality index, here denoted Q_D (Pa), is based on the engineering tensile strength s_u (Pa) and total elongation at fracture e_f (-) as

$$Q_D = s_u + K_D \cdot \log(e_f) \quad (2)$$

Here K_D (Pa) is an alloy dependent parameter to be determined empirically. Caceres used the Hollomon's equation to find a physical justification for Q_D [11] and proposed a modified quality index based on yield strength rather than total elongation at failure [12].

Tiryakioglu et al. [13] used the Kocks–Mecking strain hardening model and found that characteristics and damage of precipitates cannot explain the entire behaviour of the strain hardening rate. Structural integrity was shown to determine if fracture of cast Al–Si alloys occurs before the critical stress level to enter Stage IV, hardening is reached [14]. This led to the identification of so-called target properties, the properties that can be expected when the material is free from major structural discontinuities. A quality index was thus proposed as [7]

$$Q_T = \frac{W}{W_C} \tag{3}$$

where W is the obtained toughness, and W_C is the target value of toughness. If the target value W_C is exceeded, Q_T is defined as unity. While Q_D is a quality index with the units Pa and no upper limit, Q_T is thus an index ranging from zero to unity, indicating the fraction of the target toughness that has been reached.

In previous work [15], one of the current authors has outlined how relationships between microstructural refinement, porosity and toughness could be utilized in the design process for cast aluminium components. Using the so-called *closed chain of simulations for cast components* [16], local microstructure-based elasto-plastic material behaviour can be incorporated into load analyses using finite element analyses (FEA). By introducing the quality factor Q_T as a measure of the local material performance, it was demonstrated how local toughness can be used to determine the fracture in tensile bars as well as in complex castings. This enables a possibility to tailor the local properties and damage level by controlling the local microstructure, as demonstrated in Fig. 1a–d. It was suggested that the quality factor itself can be considered as a local parameter, which can be related to local microstructure as well as local levels of defects as porosity. It was demonstrated

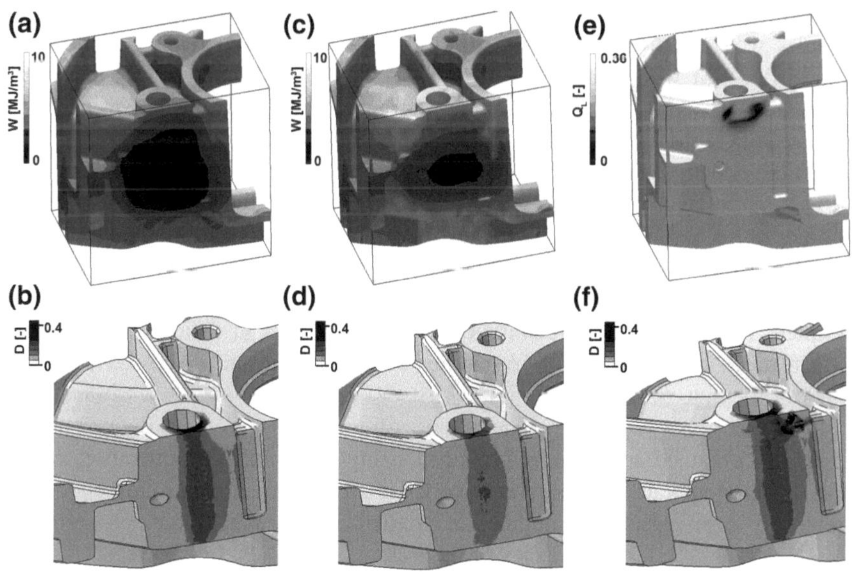

Fig. 1 Local variations in toughness and damage in an industrial casting. **a** Simulated variation in toughness **b** Corresponding local damage when subjected to load, **c** Simulated variation in toughness when a local chill is added in the mould, **d** Local damage levels has been reduced by the chill due to increased toughness **e** Locally decreased quality index due to predicted porosity **f** Increased local damage due to the decreased local toughness. Reprinted from [15] with permission from Elsevier

how such an approach would enable the prediction of damage in a complex casting subjected to load. Figure 1e–f shows how the local variations in such a local quality factor, based on the porosity predicted by a casting process simulation, can be used to predict the damage level in a complex casting containing porosity, being subjected to load. However, there is a lack of established quantitative relationships relating microstructural refinement and porosity to toughness and local variations in the quality index. If such relationships can be established, the virtual design process for robust and lightweight aluminium castings could be improved.

In the current work, a set of cast aluminium samples have been evaluated with respect to toughness, and several microstructural features and imperfections have been characterized. Statistical evaluation and regression analysis have been performed to investigate the relationships and effects of microstructure and imperfections on toughness, as well as on the quality index Q_T.

Experimental Setup

The alloy used in this experiment is a secondary alloy which meets the EN 46000 standard (equivalent to A380). The material was collected from a 8000 kg thermos after different holding times at a foundry. The chemical composition is given in Table 1. Y-shaped blocks were cast in copper and steel dies from which tensile test samples were machined. The dimensions of the samples were in accordance to ASTM B557 M with a diameter of 6 mm. A total of 25 samples were tested at a strain rate of $2 \times 10^{-4} s^{-1}$. To measure the strain a laser extensometer was used. For each set of Y-blocks, samples for the reduced pressure test were collected from which the density index, bifilm index, and porosity per cm^2 were measured.

The microstructural features measured on cross sections of the tensile test samples were area Fe-rich (polyhedral) intermetallics, total area of Fe-rich intermetallics, eutectic coarseness (average length of Si particles), secondary dendrite arm spacing (SDAS), and area porosity in the fracture surface of the tensile tested samples. The measurement techniques as well as results of measured microstructural features and basic tensile properties are published and discussed in further detail in previous work [17].

In the current work, the target value of toughness, W_C, is defined as the toughness obtained in samples produced using directional solidification with a constant cooling rate; see previous work [15, 18]. Directional solidification has previously been used in a number of studies investigating target properties of

Table 1 Chemical composition of the alloy, average wt% values

	Si	Fe	Cu	Mn	Mg	Cr	Ni	Zn	Pb	Sn	Ti	Al	SF[a]
Alloy	9.14	0.76	2.55	0.33	0.24	0.03	0.02	0.75	0.05	0.03	0.02	Bal.	1.51

[a]SF = Sludge factor; SF = %Fe + 2 × %Mn + 3 × %Cr

aluminium alloys, since it enables the production of material with high microstructural homogeneity and low amounts of defects [19]. The achieved toughness was here determined by calculating the area under the true stress–true strain curve, assuming linearity between the data points obtained from the tensile test machine. The target toughness is for each sample calculated based on its value of SDAS, using a previously established numerical relationship between SDAS and the toughness obtained in directional solidification [15].

The numerical results were evaluated numerically with Analysis of Variance (ANOVA) using the DesignExpert™ software version 10. Since in the current experiment only a limited amount of data is available with respect to the evaluated number of factors, only linear regression models have been investigated.

Results and Discussion

Table 2 shows the obtained results of the microstructural characterization, the experimental results of toughness and the investigated quality indices. Evaluating the results for toughness, a strong correlation between eutectic coarseness and toughness is identified; see Fig. 2a. Toughness decrease as the eutectic morphology becomes coarser. Toughness can also been seen to decrease with SDAS, see Fig. 2b, but the effect is not as clear and shows a larger scatter in the data.

Table 3 shows the ANOVA data for a linear regression analysis including all the measured factors. The analysis reveals, in coded factors over the investigated interval, that the effect of eutectic coarseness is the dominant effect on tensile toughness. This term also has a low p value in the ANOVA analysis, indicating that the value has a high statistical significance. The coefficient for SDAS is significantly smaller, only one-fifth of the effect of eutectic coarseness, and the statistical significance of this factor is also lower. These current results thus suggest that SDAS is not a sufficient measure to understand the tensile performance of cast aluminium alloys, and that the eutectic coarseness has a stronger correlation to mechanical behaviour and performance than SDAS.

The analysis also shows that there is no statistically significant correlation between toughness and bifilm index, number of pores or the porosity area in the fracture surface. However, there is a statistically significant influence of density index. This could may be explained by the way the density index has been measured after a reduced pressure test. The density index is thus related to the combined effect of not only porosity after solidification but also, e.g. bifilms that have been partly inflated during the reduced pressure test. The coefficient for density index in coded factors is, as for SDAS, about one-fifth of the coefficient for the eutectic coarseness, and the density index thus has a clear correlation to toughness, but the effect is smaller than the effect of eutectic coarseness.

By reducing the terms in the regression analysis to only include statistically significant terms, an equation relating toughness to density index (DI) and eutectic coarseness (EC) is identified as $W = 5.4 - 0.28 * DI - 0.04 * EC$. This equation is

Table 2 Results of microstructural characterization, toughness and quality indices

Porosity (%)	Bifilm index **(mm/cm²)**	Density index (−)	Number of pores (1/cm2)	SDAS (μm)	Eutectic coarseness (μm)	Area Fe-rich intermetallics (%)	Total area intermetallics (%)	Toughness (MJ/m3)	Q_D (MPa)	Q_T (−)
0.104	8.9	6.9	16.1	29.5	38.4	1.08	2.3	1.24	132.6	0.40
0.238	11.5	6.3	19.8	31.4	40.2	0.17	2.8	2.03	211.3	0.30
0.457	7.9	5.9	13.7	28.1	50.6	0.91	3.27	1.70	183.7	0.47
1.191	7.9	5.9	13.7	21.9	53.4	0.58	2.31	1.36	145.7	0.24
0.835	6.5	3.9	15.5	23.1	75.6	1.12	2.69	1.23	138.9	0.25
0.196	8.9	6.9	16.1	33.8	51.8	1.02	2.42	1.09	126.4	0.47
0.26	8.9	6.9	16.1	33.1	46.7	1.08	2.15	1.92	190.8	0.75
0.384	17.1	7.2	18.7	21.3	63.0	0.04	2.29	1.49	165.3	0.25
0.268	11	7.6	16.6	18.3	44.1	0.02	2.60	1.25	138.9	0.18
0.224	11.5	6.3	19.8	17	34.0	0.19	2.21	1.85	203.2	0.22
0.218	7.9	5.9	13.7	17.7	47.6	0.10	2.97	1.32	134.3	0.18
0.405	7.9	5.9	13.7	21.9	49.5	0.25	1.06	1.90	203.7	0.33
0.047	6.5	3.9	15.5	18.9	7.4	0.83	2.23	4.83	341.1	0.59
0.116	6.5	3.9	15.5	15.6	27.7	0.93	3.12	2.71	260.8	0.28
0.063	8.9	6.9	16.1	24	11.9	1.12	2.89	3.08	283.8	0.56
0.506	14.5	8.3	20	20.5	49.7	0.02	1.80	1.02	99.9	0.18
0.341	17.1	7.2	18.7	22.1	45.7	0.02	1.40	1.38	150.3	0.25
0.205	11	7.6	16.6	25.5	17.0	0.29	2.01	2.72	253.1	0.57
0.109	9.1	6.9	15.7	23.8	21.2	0.04	1.63	2.94	265.1	0.55
0.09	6.5	3.9	15.5	24	11.2	1.35	3.30	3.26	296.6	0.60
0.251	6.5	3.9	15.5	18.9	43.1	0.70	2.60	2.95	268.7	0.39
0.48	6.5	3.9	15.5	20	67.1	0.90	2.45	1.57	183.4	0.24

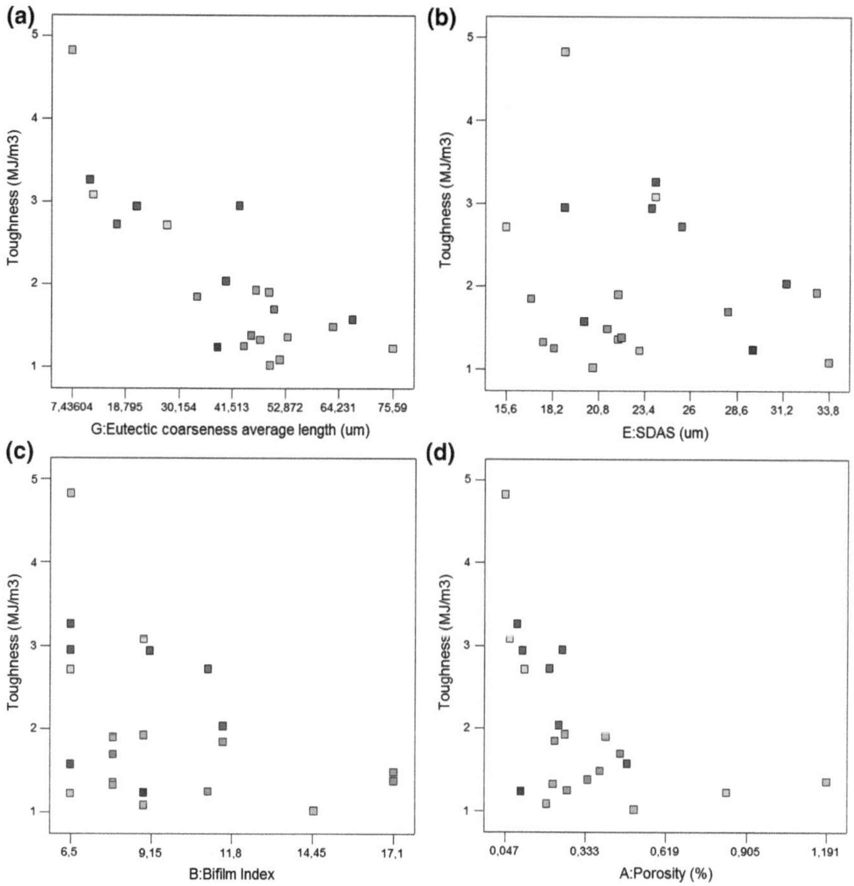

Fig. 2 Effect on toughness of **a** eutectic coarseness, **b** SDAS, **c** bifilm index, and **d** porosity. While eutectic coarseness is found to have a statistically significant effect, the other effects have low significance.

Table 3 ANOVA data for a linear model of the obtained toughness, taking into consideration all measured microstructural characteristics

Factor	p-value	Coefficient in coded factors
Bifilm index	0.4851	0.27
Density index	0.0042	−0.98
Number of pores/cm2 in RPT samples	0.6759	−0.11
Porosity, area in tensile tested samples	0.9901	−0.00370
SDAS	0.4083	0.3
Eutectic coarseness average length	<0.0001	−1.5
Area Fe-rich polyhedral intermetallics	0.604	−0.13
Total area Fe-rich intermetallics	0.2522	−0.27

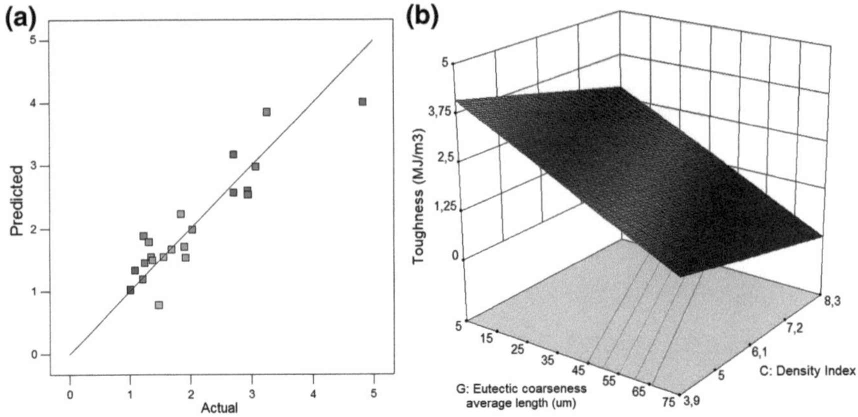

Fig. 3 **a** Predicted values of toughness versus actual, showing a reasonable agreement between the regression model and the experimental data. **b** 3D graph illustrating the regression model for toughness.

found to describe the data with a fit of $R^2 = 0.83$, thus capturing the main trends but not completely describing the experimental data. A comparison between values predicted by the model and actual measurements is shown in Fig. 3a, and a 3D plot of the model is shown in Fig. 3b.

Quality Indices

The obtained values regarding quality indices are presented in Table 2. Similar regression analyses as for toughness have been performed. The quality index Q_D is found to be mainly correlated to eutectic coarseness and density index, while there are no statistically significant effects of the other parameters. A linear regression model can be found, $Q_D = 436 - 20.6 * DI - 2.8 * EC$, which matches the experimental result with $R^2 = 0.82$. As in the case of toughness, density index and eutectic coarseness are thus the main parameters affecting this quality index. Note that for this quality index, a decreased strength and elongation (which is expected with increased microstructural coarseness) always leads to a decreased value of Q_D.

The quality index Q_T, on the other hand, is the obtained fraction of the target toughness. The average value is found to be $Q_T = 0.37$, which is quite reasonable and in line with previous results of $Q_T = 0.36$ found for an industrial HPDC casting [15]. Figure 4 shows the quality index Q_T at different eutectic coarseness. A tendency that the quality index decreases as the eutectic coarseness increases can be seen, which would indicate that a coarser microstructure is only able to reach a lower fraction of its target toughness than a fine microstructure. However, the current data is skewed since the samples with higher eutectic coarseness also

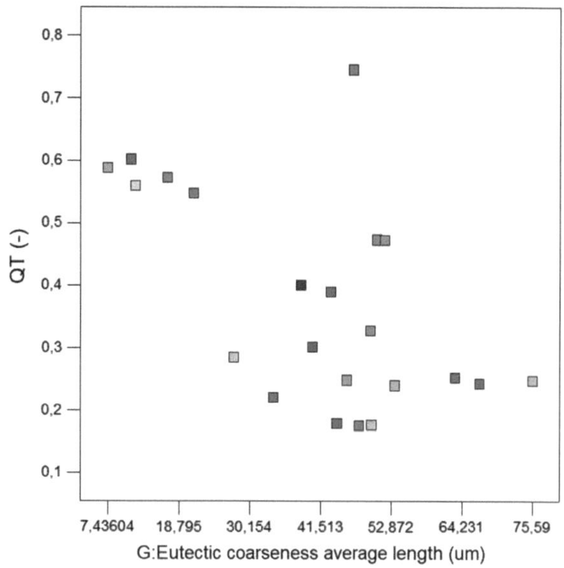

Fig. 4 Quality index QT versus eutectic coarseness.

contain a higher fraction porosity. It is thus not possible to draw any clear conclusions about the damage sensitivity of different microstructural refinements, and no statistically significant linear regression model with a reasonable coefficient of determination could be identified from the current data. One reason could be that the amount and range of data in the current work is not sufficient. Another reason could be that the method of determining the target toughness based on directional solidification, and the use of SDAS as a parameter for target toughness is not appropriate for this type of alloy; the other results in this work clearly shows that toughness has a stronger correlation to eutectic coarseness than to SDAS. It is possible that other methods and approaches to determine target toughness would led to other results. In addition, the evaluation method applied in the current work is purely empirical. Future work could focus on statistical-based evaluation methods, which is more suitable to handle this type of defect related data. By combining the previously outlined simulation methodology [15] with a statistical approach to toughness and a local quality index based on parameters predicted by casting process simulations, a new way to design more robust and reliable castings could be enabled. This could be a topic for future work in the area.

Conclusions

It is found that the toughness of the investigated aluminium alloy is highly related to the eutectic morphology, where a coarser eutectic morphology exhibits a lower toughness. The effect of the eutectic morphology is higher than the effect of SDAS and has a larger statistical significance. Density index, as a measure of the ease of porosity formation in the material, is also found to have a statistically significant effect. The quality index suggested by Tiryakioglu et al. have been evaluated, indicating that in average 37% of the materials target toughness is reached in the castings. However, the current analysis is not sufficient to establish a statistically significant relationship between the quality index and the measured microstructural parameters. Additional work is recommended in order to investigate the connections between microstructural characteristics, defects and toughness.

Acknowledgement The Swedish Knowledge Foundations are greatly acknowledged for financing the CompCAST+ research profile at Jönköping University.

References

1. Do Lee C (2007) Effects of microporosity on tensile properties of A356 aluminum alloy. Mater Sci Eng A 464:249–254
2. Lebyodkin M, Deschamps A, Bréchet Y (1997) Influence of second-phase morphology and topology on mechanical and fracture properties of Al-Si alloys. Mater Sci Eng A 234–236:481–484
3. Wang QG, Cáceres CH, Griffiths JR (2003) Damage by eutectic particle cracking in aluminum casting alloys A356/357. Metall Mater Trans A 34:2901–2912
4. Wang QG, Apelian D, Lados DA (2001) Fatigue behavior of A356/357 aluminum cast alloys. Part II-Effect of microstructural constituents. J Light Met 1:85–97
5. Campbell J (2003) Castings, 2nd edn. Elsevier Butterwort-Heinemann, Oxford
6. Alexopoulos ND, Tiryakioğlu M (2009) Relationship between fracture toughness and tensile properties of A357 cast aluminum alloy. Metall Mater Trans A 40:702–716
7. Tiryakioglu M, Staley JT, Campbell J (2004) Evaluating structural integrity of cast Al-7% Si-Mg alloys via work hardening characteristics: II. A new quality index. Mater Sci Eng A 368:231–238
8. Kuang JH, Chen YC (1996) The values of J-integral within the plastic zone. Eng Fract Mech 55:869–881
9. Tiryakioglu M, Campbell J, Alexopoulos N (2009) Quality indices for aluminum alloy castings: a critical review. Metall Mater Trans B 40:802–811
10. Drouzy M, Jacob S, Richard M (1980) Interpretation of tensile results by means of quality index and probable yield strength. AFS Int Cast Met J 5:43–50
11. Cáceres CH (1998) A rationale for the quality index of Al-Si-Mg casting alloys. Int J Cast Met Res 10:293–299
12. Cáceres CH, Makhlouf M, Apelian D, Wang L (2001) Quality index chart for different alloys and temperatures: a case study on aluminium die-casting alloys. J Light Met 1:51–59
13. Tiryakioglu M, Staley JT, Campbell J (2008) The effect of structural integrity on the tensile deformation characteristics of A206-T71 alloy castings. Mater Sci Eng A 487:383–387
14. Tiryakioglu M, Campbell J, Staley JT (2004) Evaluating structural integrity of cast Al-7% Si-Mg alloys via work hardening characteristics: 1. Concept of target properties. Mater Sci Eng A 368:205–211
15. Olofsson J (2018) Local microstructure-based material performance and damage in design and finite element simulations of cast components. J Comput Des Eng (in press)
16. Olofsson J, Svensson IL (2012) Incorporating predicted local mechanical behaviour of cast components into finite element simulations. Mater Des 34:494–500

17. Bjurenstedt A, Seifeddine S, Jarfors AEW (2015) On the complexity of the relationship between microstructure and tensile properties in cast aluminum. Int J Mod Phys B 29
18. Olofsson J, Svensson IL, Lava P, Debruyne D (2014) Characterisation and investigation of local variations in mechanical behaviour in cast aluminium using gradient solidification, Digital Image Correlation and finite element simulation. Mater Des 56:755–762
19. Seifeddine S, Johansson S, Svensson IL (2008) The influence of cooling rate and manganese content on the β-Al5Si phase formation and mechanical properties of Al-Si based alloys. Mater Sci Eng A 490:385–390

The Myth of Hydrogen Pores in Aluminum Castings

Murat Tiryakioğlu

Abstract The equilibrium solubility of hydrogen in liquid and solid aluminum is reviewed. Homogeneous and heterogeneous nucleation of hydrogen as pores are evaluated. It has been determined that homogeneous nucleation of hydrogen pores is impossible. Moreover, in situ observations of pore formation in melts with high hydrogen contents are inconsistent with heterogeneous nucleation of hydrogen pores. The only mechanism for hydrogen to contribute to pore formation is the inflation of bifilms by diffusion. Hence, in the absence of bifilms in the melt, measuring and controlling hydrogen content in the melt would be unnecessary.

Keywords Supersaturation · Fracture pressure · Microporosity · Nucleation · Growth

Introduction

Hydrogen is the only gas known to be soluble in liquid aluminum. Hydrogen absorption into liquid aluminum is due to the decomposition moisture in air as well as the oxidation of aluminum:

$$2H_2O \rightarrow 2H_2 + O_2$$

$$3H_2O + 2Al \rightarrow Al_2O_3 + 3H_2$$

The diatomic hydrogen gets dissolved in aluminum as:

$$H_2 \rightarrow 2[H]$$

There is a significant difference in the solubility of hydrogen in liquid and solid aluminum. This difference has led many researchers and practitioners to hypothe-

M. Tiryakioğlu (✉)
University of North Florida, Jacksonville, FL 32224, USA
e-mail: m.tiryakioglu@unf.edu

M. Tiryakioğlu et al. (eds.), *Shape Casting*, The Minerals, Metals & Materials Series,
https://doi.org/10.1007/978-3-030-06034-3_14

size that the rejection of hydrogen by solidifying aluminum into the mushy zone results in the local hydrogen concentration to exceed the equilibrium concentration, which results in pore formation. This hypothesis will be visited later. First, solubility of hydrogen is reviewed.

Equilibrium Solubility of Hydrogen

The equilibrium solubility of hydrogen, $S_{eq[H]}$, in aluminum was investigated in several studies [1, 2]. In most of the literature, solubility of hydrogen in liquid aluminum has been expressed in milliliters (or cubic centimeters) of diatomic hydrogen dissolved in 100 grams of aluminum, at a pressure of 1 atmosphere (101 325 Pa) and a temperature of 0 °C. To put the concentrations in this paper in context, the unit mL/100 g of Al was converted to parts per million (ppm) for hydrogen:

$$1\,\text{mL}/100\text{g of Al} = 24.1\,\text{ppm}$$

It should be noted that 1 mL/100 g of Al was mistakenly assumed to be equal to 1 ppm in several references in the literature. $S_{eq[H]}$ can be expressed in the Arrhenius form as a function of partial pressure of hydrogen, P_H, and temperature:

$$S_{eq[H]} = S_{o[H]} \cdot \sqrt{P_H} \cdot e^{\frac{-\Delta H}{RT}} \tag{1}$$

where $S_{o[H]}$ is solubility coefficient (ppm/$\sqrt{Pa}$), P_H is the partial pressure of hydrogen, ΔH is the enthalpy of solution, R is the gas constant (8.314 J/mol K), and T is temperature (K). The dependence of solubility of a diatomic gas in a liquid and solid solvent on the square root of its partial pressure is known as Sievert's law [3]. Changing pressure and temperature during experiments to determine the amount of hydrogen dissolved in aluminum is used frequently by researchers. Talbot and Anyalebechi [1] used the Sievert method in their experiments on high-purity aluminum and determined that their data are in close agreement with those by Ransley and Neufeld [4] and Opie and Grant [2]. The constants of Eq. (1) for liquid and solid aluminum as well as the equilibrium solubility of hydrogen at melting temperature, T_m, are presented in Table 1. Those for liquid aluminum have been calculated from the data from the three studies [1, 2, 4], whereas the results of Ichimura et al. [5] were used for solid aluminum.

Table 1 Constants for Eq. (1)

	Liquid	Solid
$S_{eq[H]}$ (ppm) at T_m	16.14	1.05
$S_{o[H]}$ (ppm/$\sqrt{Pa}$)	39.7	12.6
ΔH (J/mol)	51 690	63 942

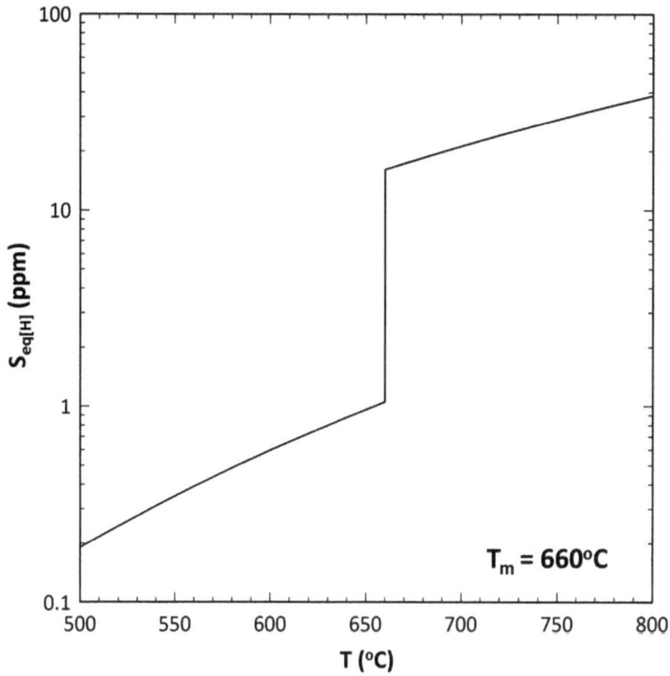

Fig. 1 Equilibrium solubility of hydrogen in liquid aluminum as a function of temperature

The change in $S_{eq[H]}$ with temperature in liquid and solid aluminum at P = 101.3 kPa (1 atm) is presented in Fig. 1. By using the numbers listed in Table 1, the ratio of $S_{eq[H]}$ in liquid to solid is calculated as 15.4.

Hydrogen Pore Nucleation

Homogeneous Nucleation

It is well known that the necessary condition for homogeneous nucleation of a pore in a solidifying liquid, i.e., the critical radius above which an embryonic pore is stable, r^*, is found by:

$$r^* = \frac{-2\sigma}{\Delta P^*} \tag{2}$$

where σ is the surface tension of the liquid (N m) and ΔP^* is the pressure differential (Pa) and is a negative number. Recently, the author [6] showed by using a combination experimental and molecular dynamics data that the fracture pressure of aluminum at the melting temperature is −4 GPa. Hence, by inserting $\Delta P^* = -4$ GPa and $\sigma = 1.03$ N m [7] in Eq. (2), r^* is calculated as 0.512 nm. To attain this pressure, the number of moles of hydrogen atoms needed in the embryonic pore is calculated as;

$$n^* = \frac{P_p . V_p}{RT_m} \tag{3}$$

where P_p and V_p are gas pressure inside the pore and volume of the pore, respectively. The embryonic pore will be spherical. By using $P_p = 4$ GPa, a positive number, n^* is found to be 2.92×10^{-22}. By multiplying this number by the Avogadro's number, it is determined that a total of 176 hydrogen atoms are needed in the embryonic pore. These numbers are discussed below.

When $P_p = 4$ GPa, whether the pore will be stable can be analyzed in two different ways; (i) by using the Sievert's law and (ii) by the effect of the pressure on the equilibrium vacancy concentration around the pore.

By using Eq. (1), the local equilibrium solubility of hydrogen in aluminum around the pore, $S_{eq[H]-p}$, can be calculated as;

$$S_{eq[H]-p} = S_{eq[H]} \cdot \sqrt{\frac{P_p}{P_H}} \tag{4}$$

Inserting $P_p = 4$ GPa and $P_H = 101.3$ kPa in Eq. (4), it is found that the equilibrium solubility of hydrogen is increased by 200 times locally. Regardless of whether the pore is surrounded by solid or liquid aluminum, the pore would be expected to collapse rather than grow because the increase in surface energy with growth needs to be overcome, whereas dissolution of hydrogen reduces the surface energy. With the same logic, the maximum pressure inside a hydrogen pore, $P_{p(max)}$ can be found as;

$$P_{p(max)} = P_H \left(\frac{S_{eq[H]-l}}{S_{eq[H]-s}} \right)^2 \tag{5}$$

where subscripts l and s mean liquid and solid, respectively. Since the ratio inside the parentheses is 15.4 and $P_H = 101.3$ kPa, $P_{p(max)}$ is approximately 24 MPa, less than 1% of the pressure needed for the homogeneous nucleation of a hydrogen pore. Therefore, homogeneous nucleation due to hydrogen supersaturation is impossible. This is in agreement with the findings of Campbell [8] and Shahani and Fredriksson [9].

Continuing with the (wrong) assumption that hydrogen will overcome the homogeneous nucleation barrier and $P_p = 4$ GPa, the effect of this pressure on the solid aluminum surrounding it will now be considered. Fukai [10] showed that under pressures of several 1–5 GPa and at high temperatures, vacancy concentrations in the order of 10^{-1} can be achieved, as opposed to the equilibrium concentration of 6×10^{-4} in solid aluminum at T_m [11]. Normally, vacancies in aluminum trap one hydrogen atom [12]. Under pressures that can create superabundant vacancies, however, they can trap up to ten hydrogen atoms [13]. Therefore, the combination of much higher concentration of vacancies and the added capacity of each vacancy to trap multiple H atoms will lead to the immediate dissolution of 176 hydrogen atoms needed in an embryonic pore. Hydrogen will subsequently remain in solid solution upon solidification and can even serve to increase strength by pinning down dislocations [14].

Oxides and Hydrogen

It has been well known since the early pioneering work on reduced pressure tests by Rooy and Fischer [15] and Brondyke and Hess [16] that the effect of hydrogen on pore formation is dependent to a great extent on the oxide content of the melt. The results of Brondyke and Hess are presented in Fig. 2, in which the volume percent of pores, f_{pore}, is plotted against hydrogen content of the melt before and after filtering. Results clearly indicate that it is the oxide, rather than the hydrogen, content that has the most effect on the formation of pores. In the absence of old oxide bifilms, it is difficult for hydrogen pores to form, which is consistent with the findings of Dispinar et al. [17] and Griffiths and Raiszadeh [18]. During filtering, coarse "old" oxides will be expected to be taken out of the melt. Much thinner "young" oxides, however, have a much higher likelihood of passing through the filter. In addition, in filtering equipment at the end of the filter, liquid metal falls as a streaming jet [19, 20] for a distance large enough to create fresh damage to the metal by surface entrainment and bubble formation. Hence, filtering is expected to eliminate most or all old oxides but pass through and even create young oxides. The slight increase in volume percent of pores at higher hydrogen contents in filtered aluminum can be attributed to smaller oxide films that have gone through the filter and fresh damage after filtering. These young oxide bifilms, possibly created during the fall after filtering, are much smaller in size and lead to microporosity in aluminum castings.

Shahani and Fredriksson [9] concluded that hydrogen pores could form only by nucleating heterogeneously on oxides, which would need approximately −2 MPa for oxide particles that are 2 μm in diameter. However, Yousefian and the author [21] found that ΔP^* is generally assumed to be −0.1 MPa (−1 atm) in modeling literature so that, a hydrogen pore would be expected to nucleate once there is hydrogen supersaturation at low solid fractions. This is in contrast with the results of Lei et al. [22] who investigated pore formation in situ during directional solidification of Al-7 wt% Si and Al-12 wt% Si alloys. They observed that pores appeared in the liquid at a distance of approximately 15 mm from the eutectic solid–liquid interface, where the hydrogen supersaturation is usually expected to be quite low and the partial pressure of hydrogen to be close to 0.1 MPa (1 atm). Murphy et al. [23] conducted in situ solidification experiments on a grain-refined Al-20 wt% Cu alloy at very low cooling rates (0.084 K/s). Their observations at a

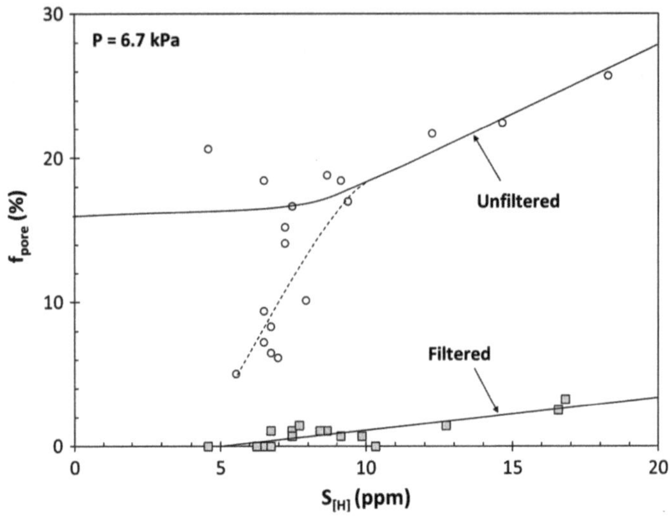

Fig. 2 Effect of filtering and hydrogen content on the volume percent of pores in an Al–Cu alloy solidified under a reduced pressure of 6.7 kPa

solid fraction of 0.13 are presented in Fig. 3. Yousefian and Tiryakioğlu found the local solid fraction to be 0.434 at the onset of pore formation for the experiments of Murphy et al. By using the equilibrium solubility of hydrogen in liquid and solid, the concentration of hydrogen was estimated to be 1.73 times the equilibrium amount. Consequently, based on Sievert's law, the partial pressure of hydrogen was calculated as 0.3 MPa, well below the number required for heterogeneous nucleation on oxides calculated by Shahani and Fredriksson. Hence, it was concluded that nucleation is bypassed during pore formation, which involves only opening of bifilms by hydrogen diffusion [21].

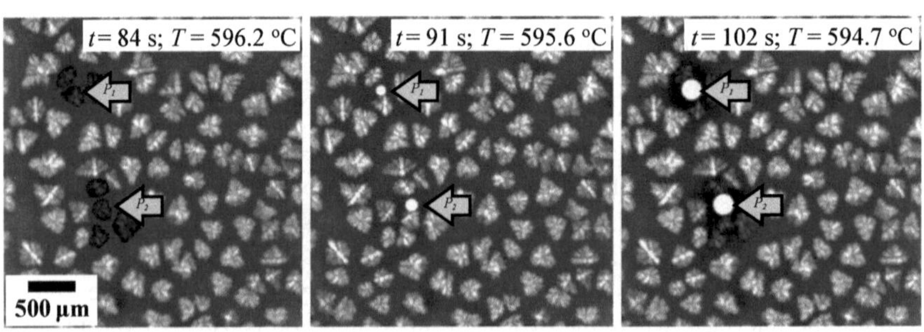

Fig. 3 The solidification of grain-refined Al-20 wt% Cu sample. The locations of pores are indicated by arrows. Grains that were affected by the nucleation and growth of pores are indicated in color (No copyright, courtesy of D. Browne)

In a recent study [24], pores in reduced pressure test samples of an A356 alloy were evaluated by scanning electron microscopy. Pores were found to be initiated by entrainment defects, i.e., oxide bifilms. However, there were many inactive bifilms between dendrites both in the vicinity and away from pores. These bifilms were visible only in x-ray maps and would remain unnoticed in routine analysis. When hydrogen content is increased, a lower number of bifilms can be expected to remain inactive. Therefore, hydrogen can be interpreted as an agent that makes preexisting damage to the metal, bifilms, visible, at least partially.

If the aluminum melt was free from bifilms, it would be unnecessary to measure hydrogen, and no degassing procedure and equipment would be needed. Hydrogen can even be used as a tool to make any damage to the metal visible. It would then be practical to dissolve maximum level of hydrogen in aluminum and yet require no pores to be formed as a condition for defining high quality in a casting quality specification, as suggested by Prof. John Campbell [25].

Conclusions

- It is impossible for hydrogen to nucleate homogeneously in liquid aluminum.
- In situ experiments in the literature on pore formation in aluminum showed that pores can form away from the solid–liquid interface and/or when the supersaturation and pressure inside the pores are low. These observations are in contrast with the heterogeneous nucleation pressures on oxide bifilms.
- Nucleation is bypassed during pore formation. Hydrogen diffuses to oxide bifilms and inflates them. Hence, hydrogen serves as an agent to make entrainment defects visible.
- Without bifilms, it would be unnecessary to measure or control hydrogen levels in the melt. It would even be practical to require maximum level of hydrogen to be dissolved in aluminum and yet require that no pores to be formed in the casting.

References

1. Talbot DEJ, Anyalebechi PN (1988) Mater Sci Technol 4:1–4
2. Opie W, Grant N (1950) JOM 2:1237–1241
3. Sieverts A (1929) Zeitschrift für Metallkunde 21:37–46
4. Ransley CEN (1948) J Inst Met 74:599–620
5. Ichimura M, Imabayashi M, Hayakawa M (1979) J Jpn Inst Met 43:876–883
6. Tiryakioğlu M, Metall Mater Trans A. (in print)
7. Lu H, Jiang Q (2005) J Phys Chem B 109:15463–15468
8. Campbell J (2006) Mater Sci Technol 22:127–145
9. Shahani H, Fredriksson H (1985) Scand J Metall 14:316–320

10. Fukai Y (2003) Phys Scr 2003:11
11. Thomas G, Willens RH (1966) Acta Metall 14:1385–1390
12. Birnbaum HK, Buckley C, Zeides F, Sirois E, Rozenak P, Spooner S, Lin JS (1997) J Alloy Compd 253–254:260–264
13. Ismer L, Park MS, Janotti A, Van de Walle CG, Phys Rev B 80 (2009)
14. Xie D, Li S, Li M, Wang Z, Gumbsch P, Sun J, Ma E, Li J, Shan Z (2016) Nat Commun 7:13341
15. Rooy E, Fischer E (1968) AFS Trans 76:237–240
16. Brondyke KJH (1964) Transaction of the American Institute of mining, Metall Eng 230:1542–1546
17. Dispinar D, Akhtar S, Nordmark A, Di Sabatino M, Arnberg L (2010) Mater Sci Eng A 527:3719–3725
18. Griffiths WD, Raiszadeh R (2009) J Mater Sci 44:3402–3407
19. Apelian D, Mutharasan R (1980) JOM 32:14–19
20. Haberl K, Schumacher P, Geier G, Stauder B (2009) Metall Mater Trans B 40:812
21. Yousefian P, Tiryakioğlu M (2018) Metall Mater Trans A 49:563–575
22. Lei Z, Hengcheng L, Ye P, Qigui W, Guoxiong S (2011) Res Dev
23. Murphy A, Browne D, Houltz Y, Mathiesen R (2016) IOP conference series: materials science and engineering, IOP Publishing, p 012067
24. Tiryakioğlu M, Yousefian P, Eason PD (2018) Metall Mater Trans A 49:5815–5822
25. Campbell J (2015) Complete casting handbook: metal casting processes, metallurgy, techniques and design. Elsevier Science

Casting Defect Analysis on Fracture Surface of 356 Aluminium Alloy

Özen Gürsoy, Eray Erzi and Derya Dispinar

Abstract Investigation of the fracture surface is an effective method in characterizing casting defects. Some data can be collected such as the population of defects and initiation of fracture by means of investigation of fracture surface via a scanning electron microscope (SEM). In this study, tensile test samples from two different 356 aluminium melts that have good and quite bad quality were poured into die and sand moulds under two different conditions: no addition and Al–Ti–B grain refiner addition. After the tensile test, casting defects in fracture surfaces are analysed and classified via SEM. The area fraction of defects (AFD) is created and correlated with bifilm index and tensile properties such as UTS and elongation.

Keywords Defect · Fracture surface · Tensile · Melt quality · Bifilm

Introduction

The presence of defects, such as bifilms, inadequate feeding, sand erosion (in sand mould casting), negatively affects the tensile properties of cast parts. Unfortunately, in many castings, particularly large-volume castings, these defects occur and unexpected out-of-service situations are encountered.

The defects like porosity in the tensile sample decrease the load bearing area (Fig. 1). The voids create a multiaxial stress centre and cause the elongation to accumulate on their near sides. The defects in fracture surface of tensile bar such as sand particles, bifilm, and inclusion can be visible on fracture surface clearly after tensile test via scanning electron microscope. Because of load bearing area

Ö. Gürsoy (✉) · E. Erzi · D. Dispinar
Metallurgical and Material Engineering Department, Istanbul University-Cerrahpasa, Istanbul, Turkey
e-mail: ozen.gursoy@gmail.com

M. Tiryakioğlu et al. (eds.), *Shape Casting*, The Minerals, Metals & Materials Series,
https://doi.org/10.1007/978-3-030-06034-3_15

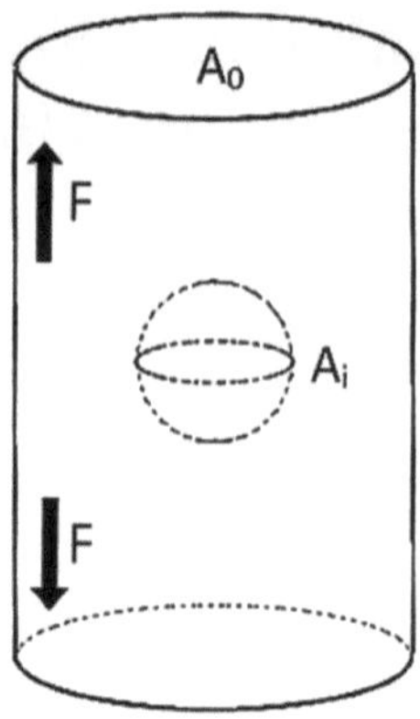

Fig. 1 Illustrated defect in the tensile sample

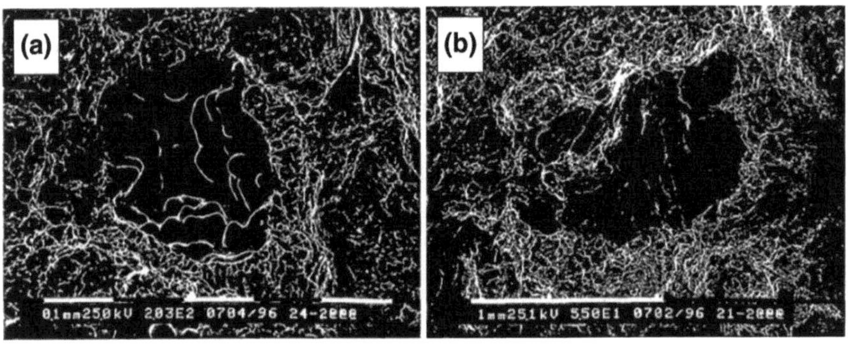

Fig. 2 SEM micrograph of clearly visible dendrites (a) and rough and dull oxide films (bifilms) [1]

decreasing depending on the presence and quantity of defects, there should be relation between area fraction of defects (AFD) and tensile values.

Caceres and Selling [1] determined two fundamental defect in tensile test bars' fracture surface of A356 aluminium alloy in their study: clearly visible dendrites and matt and rough oxides as can be seen in Fig. 2. They observed relations between either UTS values and area fraction of defects (AFD) or elongation and AFD.

Surappa et al. [2] suggested that the values of strength and elongation affected the porosity dimensions on the fracture surface, and they emphasized that decreasing the cross section causes the stress and plastic strain near to the defects. Teng et al. [3] claim that pore size should be independent of casting conditions.

Experimental Study

In this study, the relation between tensile values and fracture surface defect ratio was investigated. 356 aluminium alloys were supplied from automotive industry's scraps. The chemical composition of the 356 alloys can be seen in Table 1. The sand moulds were prepared by mixing silica sand, resin, and hardener. The charge was pre-heated at 400 °C for 1 h, and then, the casting temperature was set at 725 °C. Two different casting conditions, with (ratio of wt. 0.1%) and without Al–5Ti–1B addition, were applied. The pouring was begun 10 min after grain refinement master alloy addition. In these conditions, tensile bars are poured to analyse mechanical properties and fracture surface analysis into either die or sand moulds. After tensile tests, JEOL JSM-5600 scanning electron microscope was used in fracture surfaces examination. SEM images were analysed with Sigma Scan Pro 5 image analysis software. With reference to Fig. 1, the cross section in the smooth region, A_0, and the cross section in the defect, A_i, a simple equation was used as can be seen in (1) to calculate area fraction of defects (AFD).

$$AFD = \frac{100xA_i}{A_0} \tag{1}$$

Results and Discussion

According to tensile test results, the worst and inconsistent UTS and elongation values are observed in die casting as can be seen in Figs. 3 and 4.

Grain refiner addition cannot rely on a better result for tensile properties, and it has an adverse effect on elongation values in die cast samples. The best tensile test results are obtained distinctly in sand casting.

After the tensile tests, fracture surface analyses were carried out to five fracture surface of the tensile sample each condition. All defects in a fracture surface were included in the AFD ratio. The number of detected defect per fracture surface is 15–52. SEM micrographs of fracture surface depending on casting conditions are as can be seen in Fig. 5.

520 defects were detected in 20 fracture surfaces, and all defects in every fracture surface were examined closely. Since all defects are initiated by bifilms, a direct correlation between bifilm quantity and area fracture of defects were found in this work. A representative example is given in Fig. 6.

Table 1 Chemical composition of the 356 alloys (wt%)

Si	Fe	Mg	Ti	B	Sr
7.31	0.13	0.26	0.11	0.001	0.008

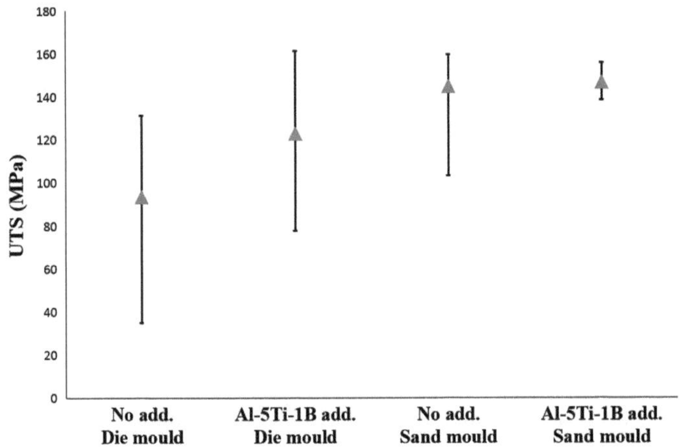

Fig. 3 UTS results of tensile tests depending on conditions

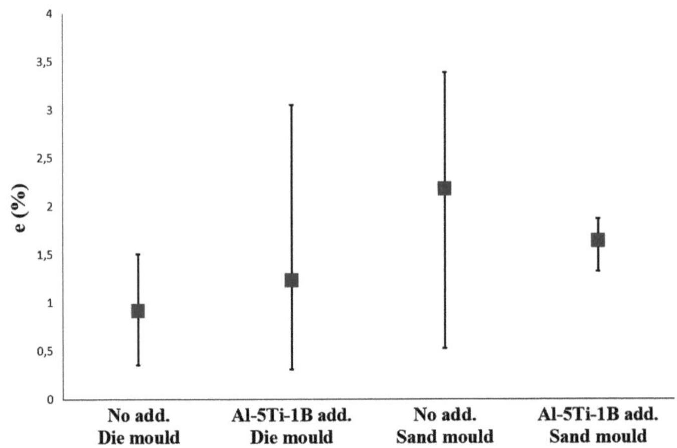

Fig. 4 Elongation results of tensile tests depending on conditions

In close examination, old oxide structures in the form of wrinkled paper as can be seen in Fig. 6 were observed in almost all of the defects as in the example Fig. 6.

Except in two cases, in all conditions, AFD ratio is followed by the linear decreasing trend for both UTS and elongation values and can be seen in Figs. 7 and 8.

AFD ratios and tensile results are almost fully alignment.

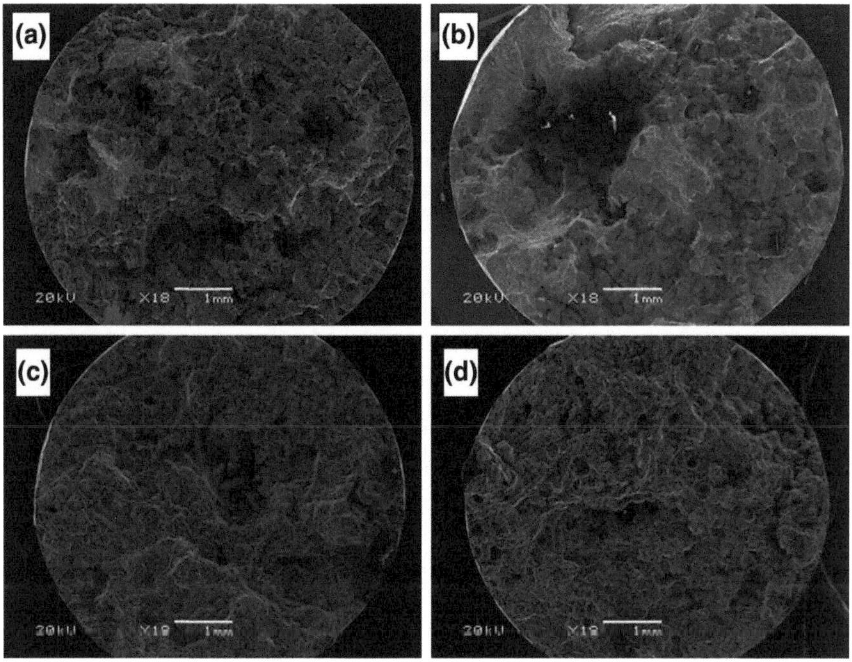

Fig. 5 SEM micrograph of fracture surfaces of different conditions (a: no grain refiner addition and die cast, b: Al-5Ti-1B added and die cast, c: no grain refiner addition and sand cast, d: Al-5Ti-1B added and sand cast)

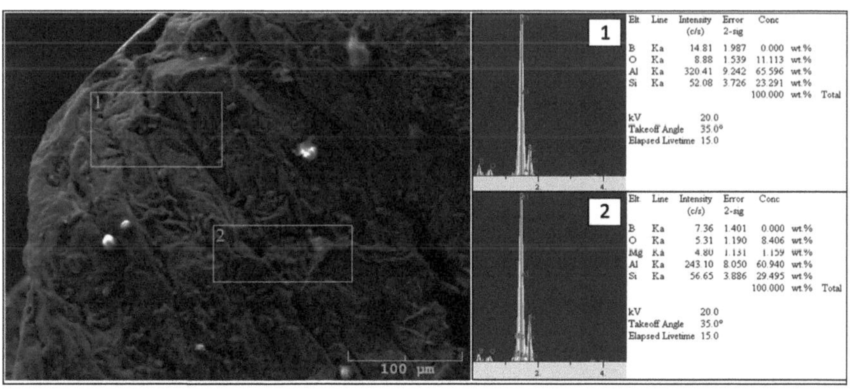

Fig. 6 SEM image of bifilm defect

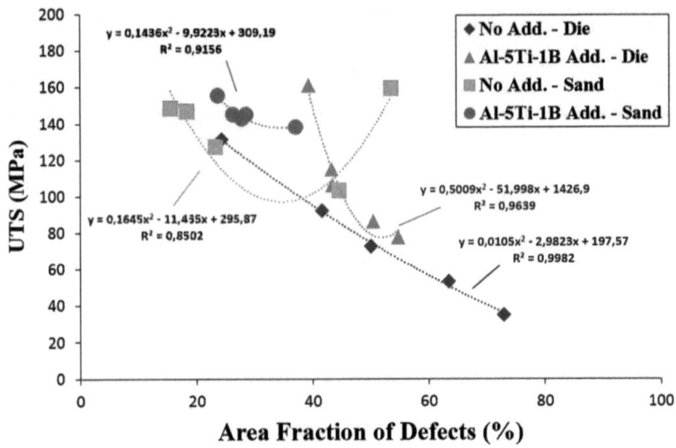

Fig. 7 Relation between UTS and area fraction of defects

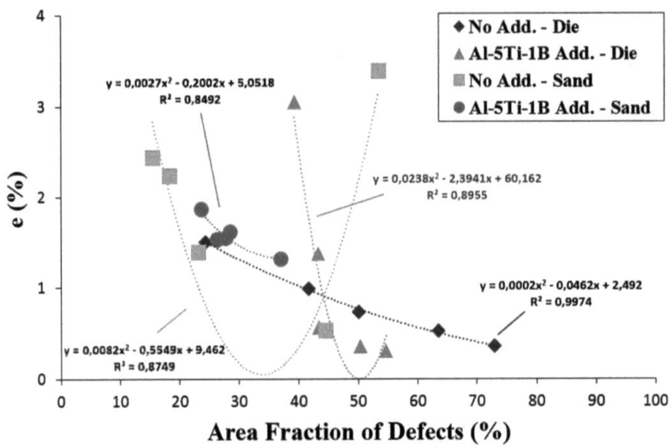

Fig. 8 Relation between elongation and area fraction of defects

Conclusion

- Area fraction of defects (AFD) ratio is a useful method for material characterization.
- There is a good relation between AFD ratio and tensile properties.
- The tensile properties increase while the AFD ratio decreases.
- Oxide structure, i.e. bifilm, is the main problem in decreasing the tensile properties.

References

1. Caceres CH, Selling BI (1996) Casting defects and the tensile properties of an AlSiMg alloy. Mater Sci Eng A 220(1–2):109–116
2. Surappa MK (1986) Effect of macro-porosity on the strength and ductility of cast Al-7 Si-0.3 Mg alloy. Scr Metall 20(9):1281–1286
3. Teng X, Mae H, Bai Y, Wierzbicki T (2009) Pore size and fracture ductility of aluminum low pressure die casting. Eng Fract Mech 76(8):983–996

Investigation of Casting Quality Change of A356 by Duration in Liquid State

Muhammet Uludağ, Mikdat Gurtaran and Derya Dispinar

Abstract This study aims to investigate how casting quality change in aluminum melt during the holding period in liquid state. 10 kg of ingot was melted in a SiC crucible by using an electrical furnace. A sample was taken every five minutes from 0 to 55 min after the alloy was melted. Reduced pressure test (RPT) samples were collected to be solidified under 80 mbar. All samples were sectioned into two parts vertically, and one of them was prepared for metallographic examination. Surface of the samples was subjected to image analysis by using via image analysis software. Bifilm index and bifilm area were analyzed in detail. As a result, it can be concluded that first 30 min of liquid metal is important, because casting quality tends to get worse after 30 min.

Keywords A356 alloy · Bifilms · Holding time · Casting quality

Introduction

Aluminum and its alloys are one of the most important engineering materials of today. Production and applications of Al increase day by day thanks to its low density and high strength ratio [1 3]. Improving of properties of these alloys creates a working area for aircraft and automobile industries. Casting technique used in production of Al parts is the most preferred method because of its superior properties such as the possibility of producing complex geometry, economy, and the suitability for mass production [4, 5]. It is very important to optimize each stage of casting process in order to produce quality material. The casting quality must be

M. Uludağ (✉) · M. Gurtaran
Metallurgical and Materials Engineering, Bursa Technical University, Bursa 16310, Turkey
e-mail: dr.uludagm@gmail.com

M. Gurtaran
e-mail: mikdat24@hotmail.com

D. Dispinar
Metallurgical and Materials Engineering, Istanbul University, Istanbul 34320, Turkey

M. Tiryakioğlu et al. (eds.), *Shape Casting*, The Minerals, Metals & Materials Series,
https://doi.org/10.1007/978-3-030-06034-3_16

high for the production of materials with high mechanical properties. It is affected by ingot quality and degassing operations. Ingot quality is one of the most important factors because a clean ingot provides defect-free cast parts.

It has been recommended to start with a clean liquid metal [6, 7]. It is explained by Campbell [8, 9] that oxide films formed on the surface of liquid metal are disturbed during casting and submerge into the melt by folding. This mechanism forms bifilms (double oxide films). Bifilms decrease liquid metal quality. These double oxide films can come from ingot (during the production stage: old bifilms) or they can be formed as a result of surface turbulence during casting operations (young bifilms). It is presented by Campbell [10–12] that bifilms in the melt are the main reason of many casting defects. Casting defects affect the mechanical properties of the alloys negatively. Surface turbulence formed during casting causes to form young bifilms. For this reason, it is necessary to avoid the formation of surface turbulence in the casting process and use optimized runner systems to eliminate the surface turbulence during filling.

Porosity encountered in aluminum alloys is one of the most common casting defects. Porosity formation mechanism is explained by many researchers in the literature [13–16]. Recently, a new approach to the porosity formation mechanism is proposed by Campbell and Dispinar [17, 18]. They have been claimed that shrinkage and gas are effective on the porosity formation but they are not main effect, they only have trigger effect. It has been shown that bifilms play a vital role on the porosity formation as the initiator of porosity [19, 20].

Bifilms in the liquid metal can be removed by degassing and addition of grain refiner as master alloys such as AlTi5B1 [21–23]. In degassing process, an inert gas (usually Ar), which does not react with the liquid metal, is purged and this gas takes along bifilms when it moves to the top of the crucible. As a result of this process, a cleaner liquid metal is obtained. Another technique to remove bifilms is the addition of Ti master alloys [24–26]. Titanium which has higher density than the density of liquid metal is added into the melt as master alloy at a certain temperature. When Ti moves to the bottom of the crucible because of its higher density, it takes along bifilms to the bottom of the crucible. Therefore, a much higher quality casting can be achieved. Ti master alloys are not only used as grain refiner but also bifilms cleaner.

This study aims to investigate the change in melt quality during holding of the melt at long durations.

Experimental Procedure

In this study, A356 alloy was used. Chemical composition of the alloy is presented in Table 1.

Table 1 Chemical composition of A356 alloy

Alloy	Si	Fe	Cu	Mn	Mg	Zn	Ti	Al
A356	6.80	0.19	0.003	0.001	0.30	0.011	0.108	Rem.

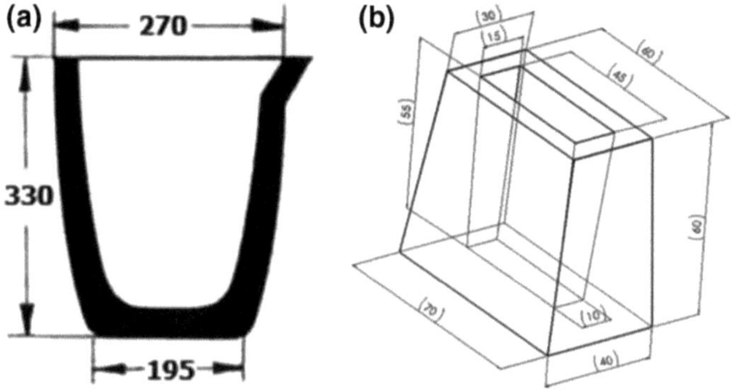

Fig. 1 Dimensions of; **a** SiC pot, **b** RPT mold

Primary A356 alloy was melted in SiC crucible in resistance furnace (Fig. 1a) which has 22 kg capacity. Melting process was carried out at 740 °C temperature, and Ti was added into the melt as AlTi5B1 master alloy at the same temperature.

Reduced pressure test (RPT) samples were collected with 5 min intervals after AlTi5B1 addition (Fig. 1b). SigmaScan image analysis software was used to determine the bifilm index and bifilm area. These results were examined statistically in Minitab.

Results and Discussion

Bifilm has been shown as the main factor that determines the casting defects. Bifilms can be quantified by RPT test.

Representative images of six RPT samples obtained at different holding time after Ti addition are given in Fig. 2. As can be seen in the figure, the number and the size of pores are quite high at 0 and 5th min. When the results are examined for from 0 to 30 min holding time, it can be said that there is a decrease in the number of bifilms depending on the holding time. After 5 min of holding, there are differences in bifilm number and size.

Bifilm analyses based on different holding time are plotted and presented in Fig. 3. It is clearly seen that bifilm index value is quite high at 0 min holding time. Although there is a decreasing in bifilm index up to 30 min of holding time, an increase in bifilm index is seen after 30 min of holding time. It is believed that the effect of Ti addition on bifilms disappears at this duration. Bifilm index value increases consistently after 30 min of holding time. The number of bifilms and the total bifilm area exhibit the same behavior as the bifilm index. While bifilm number is maximum at 0 min holding time, it is minimum at 30 min. It is seen that total bifilm area is the highest at 50 min holding time. When all results given in Fig. 3 are examined, it is understood that there is a relationship between bifilm index, total

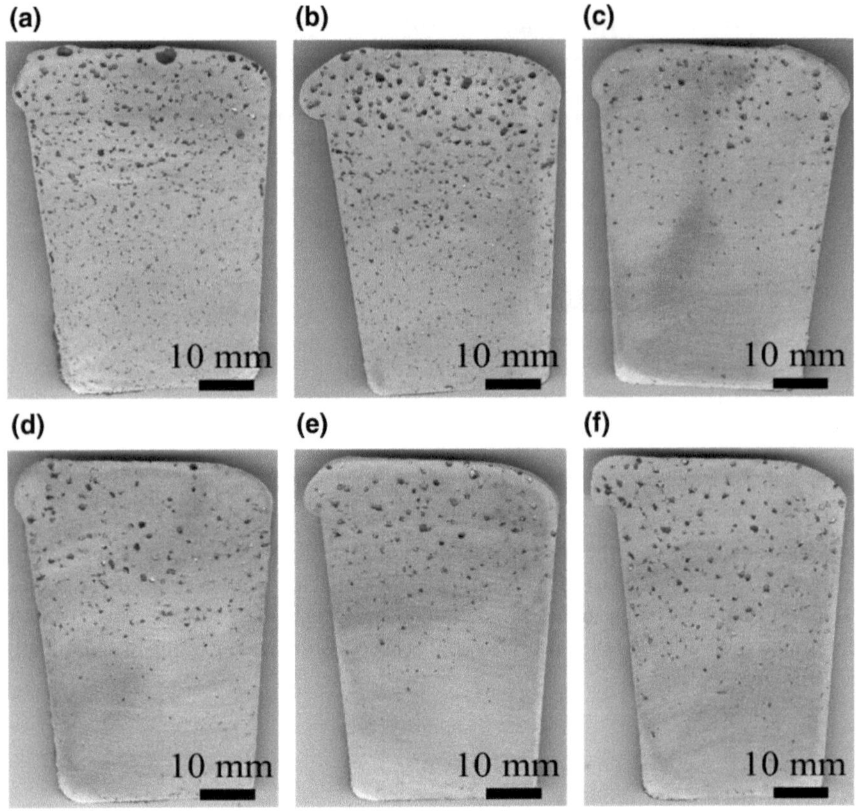

Fig. 2 Representative images of RPT for effect of holding time; **a** 0 min., **b** 5 min., **c** 10 min., **d** 15 min., **e** 20 min., and **f** 25 min

bifilm area and bifilm number; and they support each other. Figure 4 shows bifilm length depending on holding time. Bifilm length has a large scatter at 5 min hold, and it is the lowest at 25 min hold. When it is examined in terms of stability, sampling at 25 min holding time is better than others. It can be said that average bifilm length of the samples is nearly similar to each other.

Bifilm area, which is one of the calculations obtained from RPT samples depending on different holding time is presented in Fig. 5. In general, it is seen that there is no significant difference in average bifilm area for all results. However, the results of the samples are dispersed in very large scatter, especially for at 30 and 40 min. holding time.

Lognormal distributions, which helps to determine the stability of bifilm, of bifilm length and area depending on different holding time is given in Fig. 6. When Fig. 6 is examined, it is understood that bifilm length is in narrow scatter and frequency is the highest at 0 min. holding time. It can be said that average bifilm length is similar for at 5 and 10 min holding time and for at 15 and 25 min. holding

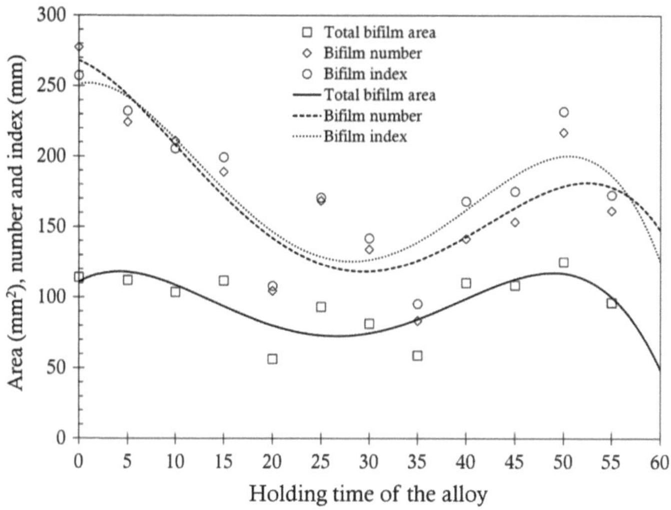

Fig. 3 Bifilm area, number and index depending on holding time

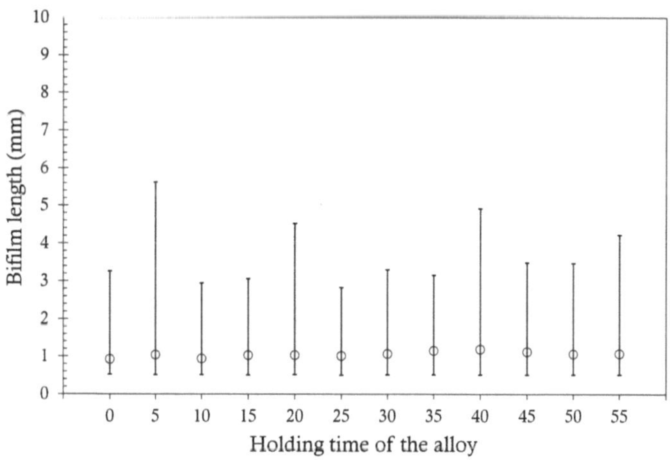

Fig. 4 Distribution of bifilm length

time. The largest scatter is seen at 20 min. holding time. It is clear that bifilm length shows varieties at low frequency and large scatter. If it is evaluated in terms of stability of the distribution of bifilm length, it is said that while the most stable result is obtained at 0 min. holding time; the most unstable result is seen at 20 min. holding time. The lognormal distribution of bifilm area shows a similar character to bifilm length. Bifilm area occurs in narrow scale at 0 min. holding time, that means 0 min holding time is optimum to achieve stable bifilm area.

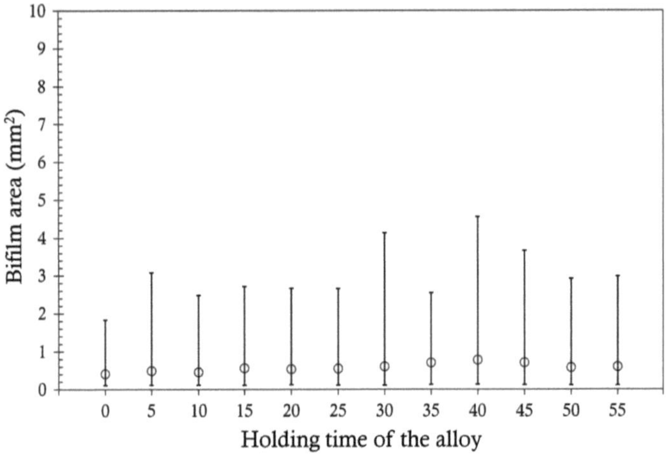

Fig. 5 Distribution of bifilm area

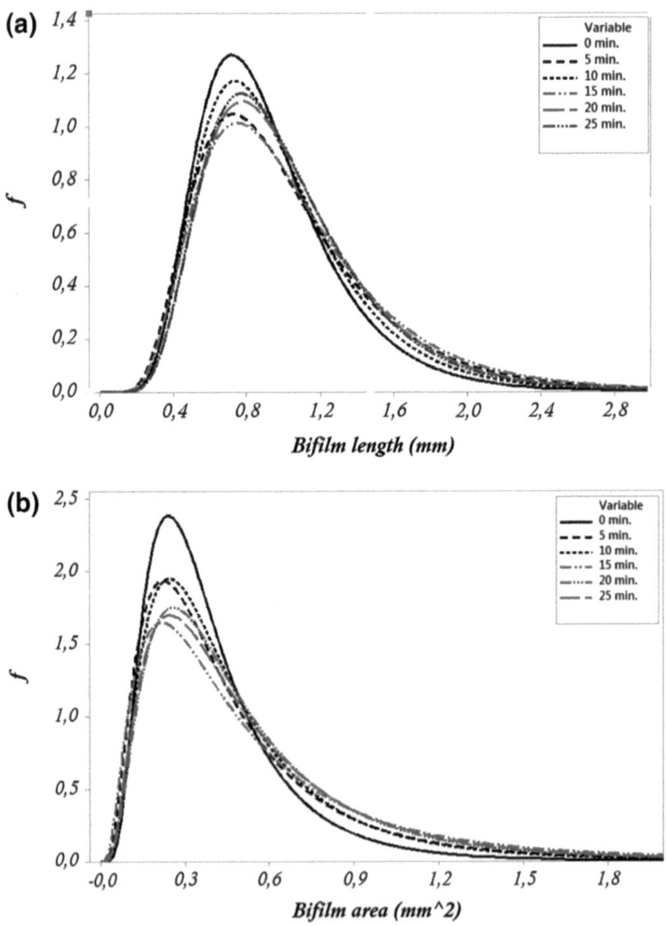

Fig. 6 Lognormal distributions of **a** bifilm length, **b** bifilm area depending on holding time

Conclusion

This study has been carried out to explain the effect of holding time on bifilms after Ti additions. The conclusions from the study may be listed as follows:

1. Ti grain refiner added into the melt as master alloys takes along bifilms when it moves to the bottom of the crucible because of its high density.
2. Bifilms are affected by Ti addition depending on holding time. While this effect is maximum at 30 min holding time, it decreases after at this time.

References

1. Campbell J (2003) Castings. Elsevier
2. Cole G, Sherman A (1995) Light weight materials for automotive applications. Mater Charact 35(1):3–9
3. Miller W et al (2000) Recent development in aluminium alloys for the automotive industry. Mater Sci Eng A 280(1):37–49
4. Rooy EL (2004) Aluminum alloy castings. Properties, Processes, and Applications 1:14–15
5. Zolotorevsky VS, Belov NA, Glazoff MV (2007) Casting aluminum alloys, vol 12. Elsevier Amsterdam
6. Dispinar D, Campbell J (2007) Effect of casting conditions on aluminium metal quality. J Mater Process Technol 182(1–3):405–410
7. Cáceres C (1998) A rationale for the quality index of Al-Si-Mg casting alloys. Int J Cast Met Res 10(5):293–299
8. Campbell J. (2015) Complete casting handbook: metal casting processes, metallurgy, techniques and design. Butterworth-Heinemann
9. Campbell J (2006) An overview of the effects of bifilms on the structure and properties of cast alloys. Metall Mater Trans B 37(6):857–863
10. Campbell J (2016) The consolidation of metals: the origin of bifilms. J Mater Sci 51(1):96–106
11. Campbell J (2006) Entrainment defects. Mater Sci Technol 22(2):127–145
12. Campbell J (2012) Stop pouring, start casting. Int J Metalcast 6(3):7–18
13. Kubo K, Pehlke RD (1985) Mathematical modeling of porosity formation in solidification. Metall Trans B 16(2):359–366
14. Lee P, Chirazi A, See D (2001) Modeling microporosity in aluminum–silicon alloys: a review. J Light Met 1(1):15–30
15. Anson J, Gruzleski J (1999) The quantitative discrimination between shrinkage and gas microporosity in cast aluminum alloys using spatial data analysis. Mater Charact 43(5): 319–335
16. Adler L, Rose JH, Mobley C (1986) Ultrasonic method to determine gas porosity in aluminum alloy castings: theory and experiment. J Appl Phys 59(2):336–347
17. Uludağ M, Dişpinar D (2017) Assessment of mechanism of pore formation in directionally solidified A356 alloy. Arch Foundry Eng 17(1):157–162
18. Uludağ M et al (2018) Change in porosity of A356 by holding time and its effect on mechanical properties. J Mater Eng Perform 1–11
19. Dispinar D, Campbell J (2004) Critical assessment of reduced pressure test. Part 1: porosity phenomena. Int J Cast Met Res 17(5):280–286

20. Dispinar D, Campbell J (2011) Porosity, hydrogen and bifilm content in Al alloy castings. Mater Sci Eng A 528(10–11):3860–3865
21. Dispinar D et al (2010) Degassing, hydrogen and porosity phenomena in A356. Mater Sci Eng A 527(16–17):3719–3725
22. Mostafaei M et al (2016) Evaluation of the effects of rotary degassing process variables on the quality of A357 aluminum alloy castings. Metall Mater Trans B 47(6):3469–3475
23. El-Sayed M, Griffiths W (2014) Hydrogen, bifilms and mechanical properties of Al castings. Int J Cast Met Res 27(5):282–287
24. Felberbaum M, Dahle A (2011) Modification and grain refinement of eutectics to improve performance of Al-Si castings. In: Light metals 2011. Springer, p 815–820
25. Zuo Y et al (2011) Refining grain structure and porosity of an aluminium alloy with intensive melt shearing. Scripta Mater 64(2):209–212
26. Campbell J, Tiryakioğlu M (2010) Review of effect of P and Sr on modification and porosity development in Al–Si alloys. Mater Sci Technol 26(3):262–268

Characterization of the Effect of Sr and Ti on Liquid Quality in Al8Si3Cu

Muhammet Uludağ, Derya Dispinar and Murat Tiryakioğlu

Abstract Al8Si3Cu alloy was studied to investigate the effect of casting conditions on melt quality of the molten metal. While Sr and Ti master alloys were used as additional parameters, holding time was performed for effect of held liquid. Experimental study was tried before and after degassing. The reduced pressure test machine was used to produce samples which are five samples for each parameter. Samples were prepared by metallographic methods to be analyzed with via digital image processing. Pores on the surface of the samples were measured in number density and distribution of them was examined statistically. Results showed that while Sr addition and holding time increased casting defects by increasing porosity, degassing decreased the number density of defects, as expected. Therefore, it is a vital result that degassing process must be applied to every casting. In addition, Ti addition was found to have a slight positive effect to reduce porosity, but it was almost negligible when compared to that of degassing process.

Keywords Al8si3cu alloy · Sr modification · Grain refinement · Liquid quality

Introduction

Bifilm is a casting defect mostly observed light alloys such as aluminum and its alloys [1]. It is formed by folding oxides from both used ingot (old bifilm) and casting process (young bifilm). This bifilms which is present in the structure as a double-layer oxide film is acceptable as the main cause of many casting defects,

M. Uludağ (✉)
Metallurgical and Materials Engineering, Bursa Technical University, 16310 Bursa, Turkey
e-mail: dr.uludagm@gmail.com

D. Dispinar
Metallurgical and Materials Engineering, Istanbul University, Istanbul 34320, Turkey

M. Tiryakioğlu
University of North Florida, Jacksonville, FL 32224, USA

M. Tiryakioğlu et al. (eds.), *Shape Casting*, The Minerals, Metals & Materials Series,
https://doi.org/10.1007/978-3-030-06034-3_17

in particular porosity and hot tear [2–4]. It was recommended by Campbell [5] that starting with a clean ingot and minimised turbulence with proper mold design are the vital parameters for preventing bifilms formation. Since the molten aluminum has a high temperature, an oxide film layer is formed on the surface. It is necessary that this oxide film layer should not be broken with any metallurgical process (i.e., mixing) at any level of casting and the oxide films should not be included in the melt [6, 7]. It was known that the mold design is the most fundamental effect for preventing the formation of turbulence. Additionally, avoiding sharp edges when the mold is designing and critical velocity of 0.5 m/s is not exceeded during mold filling. The melt included bifilms can be cleaned by filter system during melt transferring from crucible to mold cavity.

Master alloys such as Sr, Ti are added to a melt so that the morphology of microstructure changes for more regular, and thanks to these changes, mechanical properties of the alloy can be increased [8–11]. However, while Sr additions regulate eutectic morphology, increase porosity formation in Al alloys [12]. It was advised that Sr addition should not be done due to its negative effect on porosity although it has beneficial effects on microstructure [13, 14]. Campbell [15] informed that there is an interaction between bifilms and Sr addition in porosity formation. Iwahori [16] studied on the effect of Sr addition on pores in an Al alloy and found similar results with Campbell. There is a relationship between Sr modification and pore numbers in the melt. On the other hand, porosity formation depends on not only Sr addition but also oxide content in the melt. High oxide content in the melt must be available for a serious Sr interaction with bifilms. On the other hand, Ti addition in form Al-5Ti-0,2B affects porosity negatively by increasing size and number of pores in Al-Cu alloy [17]. The results of another study [18] which is on the effect of Ti addition on porosity in Al-20%Mg alloy are similar to those of Ref. [17], in which porosity was found to increase as Ti amount increased in a melt. Lee et al. [19] investigated the effect of melt treatments including modification by Sr addition and grain refinement with Ti as well as degassing on the porosity and tensile properties in A356 alloy castings. They determined that Sr and Ti additions increased porosity. However, degassing, with or without other melt treatments, significantly decreased porosity and consequently improved tensile properties. Similar results were found by Haberl et al. [20] who reported almost 50% reduction in the number density of pores in RPT samples after degassing. Although it is well known that degassing reduces the hydrogen content, the main contribution of degassing is considered to be the floatation of bifilms to the melt surface [19, 21, 22].

In this present study, the effect of Al-Ti5B1 grain refiner and Sr modifier on porosity formation was investigated under before and after degassing conditions. All results were analyzed statistically taking into account holding time of the melt from 10 min to 30 min.

Experimental Details

The alloy, Al8Si3Cu, was studied as secondary by providing as-cast ingots. Table 1 shows the chemical composition of Al8Si3Cu that was used in the study. Alloy ingots were cut to be charged into a SiC crucible by using a resistance furnace. In non-degassed melts, grain refinement and modification additions were made once the melt temperature had reached 740 °C. Master alloys used in this work were Al-Ti5B1 and AlSr15 with the chemical compositions given in Table 2. Al-Ti5-B1 was added into liquid to obtain a concentration 10 ppm Ti. AlSr15 alloy was added the melt to obtain a concentration 30 ppm. Castings and sample collection were started after 10 min of holding period.

A T-type graphite lance was preferred to achieve degassing process. 740 °C of melt temperature and 20 min duration of the process to complete degassing were carried out. After degassing, AlSr15 and/or AlTi5B1 were added into the molten metal. RPT samples were obtained using a mold that can take two samples at the same time [23]. Subsequent samples were collected in 5 min intervals up to 30 min. All RPT samples were cut into two vertically (cross section) and one half was prepared with sandpaper for image analysis. In this process, 60, 180, 400, and 600 grid sandpapers were used in sequence. Subsequently, all RPT samples were scanned with 600 dpi resolution. Digital image analysis was conducted on each specimen by following the guidelines provided by Dispinar and Campbell [24] and using SigmaScan software.

Result and Discussion

RPT samples were characterized by using an image analysis software. Results are presented in Fig. 1, which shows the cross sections of RPT samples before and after degassing trials after melt was held for 10 min. The effect of degassing process is seen in terms of amount of porosity. Pore properties such as size, number and amount are decreased after degassing. It can be said that during degassing process,

Table 1 Chemical composition (in wt%) of the Al8Si3Cu alloy used in the study

Alloy	Si	Fe	Cu	Mn	Mg	Zn	Ti	Al
Al8Si3Cu	8.14	0.64	3.12	0.44	0.22	0.49	0.02	Rem.

Table 2 Chemical composition (in wt%) of the master alloys

Master alloys	Ti	Sr	B	Fe	Si	Ca	Al
AlSr15	–	% 14–15	–	≤ % 0.2	≤ % 0.2	≤ % 0.2	Rem.
AlTi5B1	% 5	–	% 1	≤ % 0.2	≤ % 0.2	–	Rem.

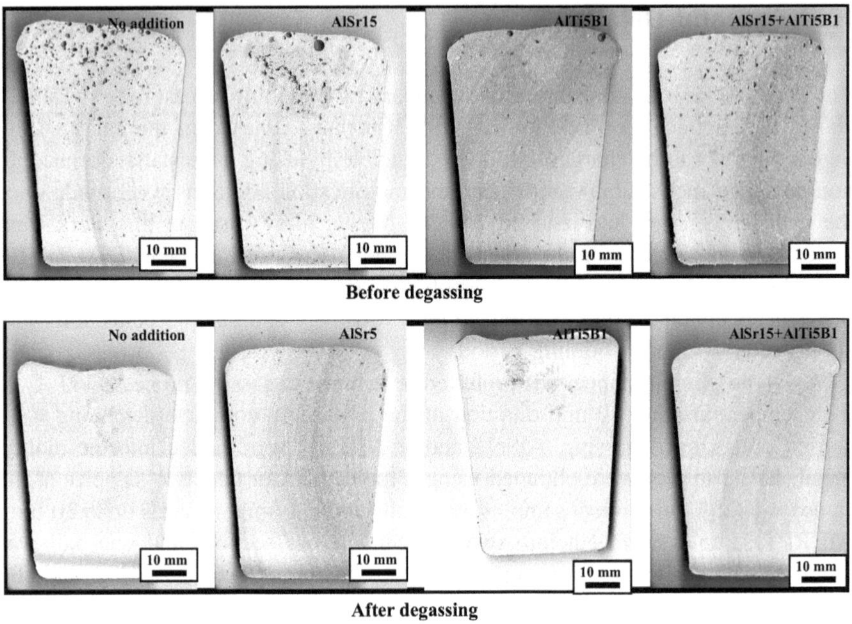

Fig. 1 Cross sections of RPT samples for all parameters after 10 min of holding time

bifilms were moved from bottom to top by rising bubbles. It is also seen that AlTi5B1 addition has minimum porosity than other parameter for before degassing. AlSr15 addition made pores bigger before degassing. Top of the RPT mold is the point that solidified lastly. The alloy used in this study has three different phases that are α-Al dendrites, Al–Si eutectic, and Al-Cu eutectic. It is known that eutectic phases are intensified at the point where solidification is completed. Because all pores were found near the top surface of the samples, either Al–Si and Al-Cu eutectic phases pushed bifilms from bottom to top in RPT mold or, alternatively, bifilms where pores grew floated upward. It is also notable that AlSr15 addition produced larger and more spherical porosities. There is also evidence that the AlSr15 addition damaged melt quality when made after degassing process.

The results of digital image processing were analyzed using two different data: number density of pores (N) and pore sizes. These two types of data are interpreted to be independent from each other; number density is a measure of how many heterogeneous nucleation sites were available during solidification. Pore sizes and their statistical distribution imply the driving force for the growth of pores, i.e., hydrogen content.

Number density data for all conditions as a function of holding time are presented in Figs. 2 and 3. Number density of pores decreases dramatically from 10 to almost zero after degassing for the non-treated alloy for each holding times. The same scenario applies to Ti. However, when Sr is involved, number density varies between 0 and 5 cm^{-2} even after degassing. On average, there is approximately

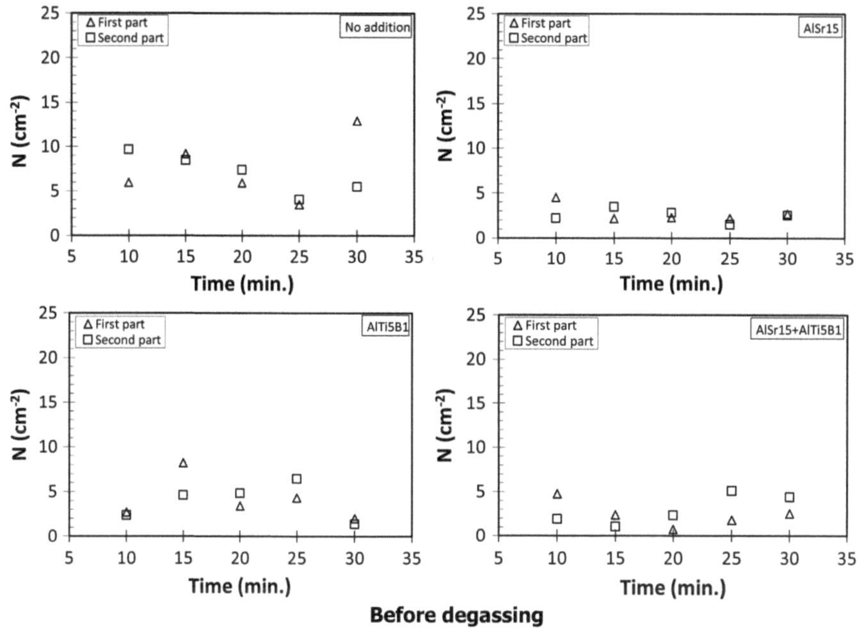

Fig. 2 Change in number density for all parameters with holding time under before degassing condition

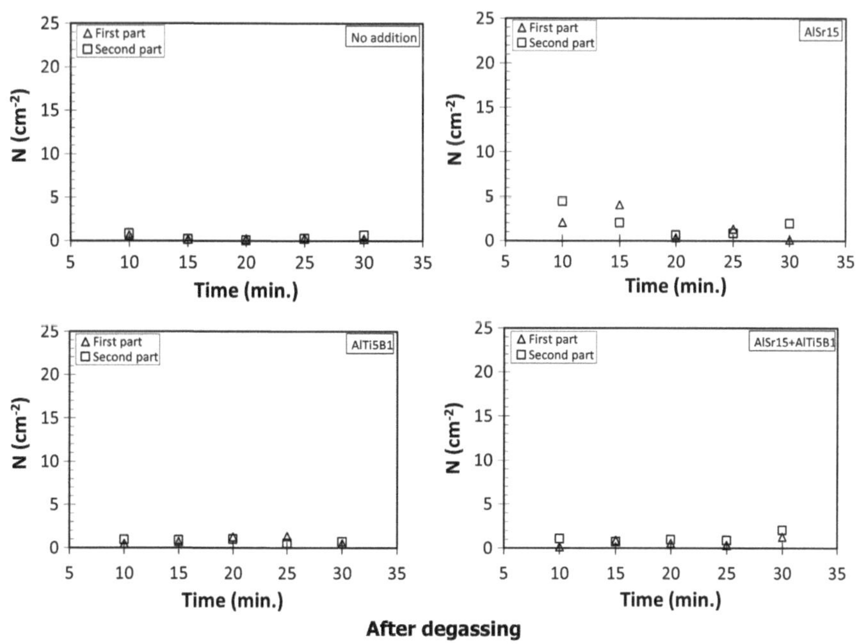

Fig. 3 Change in number density for all parameters with holding time under after degassing condition

70% decrease in the number density of pores when the melt degassed. The highest decrease was observed in the non-treated melt and the lowest change was observed for the Ti grain refiners. For the degassed melts, when 10 and 30 min is compared, the number density decreases for each addition except Ti grain refinement.

As discussed above, degassing process had a beneficial effect on porosity formation. It can be said that number density of the casting defects can be decreased by degassing for all parameters of the study (Figs. 2 and 3). If AlTi5B1 addition, AlSr15 addition, and holding time are evaluated together, there is only one process to clean defects and improve quality, which is degassing. It can be advised that to obtain high melt quality, avoiding melt addition, and apply degassing process to the melt before pouring.

Distribution of pore sizes for each parameter tested in this work is given in Figs. 4 and 5, which show fitted lognormal probability density plots to pore size data. For each case (before and after degassing), pore size distribution is the lowest for the non-treated alloy regardless of the holding time of the melt. For the Ti grain refined alloy, there is no effect of degassing, and pore size distribution does not change.

For non-degassed melts, pore sizes got smaller with holding time for all conditions, as evidenced by the shift in the distribution to left with increasing times. Assuming that the size distribution of pores is controlled by hydrogen content of the melt, it can be concluded that natural degassing takes place with time in non-degassed melts. Degassed melts, however, do not benefit from holding time to decrease the size of their pores, with the only exception of the “AlSr15 + AlTi5B1”

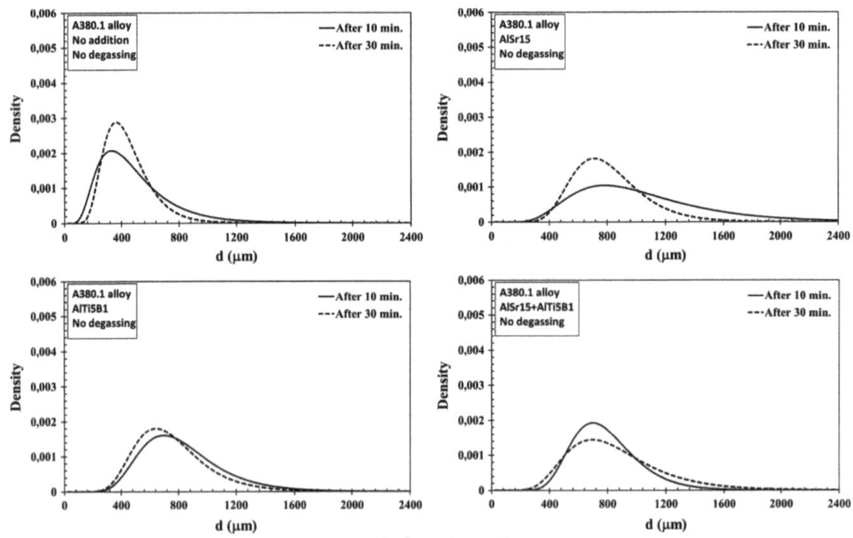

Fig. 4 Lognormal distributions of pore sizes which compare the effect of holding time for the no degassed condition

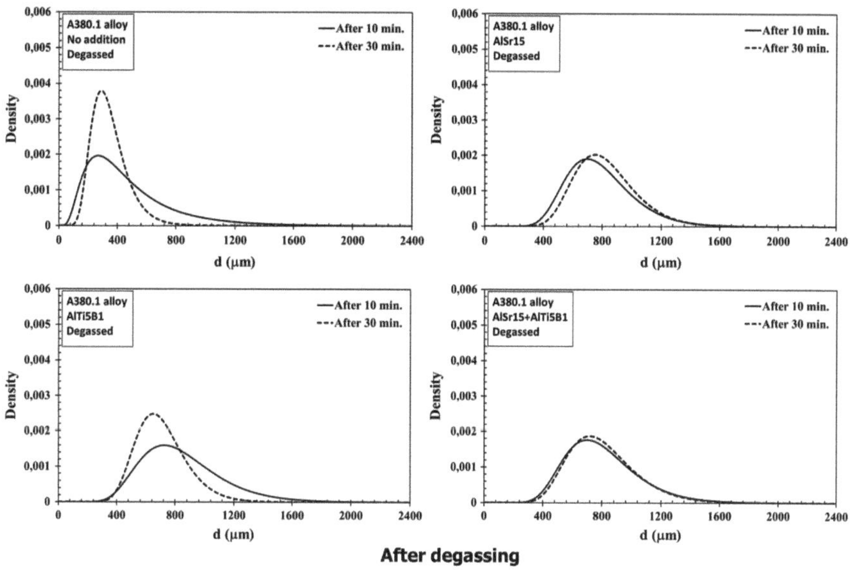

Fig. 5 Lognormal distributions of pore sizes which compare the effect of holding time for the degassed condition

condition, which behaves like the non-degassed melts. Because degassing, in most cases, has reduced the hydrogen content to such low levels that natural degassing with time barely occurs.

The anomalous behavior of degassed melts with the "AlSr15 + AlTi5B1" addition can be explained by the formation $SrO{\cdot}Al_2O_3$ spinel and its fracture to smaller pieces, as explained above. Campbell [1, 2] stated that Ti nucleates heterogeneously on bifilms and causes them to sediment to the bottom of the crucible. Hence, the combined Sr-Ti addition results in smaller bifilms, fractured due to Sr, to sediment more effectively after the nucleation of Ti on them.

Conclusions

Porosity formation that is one of the most dangerous casting defects formed by bifilms was studied for Al8Si3Cu alloy with grain refiner, modification, and holding time parameters under before and after degassing conditions. The most important and beneficial results for readers are summarized as follows:

1. AlSr15 addition, whether it adds to molten metal single or with AlTi5B1, affects pore formation negatively both before and after degassing. However, AlTi5B1 has a positive effect on porosity formation to decrease it.

2. Degassing is the most suitable process to reduce casting defects and improve melt quality.
3. Not only pore amount but also size, distribution, and shape of pores are gotten worse by addition AlSr15, holding time of the melt, and in the absence of degassing.

References

1. Campbell J (2003) Castings. Elsevier
2. Campbell J (2006) Entrainment defects. Mater Sci Technol 22(2):127–145
3. Knott J (2006) Commentaries on'Entrainment defects' by J. Campbell. Mater Sci Technol 22 (8):999–1008
4. Mi J et al (2003) Entrained oxide films in TiAl castings. Intermetallics 11(4):377–385
5. Campbell J (2015) Complete casting handbook: metal casting processes, metallurgy, techniques and design. Butterworth-Heinemann
6. Dispinar D, Campbell J (2007) Effect of casting conditions on aluminium metal quality. J Mater Process Technol 182(1–3):405–410
7. Dispinar D, Kvithyld A, Nordmark A (2011) Quality assesment of recycled aluminium. In: Light metals 2011. Springer, pp 731–735
8. Drouzy M, Jacob S, Richard M (1980) Interpretation of tensile results by means of quality index and probable yield strength-application to Al-Si7 Mg foundry alloys-France. Int Cast Met J 5(2):43–50
9. Gruzleski JE, Closset BM (1990) The treatment of liquid aluminum-silicon alloys. American Foundrymen's Society, Incorporated
10. Harris R, Lipson S, Rosenthal H (1956) Tensile properties of aluminum–silicon magnesium alloys and the effect of sodium modification. AFS Trans 64:470–481
11. Shivkumar S et al (1989) An experimental study to optimize the heat treatment of A356 alloy. AFS Trans 97:791–810
12. Fang Q, Granger D (1989) Porosity formation in modified and unmodified A356 alloy castings. AFS Trans 97(10):989–1000
13. Brosnan M, Shivkumar S (1996) Elevated-temperature tensile properties and fracture behavior of A 356 castings. American Foundrymen's Society, Inc (USA), pp 727–737
14. Gundlach R et al (1994) Thermal fatigue resistance of hypoeutectic aluminum-silicon casting alloys (94–141). Trans Am Foundrymen's Soc 102:205–224
15. Campbell J, Tiryakioğlu M (2010) Review of effect of P and Sr on modification and porosity development in Al–Si alloys. Mater Sci Technol 26(3):262–268
16. Iwahori H et al (1990) Occurring behaviour of porosity and feeding capabilities of sodium and strontium-modified Al–Si alloys. AFS Trans 94:167–173
17. Schaffer PL, Dahle AK (2005) Settling behaviour of different grain refiners in aluminium. Mater Sci Eng, A 413:373–378
18. Anyalebechi P (2013) Hydrogen-induced gas porosity formation in Al-4.5 wt% Cu-1.4 wt % Mg alloy. J Mater Sci 48(15): 5342–5353
19. Lee J-M et al (2003) Effects of melt treatments on microstructures and mechanical properties of A357 alloy. J Korea Foundry Soc 23(2):69–76
20. Haberl K et al (2009) Characterization of the melt quality and impurity content of an LM25 alloy. Metall Mater Trans B 40(6):812

21. Liu S-G et al (2015) Characteristics of mold filling and entrainment of oxide film in low pressure casting of A356 alloy. Mater Sci Eng, A 626:159–164
22. Eskin D et al (2015) Ultrasonic degassing of aluminium alloys: basic studies and practical implementation. Mater Sci Technol 31(1):79–84
23. Dispinar D, Campbell J (2011) Porosity, hydrogen and bifilm content in Al alloy castings. Mater Sci Eng A 528(10):3860–3865
24. Dispinar D, Campbell J (2013) Reduced pressure test (RPT) for bifilm assessment. In: Shape casting: 5th international symposium 2014. Wiley

Part III
Process Innovation and Modelling

The Nemak Cosworth Casting Process Latest Generation

Glenn Byczynski and Robert Mackay

Abstract This current paper provides an update to the well-known Cosworth Casting process. This paper revisits results first reported in 2009 and compares them with the latest generation of innovations that have further improved casting performance for 2019 model year engines. Traditionally only the main bearing saddle areas of the cylinder block casting were targeted for enhanced solidification via integral chilling elements. In this latest evolution, mechanical properties of other areas of the casting have further been enhanced and result in a complete solution for high-performance diesel and gasoline cylinder block castings.

Keywords Cosworth · Precision sand · Cylinder block

Background

The Cosworth process has a worldwide reputation for its ability to produce high-quality, dimensionally accurate aluminum castings. The process was crafted from basic principles with casting quality as its main focus. It has been uniquely employed to produce high volume automotive parts since the launch of Nemak Canada in 1992 (originally Ford Windsor Aluminum Plant). Originally applied for both cylinder heads and cylinder blocks, the process has evolved through two major process-driven stages in 2006 [1, 2] and in 2018. The innovation discussed in this paper describes the 2018 evolution as it is a stepwise further improvement of the 2006 Cosworth process. It now includes several chilling elements that increase regional cooling rates and thereby push casting performance to new levels in new areas of the cylinder block casting.

The major growth in cast aluminum cylinder blocks was seen in the late 1990s and continued steadily through the next 15 years of the new millennium [3]. The main factor was the continuing trend in vehicle light weighting and focus on fuel

G. Byczynski (✉) · R. Mackay
Nemak USA, Mexico, Canada
e-mail: glenn.byczynski@nemak.com

M. Tiryakioğlu et al. (eds.), *Shape Casting*, The Minerals, Metals & Materials Series,
https://doi.org/10.1007/978-3-030-06034-3_18

economy. As vehicle performance increases and engine displacement decreases, the net effect is increased specific engine power (i.e. HP/L). Where engines in the >100HP/L were rare and limited to high-performance vehicles in the mid-2000s, they are more commonplace and are in high demand in today's gasoline direct injection and turbocharged engines. This realm of high-specific power engines places greater demands on thermal management and the static and dynamic mechanical properties of engine castings. Therefore, innovative cooling and high mechanical properties are required. In addition, high-performance diesel applications place similar needs for performance materials. These are addressed with the latest embodiment of the Cosworth process for cylinder blocks.

Nemak Cosworth Latest Generation

Two major improvements were put into place for the latest generation of Cosworth process for cylinder blocks. These include *cylinder bore chills* and fine *interbore cooling channels* [4]. These new features address the latest industry requirements of improved thermal management and superior material performance in areas other than the main bearing saddle (i.e. cylinder head bolt boss columns). The net effect of cylinder bore chills is to drastically increase the localized solidification rate reducing secondary dendrite arm spacing (SDAS) and *minimizing the ability for oxide films to unfurl and produce porosity* [5, 6]. The additional benefit is an extremely homogeneous bore surface that after machining is suitable for either cylinder liner insertion (via the slip fit process) or any of the latest bore coating technologies that require good quality substrates.

Figure 1 shows and compares the main elements of the three major embodiments of the Nemak Cosworth process.

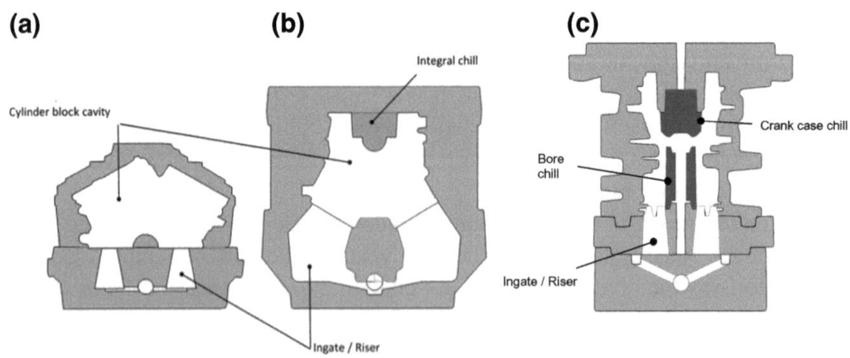

Fig. 1 Schematic showing comparison of **a** Original Cosworth "Crankcase" gating of a V6 block **b** Nemak Cosworth innovation "Head deck" gating of V6 **c** Latest generation in-line configuration (all schematics shown in mould filling orientation)

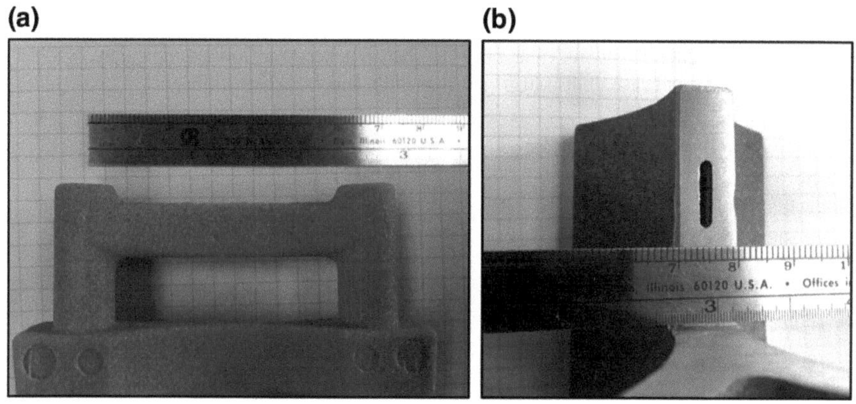

Fig. 2 Novel interbore cooling shell core **a** core profile **b** result channel (sectioned casting) resulting in a 2.0 × 11.0 mm passage between bores (narrowest dimension)

Comparing the latest generation with prior embodiments, one will note several similarities. These common features represent the fundamentals of the Cosworth process. Characteristics such as a room temperature precision sand mould, bottom filling/gating, controlled quiescent mould metal filling and directional solidification due to mould rollover (180° after filling) are maintained. These are all in agreement with the original tenets of the Cosworth process as laid out in the 1980s.

The alloy remains unchanged in the latest two variants, and applied heat treatments (T7) are also similar in the latest two Cosworth processes discussed albeit with minor necessary adjustments to artificial ageing temperatures in order to optimize the mechanical properties for each casting part number. The localized enhancements include room temperature chilling elements located both in the main bearing saddle and cylinder bore areas and a novel interbore core formed using the shell core process that introduces a narrow cooling channel between the cylinder bores near the head deck (Fig. 2).

The core can be connected to the water jacket in a variety of configurations—most commonly drilled-through during machining as to force cooling through one direction. This "cored-in" feature is superior to cross drilled designs as the cooling channel can be placed closer to the head deck where cooling is most needed. Alternate feasible designs include double or even triple channel configurations.

Discussion

Mechanical property tests were performed according to customer specifications including sample location. Though the compared castings were not identical, the sample locations and testing procedures are considered comparable. A summary of the test results from mechanical test bars taken from cylinder bolt bosses and main

Table 1 Mechanical property comparison (average values; sample size of 30 per condition)

Average values (n = 30)	2006 Cosworth (integral crankcase chills only)	Latest generation (room temperature cylinder chills and crankcase chills)
Bolt boss location		
Plastic elongation (%)	0.42	0.50
Yield strength (MPa)	**183.2**	**271.3**
Ultimate strength (MPa)	**202.4**	**294.1**
Main bearing saddle location		
Plastic elongation (%)	3.1	2.7
Yield strength (MPa)	303.0	299.2
Ultimate strength (MPa)	323.1	352.2

bearing saddles (30 samples each location and process) are shown in Table 1. As expected the largest improvements are seen from the strengths in the bolt boss areas immediately adjacent to the cylinder bore chills. Average yield strength was improved by a significant 48%, and similarly, average ultimate tensile strength was increased by 45%.

The average percent elongation to fracture in both areas of the casting remains largely unaffected. This is partly due to the secondary nature and iron phases present in the AlSi8Cu3 (0.6 Fe max) alloy and partly due to the fact that presence and effect of oxide bifilms on ductility are unaffected by solidification rate.

By evaluating all the data in the form of Weibull curves, one gets a more complete picture of the distribution of the data. Weibull statistics can be used to study fracture data, and the coefficient of determination (R^2) of over 0.96 in all cases here indicates a good fit to the Weibull distribution. The reader is referred to other publications for a more complete explanation [7–9].

The differences between bolt boss properties are further highlighted in the Weibull plots of Fig. 3. The most striking difference is the apparent transferal of the plots to the right indicating a complete population shift (no overlap of x values). This is of course related to the 45% increase in average value and is a direct result of the application of cylinder bore chills. A more subtle gauge of improvement is the Weibull modulus (***m*** or slope of the fitted line) where an approximately 25% improvement in the consistency of the data can be gauged (34 vs. 27) again related to the higher solidification rates involved.

When considering Fig. 4. Weibull plot of the main bearing saddle UTS where the Weibull moduli of 84.6 and 75.9, respectively, indicate a high level of

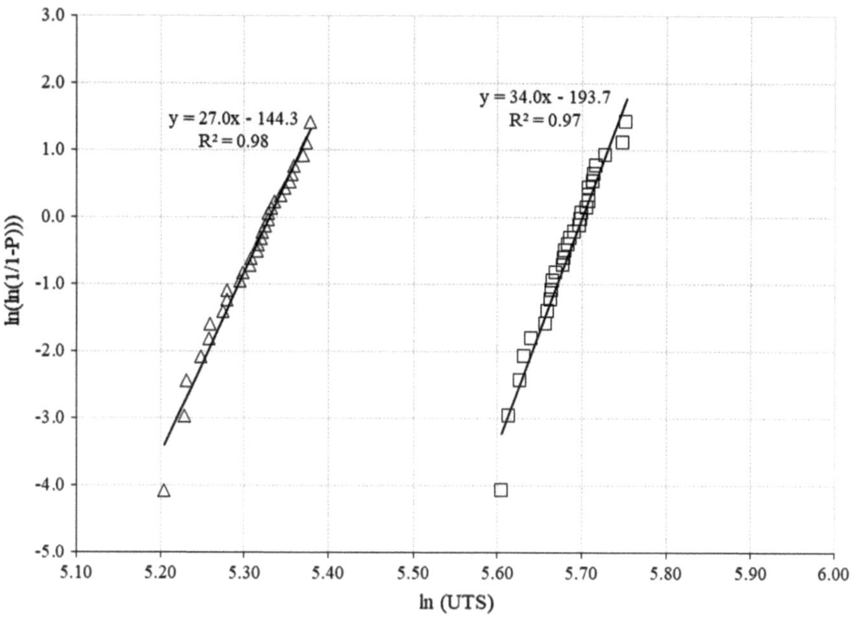

Fig. 3 Weibull probability probability plot of UTS comparing **bolt boss** properties of Cosworth 2006 (Δ) and latest generation with chilled bores (□)

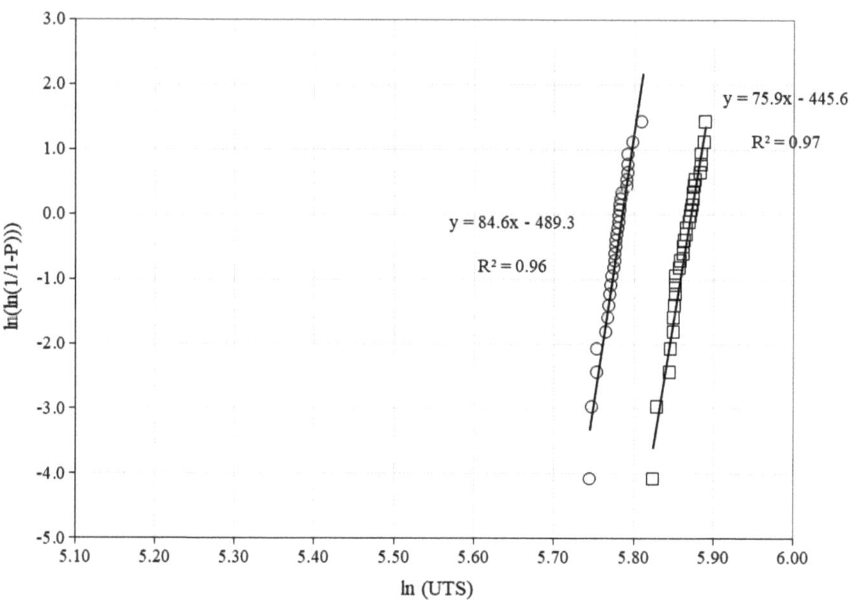

Fig. 4 Weibull probability plot of UTS comparing **main bearing saddle** properties of Cosworth 2006 (○) and latest generation (□)

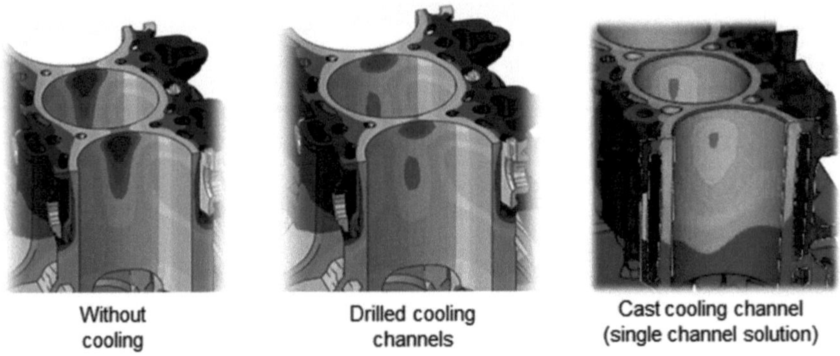

Fig. 5 Thermal simulations showing temperature profile of cylinder block interbore areas with different cooling channel designs (red colour indicating higher temperature)

capability and consistency in the data. This is commensurate with the powerful chilling effects of the cast iron in-mould chills and the fundamental principles of the Cosworth process.

Improvements in engine cooling are also predicted due to the new interbore core feature. Simulations show that the addition of the interbore core is superior at reducing temperature hotspots located near the head deck when compared with no cooling channel and cross drilled alternatives (Fig. 5).

Conclusions

- The latest embodiment of the Cosworth process has further improved cylinder block casting properties by incorporating room temperature chills in the cylinder bore areas. This has dramatically increased local solidification and has resulted in significantly enhanced yield and ultimate strengths of the bolt boss areas without much change in tensile elongation. This is consistent with the notion of suppression of bifilm unfurling and minimization of defects in the cast structure.
- An innovative interbore cooling sand core has been presented as an improvement for thermal management and is now offered as a complementary technology to the Cosworth process at Nemak
- These latest improvements have kept pace with the changing needs of the automotive industry and are able to offer innovative light-weighting solutions for the demands of future high-performance gasoline and diesel engine designs.

References

1. Byczynski G, Mackay R (2009) The nemak cosworth casting process–innovation. In: Shape casting: third international symposium, The Minerals, Metals and Materials Society, pp 199–206. ISBN: 978-0-87339-734-6
2. Byczynski G, Fritzke J (2015) Novel mold filling technology for precision sand castings in aluminum. In: AFS transactions, 119th metal casting congress, Columbus OH
3. http://www.drivealuminum.org/wp-content/uploads/2017/10/Ducker-Public_FINAL.pdf
4. Diel et al (2017) USA Patent application, US 2017/0173670 A1
5. Campbell J (2011) Complete casting handbook 2011, January 2011. Elsevier. ISBN: 13:978-1-85617-809-0
6. Byczynski GE, Campbell J (2002) The effects of oxide film on the strength and reliability of 319 alloy castings. In: 2nd International technology symposium. ASM International, USA, pp 65–74
7. Weibull W (1951) A statistical distribution of wide applicability. J Appl Mech 18:293–297
8. Byczynski G, Mackay R (2011) The use of the weibull statistical method to assess the reliability of cast aluminum engine blocks made from different casting processes. In: Shape casting: the fourth international symposium, The Minerals, Metals and Materials Society. ISBN: 978-1-11802-937-4
9. Tiryakioğlu M, Hudak D (2011) Metall Mater Trans B 41:1130–1135

Campbellology for Runner System Design

Fu-Yuan Hsu

Abstract The runner systems based on the Campbellology, the 10 casting rules of Professor John Campbell, were designed to achieve the bifilm-defect free casts during the filling process in gravity casting. In this study, three main methods were utilized to control the bottom-gating velocity under the critical velocity (the rule no. 2). In these methods, the kinetic energy form of liquid metal (with high velocity) was transformed to the static pressure form (under the critical velocity) in the bottom of the gating systems. The total head height with low energy loss was maintained in these runner designs. A continuous and positive static pressure of liquid metal against to the wall of mold cavity from the bottom to the top of cast was accomplished throughout the filling. Many design schemes to avoid the filling defects, such as oxide entrapment, confluence weld, bubble entrainment, and sand core blow defects, were also suggested in these casting processes.

Keywords Runner system design · Critical ingate velocity · Bifilm defect · Gravity casting

Introduction

In the number 2 of the 10 casting rules by Professor Campbell [1a], it addressed that the bottom ingate velocity should be controlled under a critical value, e.g., 0.5 m/s for many liquid metals, to avoid the bifilm defect formation during filling in gravity casting. This critical velocity, or the inertial force, can be obtained by balancing its opposing forces, gravity, and surface tension. The dimensionless number **Hs** for characterizing the phenomena of axisymmetric hydraulic jump, as the liquid metal in vertical direction impacts on the horizontal surface, is therefore [2],

F.-Y. Hsu (✉)
Department of Materials Science and Engineering, National United University, No.1 Lein-Da, Kung-Ching Li, MiaoLi, 36003 Taiwan, Republic of China
e-mail: willyhsu@nuu.edu.tw

M. Tiryakioğlu et al. (eds.), *Shape Casting*, The Minerals, Metals & Materials Series,
https://doi.org/10.1007/978-3-030-06034-3_19

$$\mathbf{Hs} \equiv \frac{Inertial\, pressure(P_i)}{Gravitational pressure(P_g) + Surface pressure(P_s)} = \frac{\rho \cdot V^2}{(\rho \cdot g \cdot H) + (4T/H)} \quad (1)$$

Here, ρ is the density of liquid metal, V is the velocity, g is gravity, H is the height of the hydraulic jump, and T is surface tension. If the **Hs** number is equal to or less than 1, the transition happens. In other words, if the total effect of the gravitational pressure (P_g) and surface pressure (P_s) become dominant, the hydraulic jump starts. If aluminum alloy is considered in this equation, the constants of density ρ, gravitational acceleration g, and surface tension T become 2430 kg m^{-3}, 9.81 m s^{-1}, and 0.8825 N s^{-1} [3], respectively. Then, this equation gives the relation between the average velocity V of the jet flow before the transition and the total height H of the jump after the transition as the transition of hydraulic jump takes place (i.e., **Hs** = 1).

If the two pressures, P_g and P_s, plotted against the height H of the jump, the lowest sum of the two pressures could be identified. It means that the minimum energy to which the system would approach and be stable. The critical height of 12 mm could be found. This critical height is also known as "glissile drop" height. At the critical height, 12 mm, the average velocity V of the jet prior to the transition could be calculated from Eq. 1 (as **Hs** = 1), and therefore, the critical velocity is 0.4886 m s^{-1} [2].

This kind of the axisymmetric flow phenomena, characterized by the Hs number, is similar to that the liquid metal emerges vertically from bottom ingate and diverge radially on horizontal surface in the bottom-gating system. Gravity and surface tension should suppress the high kinetic energy of the liquid metal emerging from bottom ingate (i.e. Hs < 1), in order to avoid the oxide on liquid surface folded into form bifilm defect.

Folding mechanism in the diverging flow condition, like axisymmetry hydraulic jump, starts as the inertial pressure is dominant (i.e., Hs > 1).

Bifilm is folded from two parts of oxide films at the transition point (e.g., Hs = 1) with the help of two currents, the reverse eddy and the emerge jet. One part of the oxide films is on the surface of the reverse eddy current next to the jet, and the other is on the surfaces of this jet at the point of energy transforming. At this transformation point, two films from two different parts of oxides are folded and bifilm is formed next to the jet.

This oxide bifilm then is entrained into metallic bulk. The bifilm with convoluted morphology at the beginning of entrainment will unfold by various actions, such as hydrogen inflation in aluminum alloy, graphite precipitation in cast iron, dendrite grains protrusion, and intermetallic compounds deposition, during solidification [1b]. Since there is no binding in between two oxide films, this unfolded bifilm serves as a non-metallic crack instituted inside the bulk. It will weaken the metal as it is subjected to a force. Eventually, the metal becomes brittle.

Resulting from turbulent filling condition, that bifilms are scattered within a casting is an evidence of a random nature in mechanical properties of casting metals. Many authors found that Weibull distribution is best to describe it [4–9].

Thus, a reliable casting should be produced if its filling condition were controlled and no bifilms exist. It implies that a high kinetic energy should be transformed back to potential energy state before liquid metal entering a casting cavity from bottom ingates. Therefore, a good runner system, which can control the ingate velocity under the critical value, is recommended by Campbellology.

Method

At elevated temperature, many liquid metals are highly oxidized by the oxygen in air. Oxide film is quickly formed on the surface of liquid metal as its surface contacts air. This film on the free surface is a good protective layer to prevent further oxidation for the metal underneath.

New oxide film on the surface of metal flow front will form a metal stream fills along with a runner system. Like blowing a bubble, the surface of an inflated bubble always expands, if a positive pressure is applied continuously. It will have no chance for the oxide film on the glowing free surface to fold (or entrain) and to form bifilm. Therefore, in whole filling process liquid metal should fill fully in the space of runners along way to cast cavity. If filling stream pushed positively against to walls of runners in all time, the surface oxide film then attaches safely on the walls. No folding event will happen until the filling is completed. This means that a good runner design in gravity casting is to provide exact size of runners for the volume required by streams filling along runners and cast cavities in a form of potential head recovered.

In converging runner, however, filling streams will be pressurized by this runner. Although the oxide films on the stream surfaces are against the walls of the mold cavity, kinetic head of stream will be transformed. As results, the stream velocity increases. This high momentum stream will be detrimental if it enters the big space of cast cavity through bottom ingate. Surface turbulence is inevitable at this great kinetic state.

In the diverging runner, such a diffuser runner, if a high momentum stream enters, the stream will go straight through this runner without filling the space left behind. In gravity casting, cast cavity is empty before the filling starts. To reduce the velocity of the filling stream, many textbooks suggest to enlarge the runner geometry based on the continuity principle in fluid dynamic. But this principle only can be applied in the cross section of a natural form of stream, not for the deliberately designed runner. If a large geometry of runner were intentionally designed, such a diverging runner, unfilled space, the empty region, will be remained in this runner.

These regions will stay empty not until the rolling back wave of streams fill them. But air in these regions will also incorporate with the high momentum stream, since negative pressure less than the atmosphere will be developed by the socking effect described by the Boyle's law. New fresh oxygen in the atmosphere will add into these low-pressure empty regions through the porous sand mold or the parting line in permanent mold. Inevitably, oxygen (or moisture) in the air will oxide the liquid metal during filling.

At high melting temperature with this negative pressure environment in these regions, the organic resin binder in sand mold will burn out quickly and moisture and carbon soot will produce. Defects, such as hydrogen and lustrous carbon (carbon film) [1c], will also contaminate liquid metal.

In these regions, bubbles sometime will be entrained into the bulk metal if the rolling back wave engulfed the air in these regions. Bubble trail defect, another kind of bifilm, is also found in the casting [1d].

Filling a big casting with a bottom ingate velocity under 0.5 m/s, suggested by Campbellology, actually is a challenge. With such a low velocity in ingate, premature solidification of liquid metal should be prevented. Thus, one of runner-designed rules is to provide a large flow rate (i.e., unit of m^3/s) but a low ingate velocity.

This can be achieved by multiple-ingate design [11]. In each bottom ingate, the inlet velocity is small (under 0.5 m/s) while the total cross-sectional areas (mm^2) of the inlet streams for all multiple ingates can be large.

Thus, principle of designing runners is to make exact shape of runner geometry to fit filling stream with a flow rate required. Using this principle, the author has proposed L-shape and multiple-gated runners to fit filling streams as turning in a right angle, elsewhere [10, 11].

Methods for achieving the ingate velocity below the critical velocity of 0.5 m/s are classified into three categories, which are applying mechanical force to control, designing runner to transform and using media to block.

In this study, three examples, which are Durville 180° tilting control (Fig. 1), diffusing runner design (Fig. 2) and foam filter usage (Fig. 3), were suggested and they also followed in the three categories, respectively. For convenient, these casting methods are named as the tilt casting, the diffuser casting, and the filter casting in the followings. Table 1 has listed some information of the runner system designs in these three castings.

The Tilt Casting

Figure 1 shows the runner system for the (Durville 180°) tilt casting. Five ingates are designed in this runner system. L-shape runners and multiple-gating are also included. A tilt caster with a rotating stepper motor was used for controlling the tilt angle at each time step in order to follow the procedures of angular velocity, derived from the modeling results. In this caster, the rotating axis is the cross arrows as shown in Fig. 1d and e. The ingate velocity and the flow rate are all controlled by the caster.

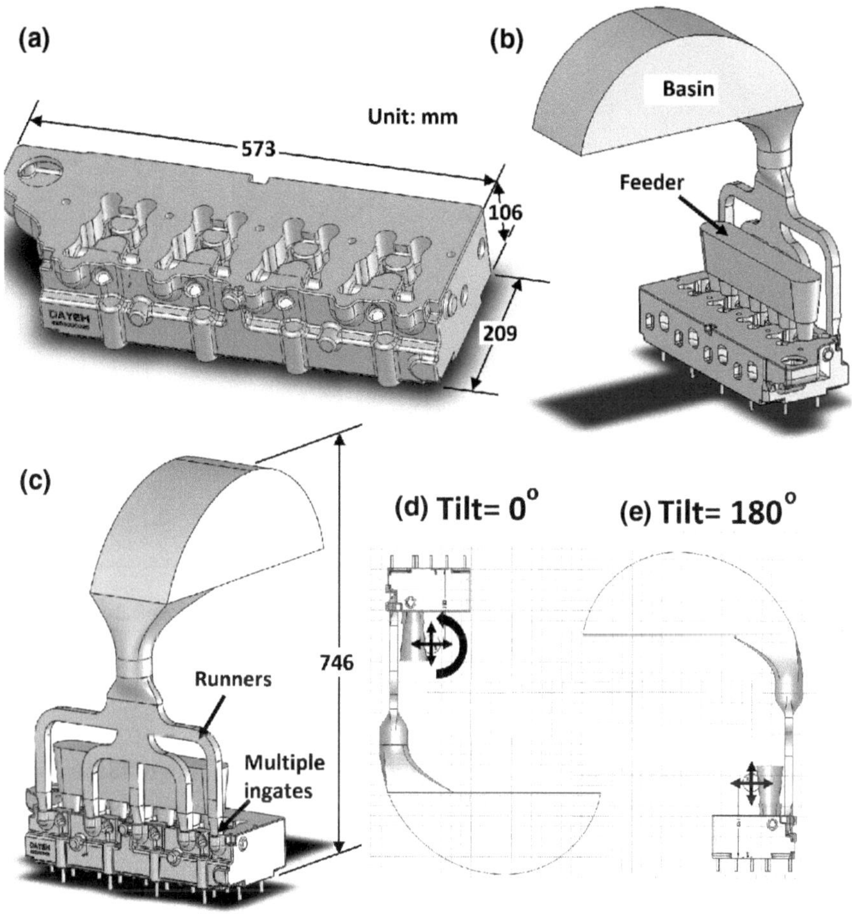

Fig. 1 (Durville 180°) tilt casting: an example of casting the 17.9 kg aluminum diesel engine head

In this caster, resin binder sand molds are used and automated production line can be achieved. Aluminum alloy, A356 (Al–7%Si–Mg), was melted at 700 ± 20 °C. A removable refractory basin (e.g., Fig. 1b) can collect the liquid aluminum by gently rotating mechanism. Once the basin filled with the volume required by the casting, it can attach with ceramic seal on the sand mold ready in the caster. The caster will start the tilting, and the tilting procedure by design will last approximately 7.0 s.

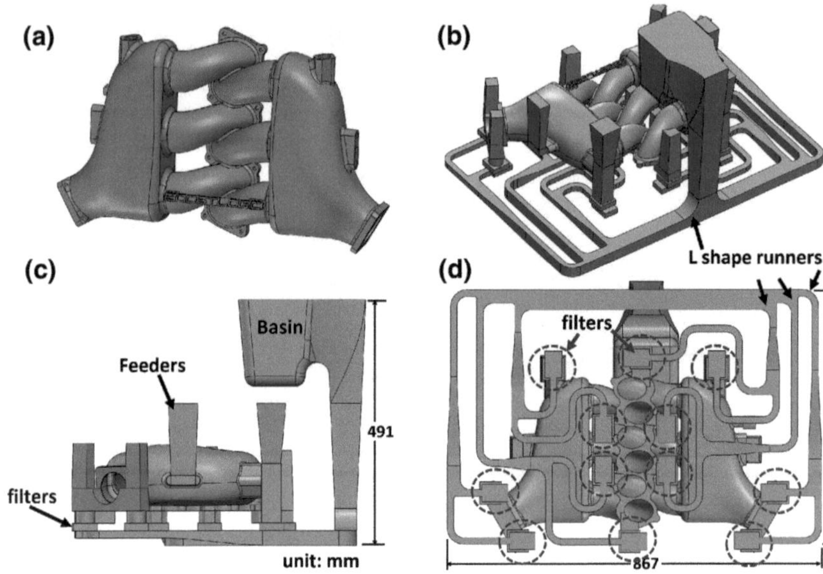

Fig. 2 Diffuser casting: an example of gravity casting 130 kg aluminum chemical valve

The Diffuser Casting

Figure 2 presents the runner system which is the combination of 6 diffusers for the 130 kg gravity casting. The geometry of a diffuser has been introduced by the author elsewhere [12]. In this runner system, eight ingates are connected to the diffusers as a multiple-gating system. The other three bypass runners are linked next to the ingates (e.g., Fig. 2b and d). Various L-shape runners are also used for transferring and separating the mainstream into multiple runners. Exact size of L-shape runners for the streams is designed based on the principle previously described.

To provide 188 kg of liquid metal at once for pouring, a bottom-feeding basin with a stopper is designed. After this basin collects the clean aluminum melt, it can be transferred to the mold for pouring. The bottom outlet of the basin can attach to the entrance of the sprue, and the stopper can be lifted to begin filling. The filling takes approximately 35 s.

The Filter Casting

Figure 3 demonstrates the runner system where 12 filters are included. A silicon-carbide ceramic 10 ppi (pore per inch) foam filter with dimensions of 50 mm × 50 mm × 15 mm is applied to this aluminum gravity casting.

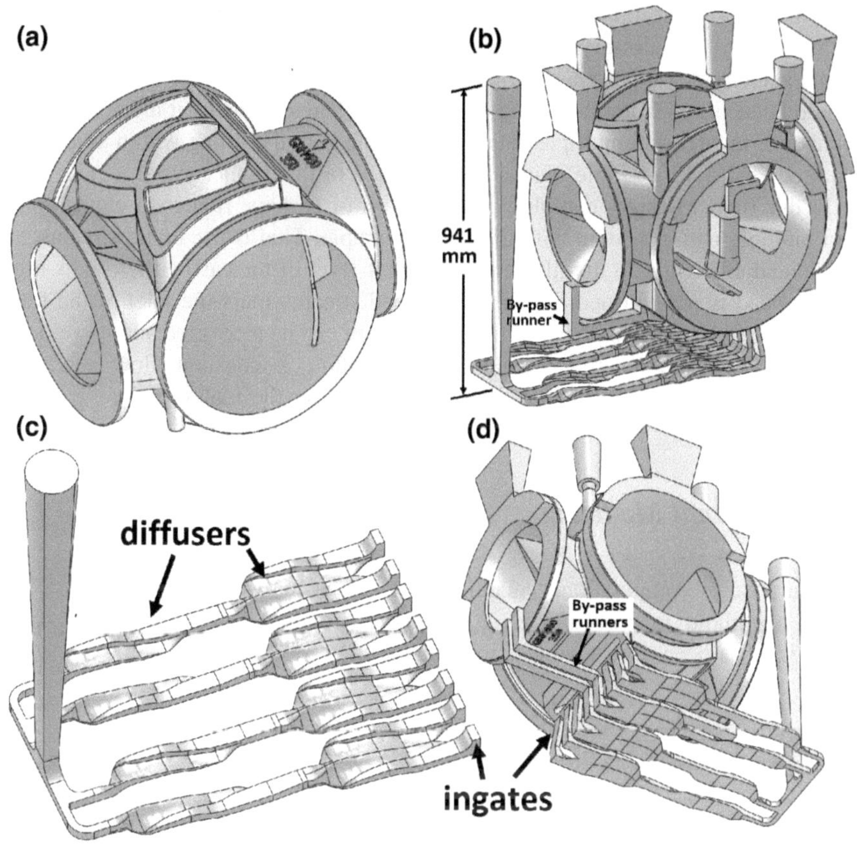

Fig. 3 Filter casting: an example of casting the 5.7 kg aluminum intake manifold

Table 1 Information of runner design for the three castings in this study

	Total weight of casting with runner system (kg)	Total volume of casting with runner system (mm^3)	Cross-sectional area of ingates	Number of ingates
Tilt casting	22.4	9 236 494	1774	5
Diffuser casting	187.6	77 203 352	6477	8
Filter casting	24.0	9 887 802	16 710	12

Orientation of a foam filter installed in a runner is also suggested by the author, elsewhere [13]. Like the principle explained previously, a high flow rate within a minimum ingate velocity can be assured by this kind of installations.

In this study, the intake manifold casting is a thin section with 3.5 mm. That high flow rate is necessary to prevent the filling to stop prematurely. The bottom ingate velocity below 0.5 m/s is controlled by the foam filter. A large flow rate thus should be acquired from increasing cross-sectional area of the total ingates. A large number of ingates or a multiple-gating system can help the filling within a flow rate required. Thus, 12 multiple ingates are designed in this runner system.

In this runner design, complex combinations of L-shape runners are assembled into multiple-gating system with the same principle. That is the liquid streams are distributed efficiently to the filters at the lowest level of the casting before its high kinetic head is transformed. As the high-speed stream enters the inlet of the filter installation, the positional head of this stream then can be recovered and its velocity decreases below the critical value. To maintain a constant flow rate, the cross-sectional area of the outlet can be enlarged after the filters.

Computational Modeling

The purpose of computational modeling is to assist the runner design before initiating the mold manufacturing. A computational fluid dynamics (CFD) code, Flow-3D, has been used for these investigations.

Result

Figure 4 shows some pictures of the tilt casting experiment. The filling of the casting lasts 7.0 s. In Fig. 5, the filling process at various time frames in velocity scale ranging from 0.0 to 0.5 (m/s) is illustrated. At 3.12 s, the velocity of the filling stream at the ingates of the casting is less than 0.5 (m/s) as the color of the stream is not red. After filling the ingate areas, the rotating caster increasing its tilt angular velocity without generating the surface turbulence in the filling front (e.g., the yellow and blue colors at 4.64 s in Fig. 5). At this time, the ingate velocity is over 0.5 m/s (i.e., the red color). Moreover, in Fig. 6, pressure contours of the filling streams before and after entering the ingates are shown, respectively. The pressure difference of the stream has no sign of negative pressure developed, and its lowest pressure is in green color (i.e., a positive pressure).

Some pictures of the diffuser casting are shown in Fig. 7. The filling time of this casting is 32 s. Figure 8 shows the various time frames of the temperature contour, and there is no free turbulence found in the modeling results. The velocity of the filling stream next to the ingated regions is under 0.5 m/s as shown the colors other than red (e.g., in the velocity scale of Fig. 9a). At the same time frame, in Fig. 9b, the pressure contour also shows positive pressures of the filling fronts, where the lowest pressure is in green color.

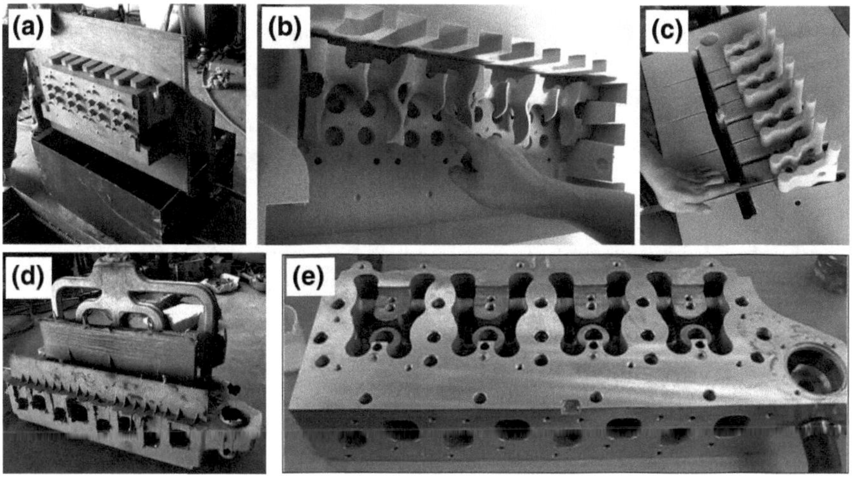

Fig. 4 Tilt casting pictures: **a** the pattern for sand molding **b** water jacket assembled **c** venting design **d** after casting **e** after machining

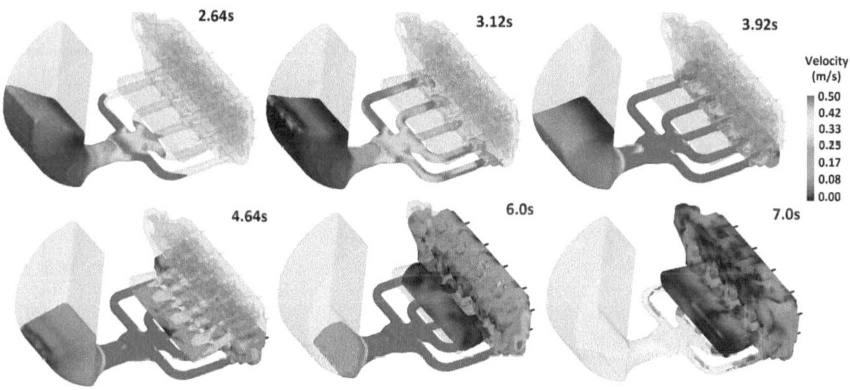

Fig. 5 Modeling results of the tilt casting in velocity contour at various time frames

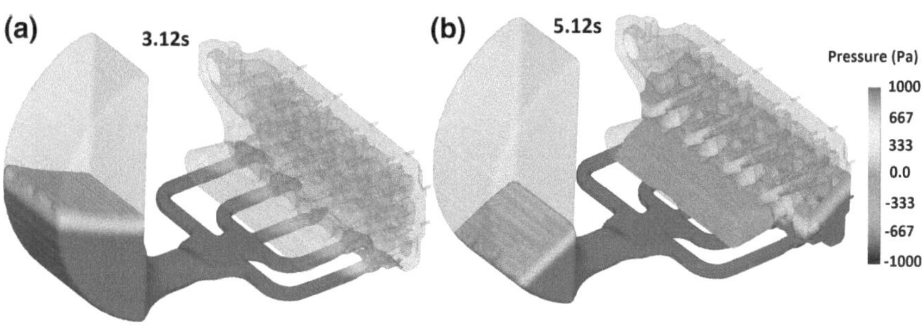

Fig. 6 Pressure contours before (**a**) and after (**b**) the ingates. Noted the green color is the pressure difference of 0.0 (Pa). (Color figure online)

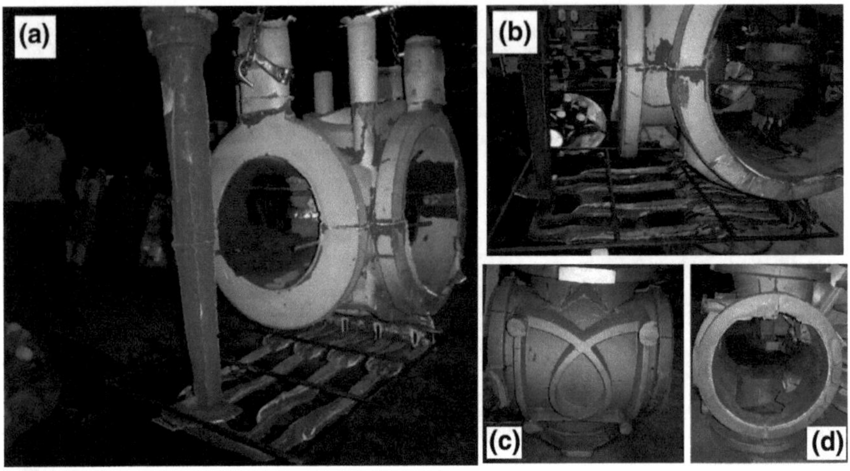

Fig. 7 Diffuser casting pictures: **a** after casting **b** the diffusers **c** the top view and **d** the front view

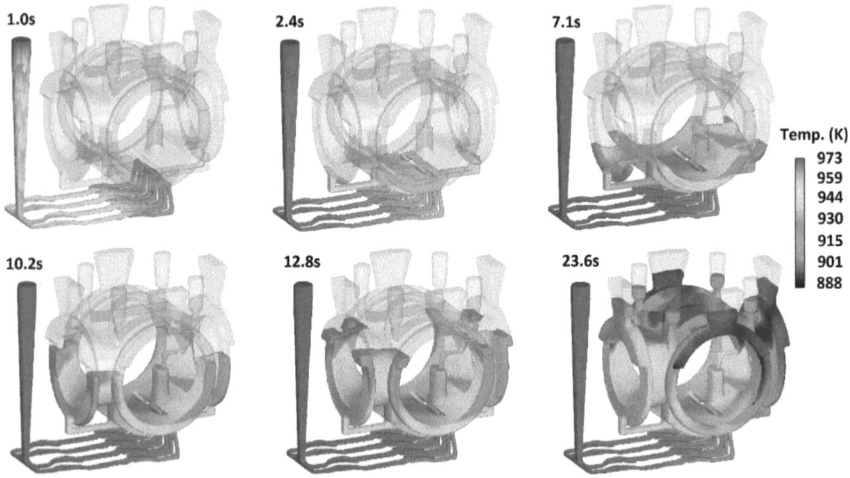

Fig. 8 Modeling results of the diffuser casting in temperature contour at various time frames

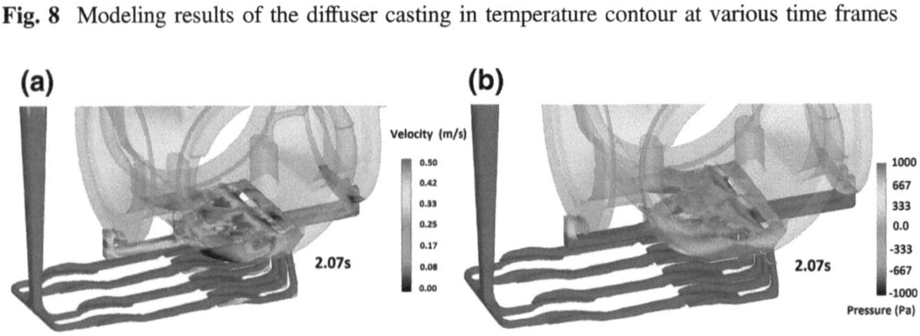

Fig. 9 Velocity and pressure contours in the ingate regions at the same time frame

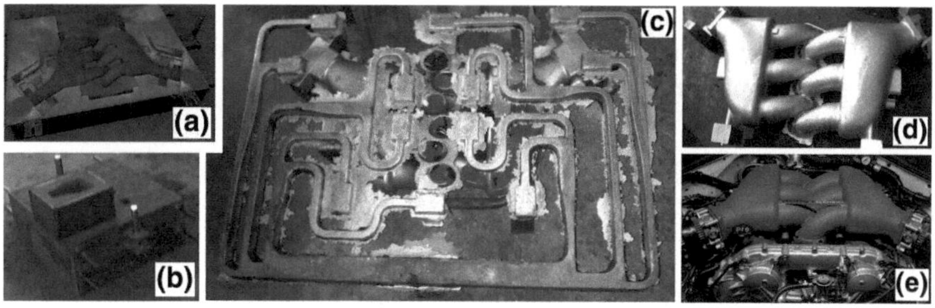

Fig. 10 Filter casting pictures: **a** the green sand mold and cores **b** the assembly **c** the bottom view of the casting **d** the top view and **e** the mounting

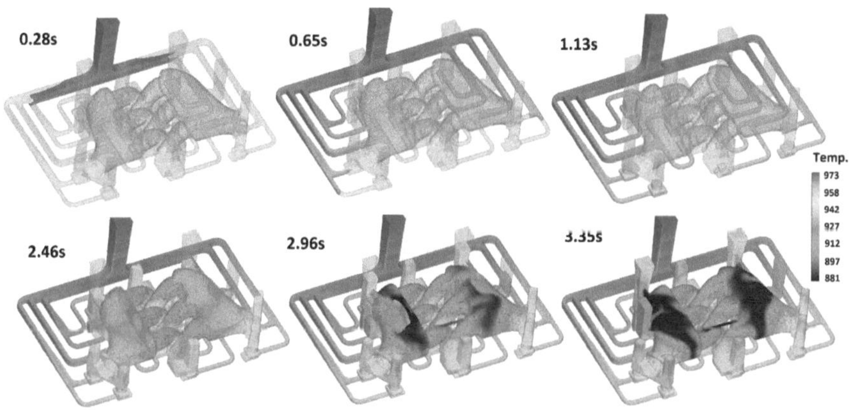

Fig. 11 Modeling results of the filter casting in temperature contour at various time frames

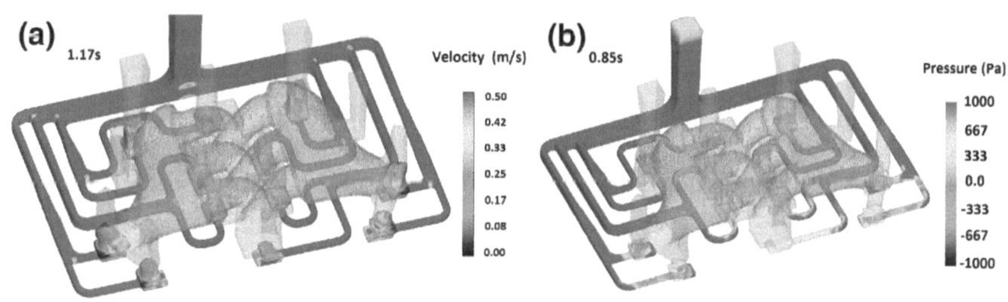

Fig. 12 Velocity and pressure contours in the ingate regions at similar time frames. (Color figure online)

Figure 10 presents some pictures of the filter casting. In this aluminum gravity casting, the filling takes 3.35 s. At 1.13 s, the filling streams enter the sites of the filter installations as shown in Fig. 11. The stream velocity in the ingate areas is below 0.5 m/s (e.g., colors other than red appeared in Fig. 12a). Positive pressures of the filling streams along the multiple runners to the ingates are revealed in Fig. 12b. The lowest pressure of the filling contour is in the green range, and there is no negative pressure in the filling front.

Discussion

All three casting methods success the fillings from the bottom to top of castings in a quiescent manner, and no surface turbulence is found in all casting processes. From the evidence in modeling results, positive pressures (i.e., greater than the atmosphere) are achieved as filling along runners to cast cavities. The potential head (or elevation head), described by Bernoulli's equation, is recovered from the kinetic head (or the velocity head), which may be developed at the base of the sprue by gravity.

In all casting methods, the averaged ingate velocity and the flow rate are listed in Table 2. For the tilt, diffuser and filter castings, their averaged ingate velocities are, respectively, 0.74, 0.37, and 0.18 (m/s), which are calculated among total volume of casting, cross-sectional area of ingates and filling time. In the case of the tilt casting, mechanical force can be artificially applied to increase the flow rate of the ingates after entering them. In contrast, that will decrease gradually such the fillings by gravity in the diffuser and filter castings, since the liquid level in basin will drop after pouring stops. Thus, the ingate velocity in the tilt casting over the critical velocity can be found although it is below this critical value at the time of entering ingates. This implies that developing of surface turbulence is not solely considered by the velocity head in ingates. Others, like potential recovery and flow rate, should also be involved.

If the potential head is recovered from the velocity head, hydrostatic pressure (i.e., the positive pressure) will take place in the cast cavities. The filling stream with positive pressures against to the walls along runners and cavities can be achieved. Flow rate has also been considered as an energy to elevate liquid metal to top of casting. A higher flow rate, a shorter filling time. It means that it is more efficient to fill the cavities in altitude. The defects, such as sand core blowing, confluence welding, and sand loosening, due to negative pressures, then can easily be avoided.

Table 2 Calculations from the modeling results and data in Table 1

	Filling time (s)	Averaged ingate velocity (m/s)	Flow rate (kg/s)
Tilt casting	7.0	0.74	3.2
Diffuser casting	32	0.37	5.9
Filter casting	3.35	0.18	7.2

Conclusion

1. An ingate velocity under 0.5 (m/s) is achieved by three casting methods. Flow rate in the tilt, diffuser, and filter castings, receptively, are 3.2, 5.9, and 7.2 (kg/s). A higher flow rate with a minimum ingate velocity is required in runner system design to prevent the defects derived from negative pressures.
2. The size of runner should be equal to that of filling streams in order to fit them. If oxides films on filling front can safely be against to the walls of runners, no surface turbulence will be created and, thus, bifilm entrainment could be avoided.
3. Other than the minimized velocity head, the potential head recovery and flow rate enlargement should be added in the principle of designing a good runner system.

References

1. Campbell J (2011) Complete casting handbook, metal casting process, metallurgy, techniques and design, 1st edn. p 614 [1a], pp 24–42 [1b], p 327 [1c], pp 67–72 [1d]
2. Hsu F-Y (2016) Bifilm defect formation in hydraulic jump of liquid aluminum. Metall Mater Trans B 47(3):1634–1648
3. Smithells CJ, Brandes EA (1976) Metal reference book, 5th edn. Butterworths
4. Green NR, Campbell J (1993) Statistical distributions of fracture strengths of cast Al–7Si–Mg. Mater Sci Eng, A 173:261–266
5. Green NR, Campbell J (1993) Statistical distributions of fracture strengths of cast Al–7Si–Mg. Mater Sci Eng, A 173:261–266
6. Green NR, Campbell J (1994), Influence of oxide film filling defects on the strength of Al–7Si–Mg alloy castings. Trans AFS, pp 341–347
7. Tiryakioglu M, Campbell J (2010) Weibull analysis of mechanical data for castings: a guide to the interpretation of probability plots. Metall Mater Trans A 41A:3121–3129
8. Tiryakioglu M, Campbell J, Alexopoulos ND (2009) Quality indices for aluminum castings: a critical review. Metall Mater Trans B 40:802–811
9. Hsu F-Y, Yang Y-M (2012) Confluence weld in an aluminum gravity casting. J Mater Process Technol 212:825–840
10. Hsu F Y, Jolly MR, Campbell J (2007) The design of L-shaped runners for gravity casting. In. Shape casting 2nd international symposium, Proceedings of 2007 TMS (The minerals, metals & materials society) annual meeting. Orlando, FL, USA, pp 101–108
11. Hsu F-Y, Jolly MR, Campbell J (2009) A multiple-gate runner system for gravity casting. J Mater Process Technol 209:5736–5750
12. Hsu F-Y, Lin H-J (2009) A diffusing runner for gravity casting. Metall Mater Trans B 40 (6):833–842
13. Hsu F-Y, Lin H-J (2011) Foam filters used in gravity casting, Metall Mater Trans B 40 (6):1110–1117

A Solidification Model with Application to AlSi-Based Alloys

Adrian V. Catalina, Liping Xue and Charles A. Monroe

Abstract The behavior of a cast component under service conditions depends primarily on its local microstructural features, such as the amount and morphology of phases and chemical compounds precipitating during the solidification process. Therefore, developing a good understanding between the influencing factors (chemical composition of the alloy and cooling conditions) and the solidification path is of great importance for establishing the casting process conducive to desired results. Because the solidification path of commercial AlSi-based alloys is quite complex, predictive capabilities for this type of alloys have been developed and incorporated into the commercial casting simulation software FLOW-3D® CAST. Models for quantitative description of the precipitation reactions, their temperature range, precipitation rate of phases and compounds, as well as the secondary dendrite arm spacing in relation to the chemical composition and cooling conditions are presented in this chapter. Comparisons of predicted microstructure with experimental measurements on a test casting are also discussed.

Keywords AlSi alloys · Mathematical modeling · Precipitation reactions · Microstructure · Quantitative prediction

---

A. V. Catalina (✉) · L. Xue
Flow Science, Inc., Santa Fe, NM 87505, USA
e-mail: adrian.catalina@flow3d.com

L. Xue
e-mail: liping.xue@flow3d.com

C. A. Monroe
The University of Alabama at Birmingham, Birmingham, AL 35294, USA
e-mail: camonore@uab.edu

M. Tiryakioğlu et al. (eds.), *Shape Casting*, The Minerals, Metals & Materials Series,
https://doi.org/10.1007/978-3-030-06034-3_20

Introduction

The class of AlSi-based alloys is widely used by the aerospace and automotive industries due to its lightweight, good mechanical properties, and excellent thermal conductivity. A given blend of mechanical and thermo-physical properties of a cast component is determined by the microstructure resulting from the solidification process while the microstructural features are the result of the chemical composition and the cooling conditions during the liquid-to-solid transformations in the alloy. A detailed understanding of the solidification path; i.e., the precipitation sequence of the various phases and compounds is, therefore, essential when designing a casting process. The published literature provides a few empirical relationships for calculating the liquidus [1] and the AlSi eutectic temperature [2, 3] based on the initial chemical composition of hypoeutectic AlSi alloys. A silicon equivalent (Si_{eq}) concept was developed by Djurdjevic et al. [4] for calculating the liquidus temperature of hypoeutectic AlSi alloys. The method, which treats a multicomponent alloy as a pseudo-binary one, was later extended to hypereutectic alloys [5], and then a capability for predicting the onset of the AlSi and AlSiCu eutectic temperatures was added [6]. Because the developed equations make use of the initial composition of the alloy, the accuracy of predicting the liquidus temperature is quite good but degrades for lower temperature transformations. Such empirical relationships have certain usefulness to the metal caster when establishing the pouring temperature, for instance. However, having the capability of predicting the type and amount of phases and compounds that precipitate during the solidification process requires models of a much higher degree of complexity. This is the realm of CALPHAD-based software packages, such as Thermo-Calc, Pandat™, or JMatPro® which are widely used for predicting the solidification path and the thermodynamic properties of multicomponent alloys. They are extremely valuable to the industry for designing materials and/or processes. However, because they operate with complex thermodynamic databases and algorithms, the answer to the question of why an alloy follows a certain solidification path may not be immediately and clearly available to a designer.

The work presented in this paper is aimed toward acquiring a better understanding of the mechanisms that govern the solidification of AlSi-based alloys. Several empirical relationships for the critical transformation temperatures are presented, along with models for predicting the microstructure evolution during the solidification process. The JMatPro® software package was used for developing the empirical relationships that cover a large range of chemical compositions for AlSi-based alloys.

Precipitation Reactions in AlSi-Based Alloys and Descriptive Relationships

Types of Precipitation Reactions

A list of representative precipitation reactions that can occur in AlSi-based alloys is presented in Table 1. For the chemical composition of a certain alloy, some of these reactions may occur, while others may not.

The notations Al_{fcc} and Si_D used in Table 1 stand for Al solid solution (α) and Si precipitate, respectively. In addition to the reactions presented in the table, precipitation of α-Fe and β-Fe-intermetallics also occurs when Fe is present.

Descriptive Empirical Relationships

With the help of JMatPro® software, several empirical relationships have been developed for describing the onset of the precipitation reactions presented in Table 1. A few examples are provided below.

The pro-eutectic reaction R1 is characterized by the liquidus temperature (T_{liq}) that can be calculated as:

$$T_{liq} = T_m + \sum_{i=1}^{N} m_{l,i} \cdot X_i \tag{1}$$

where T_m is the melting point of pure Al, $m_{l,i}$ is the liquidus slope of the alloying element i, and X_i is the liquid concentration of i. Equation 1 is valid from the start of solidification to the onset of the eutectic transformation. The liquidus slopes for this reaction are presented in Table 2.

If during the pro-eutectic solidification stage, the liquid concentration of Mg reaches or exceeds the value Mg_{R2} which is given by:

Table 1 List of precipitation reactions for AlSi-based alloys

Reaction #	Reaction type	Observations
R1	$Liquid \rightarrow Al_{fcc}$	Pro-eutectic
R2	$Liquid \rightarrow Al_{fcc} + Mg_2Si$	Pro-eutectic
R3	$Liquid \rightarrow Al_{fcc} + Si_D$	Eutectic
R4	$Liquid \rightarrow Al_{fcc} + Si_D + Mg_2Si$	Eutectic
R5	$Liquid \rightarrow Al_{fcc} + Si_D + Al_5Cu_2Mg_8Si_6$	Eutectic
R6	$Liquid \rightarrow Al_{fcc} + Si_D + Al_2Cu$	Eutectic
R7	$Liquid \rightarrow Al_{fcc} + Si_D + Al_2Cu + Al_5Cu_2Mg_8Si_6$	Eutectic

Table 2 Relationships for calculating the liquidus slopes in AlSi-based alloys

Element	Liquidus slope, $m_{l,i}$, °C/wt%
Si	$-5.584 - 0.081 \cdot \mathrm{Si} - 9.76\mathrm{E}-4 \cdot \mathrm{Si}^2 - 0.1169 \cdot \mathrm{Cu}$ $+0.267 \cdot \mathrm{Mg} - 0.1 \cdot \mathrm{Zn} + 0.124 \cdot \mathrm{Fe}$
Cu	$-2.695 + 6.574\mathrm{E}-03 \cdot \mathrm{Cu} - 8.191\mathrm{E}-04 \cdot \mathrm{Cu}^2$
Mg	$-4.033 - 0.088 \cdot \mathrm{Mg} - 0.014 \cdot \mathrm{Cu}$
Zn	$-1.449 + 0.092 \cdot \mathrm{Zn} - 0.0395 \cdot \mathrm{Cu}$
Fe	$-2.891 + 0.09 \cdot \mathrm{Fe} - 0.1048 \cdot \mathrm{Cu}$
Mn	-1.667

$$Mg_{R2} = 12.949 - 0.734 \cdot Si - 0.199 \cdot Cu - 0.177 \cdot Zn \quad (2)$$

then Mg_2Si will start precipitating along with the Al_{fcc} phase (reaction R2 in Table 1).

The eutectic reaction R3 ($Liquid \rightarrow Al_{fcc} + Si_D$) is thermodynamically invariant in the binary AlSi system and, according to JMatPro®, is characterized by the transformation temperature $T^o_{R3} = 576.916\,°\mathrm{C}$ and a Si concentration $Si^o_{R3} = 12.535wt\%$. In multicomponent systems, however, degrees of freedom are generated by the presence of the other alloying elements. Examples of the effects of Mg and Cu on the transformation temperature and eutectic concentration of Si are shown in Fig. 1. It can be observed from the figure that both Mg and Cu shift the transformation to lower temperatures compared to that of the binary AlSi alloy. It can also be observed that Mg shifts the eutectic concentration of Si to higher values, while Cu has an opposite effect. Therefore, it can be concluded that both Mg and Cu are controlling elements for this eutectic reaction. In fact, every alloying element except Si would be a controlling element for this precipitation reaction.

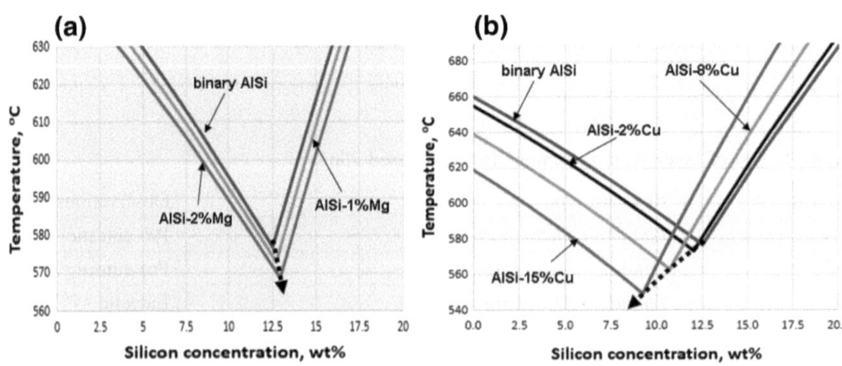

Fig. 1 Effect of Mg and Cu additions on the liquidus temperature curves and Si concentration at the eutectic point in: **a** AlSiMg alloys; **b** AlSiCu alloys. The curves were generated using the Pandat™ software and COST507 database

The temperature and Si concentration change during the eutectic transformation R3 can be described by relationships of the type:

$$T_{R3} = T^{o}_{R3} + \sum_{i=1}^{N_c} m_i^{R3} \cdot X_i \quad (3)$$

$$Si_{R3} = Si^{o}_{R3} + \sum_{i=1}^{N_c} g_i(X_i) \quad (4)$$

where N_c is the number of alloying elements other than Si, X_i is the liquid concentration of the element i, m_i^{R3} is the temperature slope of i for the eutectic transformation, and $g_i(X_i)$ is a function of X_i. Empirical relationships for m_i^{R3} and $g_i(X_i)$ of reaction R3 as well as similar functions for the other precipitation reactions presented in Table 1, including the precipitation of Fe-intermetallic compounds, have been developed. However, due to the lack of space, they will not be presented here. Note that in all the empirical relationships presented thus far the actual/current concentration values X_i at a certain time during solidification should be used. A model for the solid fraction evolution and liquid concentration change of the alloying elements during the solidification process is presented in the next section.

Solidification Model

When modeling the liquid/solid phase change during solidification of an alloy, a heat transfer equation with a source term has to be solved. In its simplest form, when the melt convection effects are neglected, this is:

$$\rho C_p \frac{\partial T}{\partial t} = \nabla(k \nabla T) + \rho \Delta H_f \frac{\partial f_s}{\partial t} \quad (5)$$

where T is temperature, t is time, ρ is density, C_p is specific heat, k is thermal conductivity, ΔH_f is the latent heat of fusion, and f_s is the solid fraction in the volume element. The solid fraction evolution term (i.e., $\partial f_s/\partial t$) in Eq. 5 can be linked with the temperature evolution ($\partial T/\partial t$) and the liquid concentration of the alloy in the volume element by making use of the Gulliver–Scheil (GS) solidification model and the thermodynamic equilibrium at the solid/liquid interface. It can, therefore, be shown that (see also [7]:

$$\frac{\partial f_s}{\partial t} = S_T \frac{\partial T}{\partial t} \quad \text{with} \tag{6}$$

$$S_T = \frac{1 - f_s}{\sum_{i=1}^{N} \bar{m}_i (1 - k_i) X_i} \quad \text{and} \quad \bar{m}_i = m_i + \frac{\partial m_i}{\partial C_{l,i}} X_i \tag{7}$$

where m_i, k_i, and X_i are the liquidus slope, the solid/liquid partition coefficient, and the liquid concentration of the alloying element i, respectively. The quantity $\bar{m}_i$ accounts for the fact that m_i is usually concentration dependent. For simplicity, the grain curvature effect was neglected in the above equations although it can be included as shown in [7]. In the numerical calculations, the quantity S_T can be calculated based on the last known values of f_s, m_i, k_i, and X_i. Then Eq. 6 is substituted into Eq. 5 from which the new temperature is calculated. Once the new temperature is known, the solid fraction evolution and the liquid concentration change during a time-step can be calculated by means of Eq. 6 and GS model, respectively.

The model presented above can be applied to any stage of solidification, i.e., primary solid solution or eutectic, provided that m_i and k_i as well as the onset conditions for each solidification stage are known. The partition coefficients between the liquid phase and Al_{fcc} phase as suggested by Sigworth [8], k_i^{α}, can be used during each stage of solidification. Some complications arise when multiple phases/compounds are precipitating concomitantly which is the case for all but reaction R1 in Table 1. During the reaction R2, for instance, the segregation of Si and Mg into the liquid can be calculated based on the partition coefficients k_i^{α} shown in [8]. At the same time, however, one has to account for the fact that some quantities of Si and Mg are consumed because of Mg_2Si precipitation. As another example, in binary AlSi systems, the partition coefficient of Si during the eutectic reaction R3 is practically equal to 1. However, the continual shift of the eutectic point during the solidification of multicomponent alloys means that the partition coefficient undergoes changes due to the segregation of the other alloying elements. This problem can be solved by considering some *apparent* or *overall* values of the partition coefficients. A model for calculating the overall partition coefficient of Si is presented below for the case of the eutectic reaction R3 occurring in a multi-component alloy. For this purpose, one may consider the GS model applied to Si and another alloying element i:

$$dSi = Si \cdot \left(1 - \bar{k}_{Si}^{R3}\right) \cdot \frac{df_s}{1 - f_s} \tag{8}$$

$$dX_i = X_i \cdot \left(1 - k_i^{\alpha}\right) \cdot \frac{df_s}{1 - f_s} \tag{9}$$

where $\bar{k}_{Si}^{R3}$ represents the apparent partition coefficient of Si. For the alloying element i, the partition coefficient k_i^{α} can be used as a reasonable approximation because the fraction of Si_D is quite small. The value of k_i^{α}, however, can be refined once the fraction of Si_D in the evolving microstructure becomes known. From Eqs. 8 and 9 one obtains:

$$\frac{dSi}{dX_i} = \frac{Si \cdot \left(1 - \bar{k}_{Si}^{R3}\right)}{X_i \cdot \left(1 - k_i^{\alpha}\right)} \tag{10}$$

and therefore:

$$\bar{k}_{Si}^{R3} = 1 - \frac{1}{Si} \cdot \left[(1 - k_i) \cdot X_i \cdot \frac{dSi}{dX_i}\right] \tag{11}$$

For a number N_c of alloying elements other than Si, Eq. 11 can be written as:

$$\bar{k}_{Si}^{R3} = 1 - \frac{1}{Si} \cdot \sum_{i=1}^{N_c} (1 - k_i) \cdot X_i \cdot \frac{dSi_{R3}}{dX_i} \tag{12}$$

where Si and X_i are the current liquid concentrations of Si and element i, respectively. Also, the quantity dSi_{R3}/dX_i can be calculated based on Eq. 4 and the empirical functions $g_i(X_i)$. With $\bar{k}_{Si}^{R3}$ now known, the phase fractions in the eutectic microstructure can be calculated from:

$$\begin{cases} f_{Si_D} + f_{\alpha} = 1 \\ f_{Si_D} \cdot Si_{Si_D} + f_{\alpha} \cdot Si_{\alpha} = \bar{k}_{Si}^{R3} \cdot Si \end{cases} \tag{13}$$

where f_{Si_D} and f_{α} are the fractions of Si_D and Al_{fcc}, respectively. Also Si_{Si_D} and Si_{α} represent the concentrations of Si in Si_D and Al_{fcc}, respectively. It follows that:

$$f_{Si_D} = \frac{\bar{k}_{Si}^{R3} \cdot Si - Si_{\alpha}}{Si_{Si_D} - Si_{\alpha}} \tag{14}$$

With $\bar{k}_{Si}^{R3} = 1$, Eq. 14 becomes the lever rule in the binary alloy.

Secondary Dendrite Arm Spacing (SDAS)

The SDAS can be calculated based on the model proposed by Rappaz and Boettinger [9], i.e.:

$$SDAS = 5.5 \cdot (M \cdot \Delta t_s)^{1/3} \tag{15}$$

where Δt_s is the solidification time and the quantity M is calculated from the chemical composition and the liquid diffusion coefficients of the alloying elements:

$$M = -\frac{\Gamma}{\sum_{i=1}^{N} m_i(1-k_i)(X_{f,i}-X_{o,i})} \cdot ln\left[\frac{\sum_{i=1}^{N} m_i(1-k_i)X_{f,i}/D_{l,i}}{\sum_{i=1}^{N} m_i(1-k_i)X_{o,i}/D_{l,i}}\right] \quad (16)$$

where Γ is the Gibbs–Thomson coefficient, $D_{l,i}$ is the liquid diffusion coefficient of element i, and $X_{o,i}$ and $X_{f,i}$ are the liquid concentrations of i at the beginning and end of solidification, respectively. The diffusion coefficients used in this work were taken from [10]. Also, in the calculated results presented in the next section, Δt_s was considered as the time between the start of solidification and the start of the eutectic transformation. The $X_{f,i}$ values were determined according to this assumption.

Results

In this section, some calculated results are, first, compared to the JMatPro® predictions. Then, the model predictions are compared against the experimental measurements performed on a test casting.

Model Versus JMatPro® Predictions

The alloy B319 with the chemical composition (wt%) of 5.9 Si, 3.28 Cu, 0.42 Mg, 0.75 Fe, 0.36 Mn, and 0.98 Zn was chosen as an example. The results obtained using the developed model as well as those from JMatPro® simulations are presented in Figs. 2, 3, and 4.

Figure 2a shows the temperature evolution with respect to the total fraction of solid. The precipitation reaction number is also plotted in this figure. It can be observed that the precipitation of primary Al_{fcc} phase (reaction R1) is briefly preceded by another event (see inset) which is the precipitation of the α Fe-intermetallic compound. This precipitation of the Fe-intermetallic is also visible in Fig. 3a, b which shows a rather abrupt decrease of Fe and Mn concentrations at the very beginning of solidification. Apart from the precipitation of Fe-intermetallic compounds, the solidification sequence of the alloy is R1→R3→R5→R7. The liquid concentrations of Si, Mg, and Cu continually increase during the reaction R1. The concentration of Si is around 10.6wt% at the onset of the eutectic reaction R3 (i.e., about 2wt% lower than in the binary AlSi system) and after that it assumes a descending trend due to the precipitation of the eutectic Si_D. Because Mg and Cu do not participate in this reaction, their concentrations continue their ascending trend until the onset of reaction R5 where the compound $Al_5Cu_2Mg_8Si_6$ starts forming.

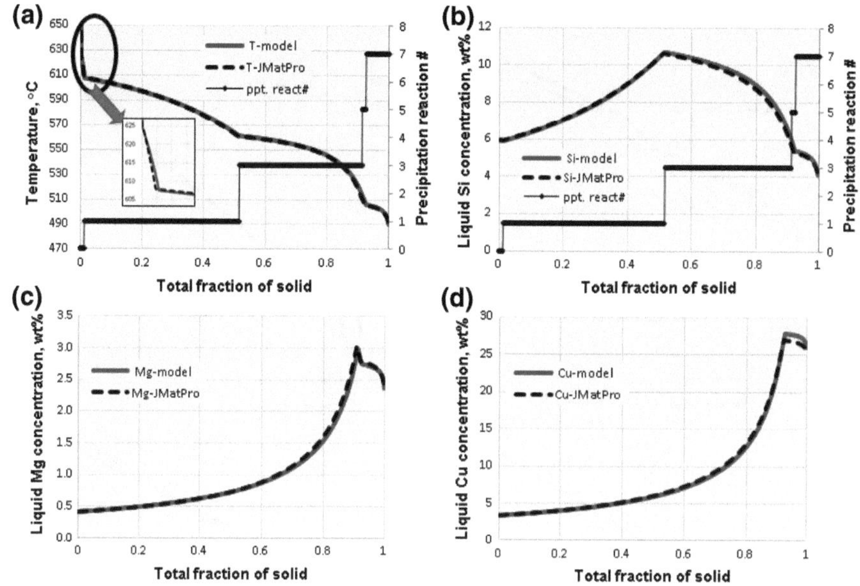

Fig. 2 Evolution of temperature (**a**) and liquid concentration of: **b** Si, **c** Mg, and **d** Cu during the solidification of B319 alloy

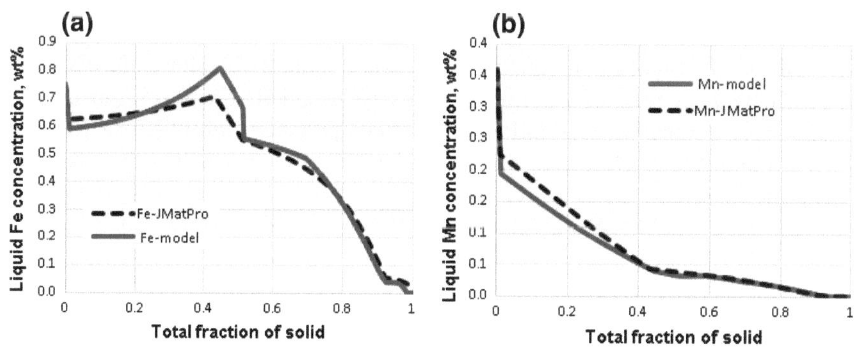

Fig. 3 Evolution of liquid concentration of: **a** Fe, **b** Mn during solidification of B319 alloy

The duration of reaction R5 is quite brief, but it can be clearly distinguished in Fig. 2c when Mg decreases rather abruptly from ~3wt% to ~2.75wt%. The concentration of Cu reaches its peak value of ~27wt% and then starts decreasing due to the precipitation of $Al_5Cu_2Mg_8Si_6$ and Al_2Cu during the reactions R5 and R7 (see Fig. 2d). Figure 3a, b shows the evolution of Fe and Mn concentrations. The continual change of these concentrations on a descending path, except for Fe during R1 reaction, suggests that the precipitation of Fe-intermetallic compounds occurs during every stage of the solidification process. This is also visible in Fig. 4a which

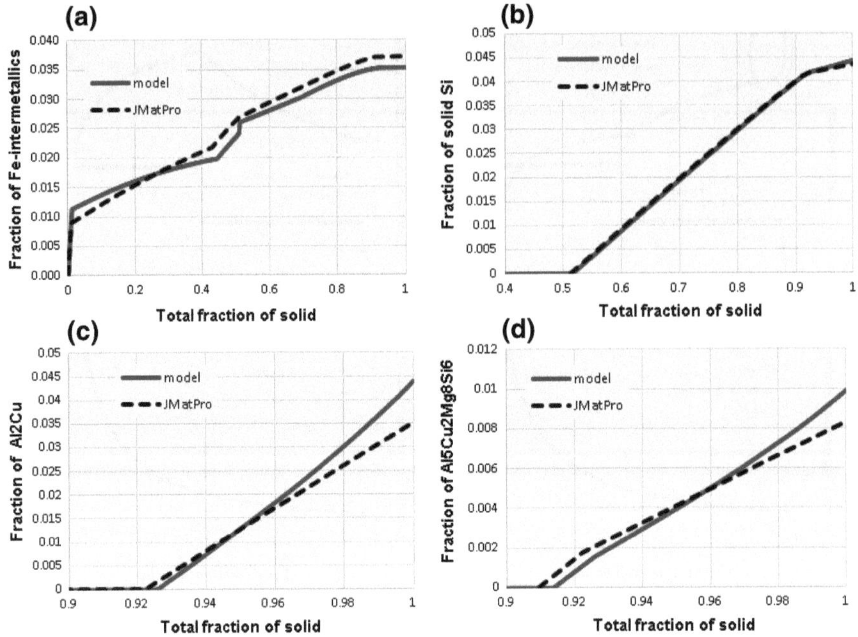

Fig. 4 Evolution of precipitates fractions **a** Fe-intermetallics, **b** Silicon, **c** Al_2Cu, **d** $Al_5Cu_2Mg_8Si_6$ during the solidification of B319 alloys

shows the evolution of Fe-intermetallics fraction. Figure 4 also shows the evolution of Si_D fraction as well as the fractions of Al_2Cu and $Al_5Cu_2Mg_8Si_6$ in the microstructure. Comparing to JMatPro®, the model predicts slightly higher fractions of Al_2Cu and $Al_5Cu_2Mg_8Si_6$ in the final microstructure, but still with reasonable accuracy.

Model Predictions Versus Experimental Measurements of SDAS

In order to further test the developed model, its predictions are compared against experimental measurements performed on a test casting having the shape of a wedge [11]. The geometry and dimensions of the casting are shown in Fig. 5. The casting was produced by means of a lost foam process, with pouring performed through a sprue (76 mm × 76 mm × 76 mm) placed at the top of the thick section. Also, a steel chill of dimensions 240 mm × 40 mm × 25 mm was placed at the bottom of the thin section in order to obtain a relatively large range of cooling rates on the height of the casting. The A319 alloy of chemical composition (wt%) 6.99Si, 0.526 Mg, 3.50Cu, 0.51Fe, 0.35Mn, 0.76 Zn was used in this experiment.

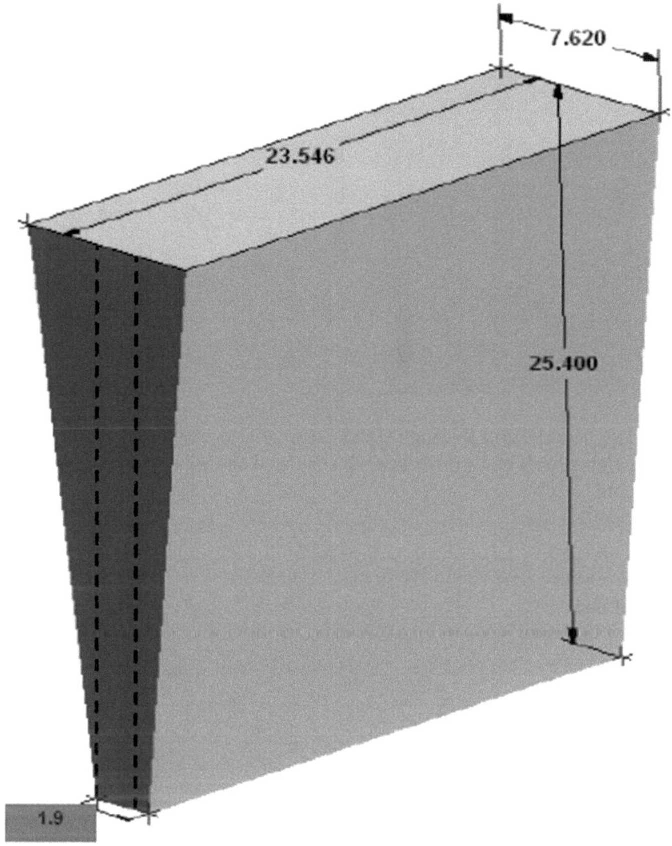

Fig. 5 Geometry and dimensions (cm) of the test casting [11]

More details of the experiment can be found in [11, 12]. SDAS measurements on the produced casting were performed at various distances on the height of the casting, starting at 12.7 mm from the chill and then at an increment of 25.4 mm.

The solidification of the test casting was simulated with the FLOW-3D® CAST software in which the developed solidification model was implemented. Empirical relationships relating the thermo-physical properties (density, specific heat, thermal conductivity, latent heat of fusion) to temperature, chemical composition, and the microstructure of the alloy have also been developed and implemented into the solidification code.

In order to assess the melt convection effect on SDAS, simulations with and without considering natural melt convection were performed. The SDAS predictions, along with the experimental measurements are shown in Fig. 6. As expected, the predicted SDAS increases with the distance from the chill. When accounting for the natural melt convection (curve 2 in Fig. 6b), the predicted SDAS is slightly larger comparing to the case of a quiescent melt (curve 1), especially in the regions

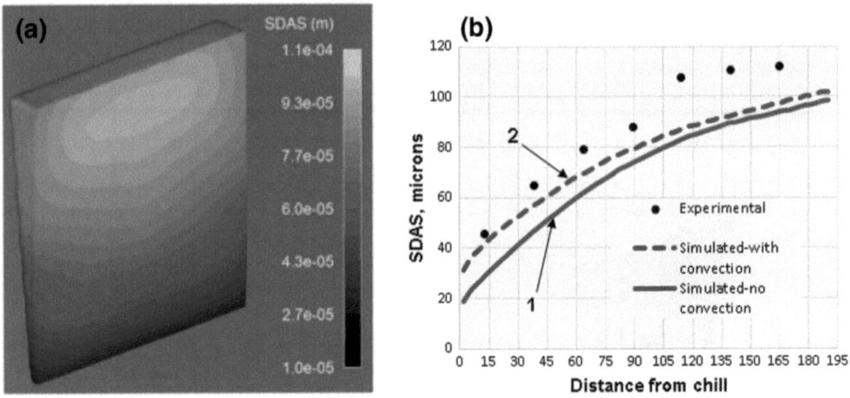

Fig. 6 Secondary dendrite arm spacing (SDAS): **a** simulated size distribution on cross section of the casting (simulation with melt convection); **b** simulated and experimental measurements Versus distance from chill

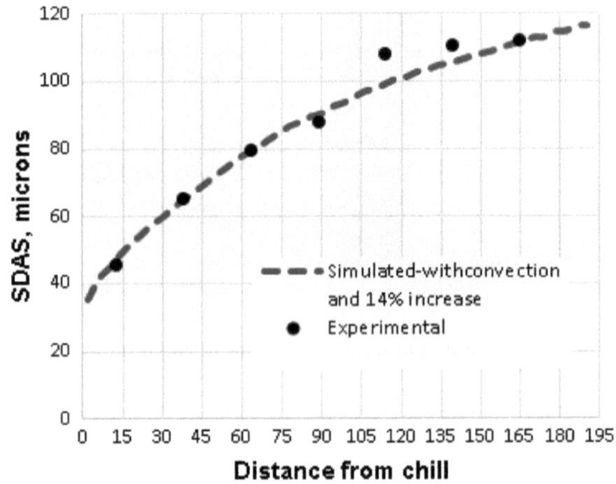

Fig. 7 Comparison between the experimental and theoretical SDAS increased by 14%

closer to the chill. Nevertheless, the values are still, but consistently, under predicted by ~14%. It can be seen in Fig. 7 that when shifting the curve 2 upwards by 14% a very good agreement with the experimental data is obtained. This shifting is equivalent to increasing the coefficient in Eq. 15 from 5.5 to 6.27. Because of the many simplifying assumptions used in developing the SDAS prediction model [9, 13], the need for this increased coefficient is not surprising. This increase, however, may not be suitable for alloys other than the one used in this study.

Conclusions

It was shown in this paper that the solidification path of AlSi-based alloys can be described using empirical relationships derived from the analysis of a large number of data generated by means of a CALPHAD-based software package. Examples of models for calculating the partition coefficients of the alloying elements and quantitative prediction of precipitating phases and chemical compounds during complex eutectic reactions have been presented. The predictions of the developed model showed a good agreement with the JMatPro® results. The simulations performed on the test casting showed that when accounting for the natural melt convection during solidification the SDAS was slightly higher comparing to that obtained for a quiescent melt. A further increase by 14% of the calculated SDAS was necessary in order to achieve a good agreement with the experimental measurements.

References

1. Drossel G (1981) Der Einflusse von Schmelzebehandlungen auf die Dichtheit von Gusskoerpern aus Al-Si-Gusslegierungen. Giessereitechnik 27(1):7–12
2. Mondolfo LF (1979) Aluminum alloys: structure and properties. Butterworth (Publishers) Inc., Boston
3. Tenekedjiev N, Mulazimoglu H, Closset B, Gruzleski J (1995) Microstructures and thermal analysis of strontium-treated aluminum-silicon alloys. American Foundrymen's Society, Inc., Des Plaines, Illinois
4. Djurdjevic MB, Kierkus WT, Byczynski GE, Sokolowski JH (1998) Calculation of liquidus temperature for Aluminum 3XX series of alloys. AFS Trans, 143–147
5. Hernandez FCR, Djurdjevic MB, Kierkus WT, Sokolowski JH (2005) Calculation of the liquidus temperature for hypo and hypereutectic aluminum silicon alloys. Mat Sci Eng A 396:271–276
6. Djurdjevic MB, Vicario I (2013) Description of hypoeutectic Al-Si-Cu alloys based on their known chemical compositions. Revista de Metalurgia 49(5):340–350
7. Catalina AV, Leon-Torres JF, Stefanescu DM, Johnson ML (2007) Prediction of shrinkage-related defects in steel castings. In: Proceedings of the 5th decenial international conference solidification processes, Sheffield, pp. 699–703
8. Sigworth GK (2014) Fundamentals of solidification in aluminum castings. Intl J Metalcast 8 (1):7–20
9. Rappaz M, Boettinger WJ (1999) On dendritic solidification of multicomponent alloys with unequal liquid diffusion coefficients. Acta Mater 47(11):3205–3219
10. Du Y, Chang YA, Huang B, Gong W, Jin Z, Xu H, Yuan Z, Liu Y, He Y, Xie F-Y (2003) Diffusion coefficients of some solutes in fcc and liquid Al: critical evaluation and correlation. Mat Sci Eng A363:140–151
11. Chintalapati P (2009) Solidification under pressure of aluminum castings. Ph.D. dissertation, The University of Alabama at Birmingham
12. Chintalapati PP, Griffin JA, Griffin RD (2007) Improved mechanical properties of lost foam cast A356 and A319 aluminum solidified under pressure. AFS Transactions, 881–897
13. Kurz W, Fisher DJ (1992) Fundamentals of solidification. Trans Tech Pub Ltd, Switzerland

Physical Modeling of Transport Phenomena in Asymmetrical Multi-strand Tundish with Retaining Wall

Wei Xiao and Yanping Bao

Abstract In multi-strand tundish, the structure is more complex and has various types compared with single/double-strand tundish and the horizontal distance between ladle shroud and each submerged nozzles is much different which leads to the inhomogeneity between each outlets. In this paper, physical simulation was conducted in a full-scale physical model of asymmetrical multi-strand tundish to study the effect of the direction of the diversion hole on the retaining wall on tundish performance. The optimized retaining wall is applied to field production, and industrial experiments were conducted to analyze the effect of inclusion removal before and after optimization. The effects were quantified by measuring residence time of fluid in the tundish. According to the results, the optimized retaining wall improved the homogeneity between each outlets and enhanced the metallurgical function of the tundish.

Keywords Tundish · Multi-strand · Physical simulation · Retaining wall

Introduction

The tundish has the function of purifying liquid steel and removing non-metallic inclusions in the process of continuous casting. It is an important process to improve the quality of steel products. The flow behavior of molten steel in the tundish has an important influence on the mixing of molten steel in the tundish, the removal of non-metallic inclusions, and the behavior of the steel–slag interface [1]. Therefore, a certain flow control device is set up in the tundish so that a reasonable flow field can be formed in the tundish, which plays an important role in achieving the metallurgical function of the tundish.

W. Xiao · Y. Bao (✉)
State Key Laboratory of Advanced Metallurgy, University of Science and Technology Beijing, Beijing, China
e-mail: YanpingBao@hotmail.com

M. Tiryakioğlu et al. (eds.), *Shape Casting*, The Minerals, Metals & Materials Series,
https://doi.org/10.1007/978-3-030-06034-3_21

The asymmetric four-flow tundish in the steel plant was studied in this paper. In multi-strand tundish, the structure is more complex and has various types compared with single/double-strand tundish. The horizontal distance between ladle shroud and each submerged nozzles is much different. The distance between the outside submerged nozzle and the ladle shroud is longer, and there is more heat loss led to easy solidification of molten steel and difficulties in casting start and continuous casting process. The distance between the inside submerged nozzle and the ladle shroud is shorter which leads to the short time of inclusion floatation, difficulty of inclusion removal, and easy to appear quality problem of billet [2–4]. Hence, many researchers have done a great deal of research on the consistency of the flow characteristics of the multi-strand tundish [5–8].

In this paper, physical simulation was adopted to study the flow field of asymmetric multi-strand tundish in the steel plant by using different flow control device to obtain the most appropriate flow field [9–16]. The optimized flow control device was applied to field production, and industrial experiments were conducted to analyze the effect of inclusion removal before and after optimization.

Analysis of Prototype Usage

There were many large inclusions in casting billets in the steel plant, and the horizontal distribution of large inclusions between different streams was uneven and much different. It still can be improved in the effect of metallurgical functions of the tundish. Long-term detection of the steel billet quality had been conducted. In the detection, size of inclusions in billets in different furnace had been tested and the control level of inclusions in tundish had been analyzed; size of inclusions in different streams had been detected, and the inhomogeneity between different streams had been analyzed. The following results are obtained:

In Fig. 1a, there are 15–25% inclusions of which size are >20 μm in the billets. In Fig. 1b, the amounts and the size of inclusions of each streams are quite different showing the inhomogeneity between different streams. Therefore, it is necessary to optimize the structure of the tundish, to improve and uniform the cleanliness of molten steel in tundish.

Experiment

Experimental Principle

On the basis of the similarity principle, the key physical phenomena in the casting process of the tundish are studied by using water to simulate the molten steel. The flow of molten steel in tundish is mainly affected by gravity, inertia force, and

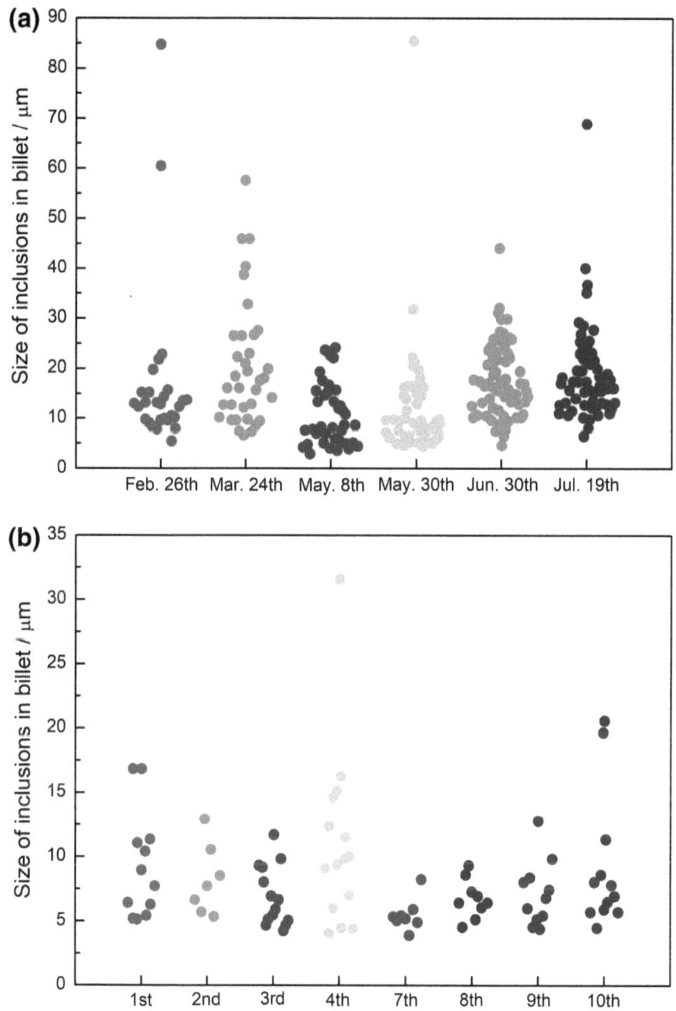

Fig. 1 Size of inclusions in billet at different time (**a**) and different outlets (**b**)

viscous force. Therefore, the Froude $\mathrm{Fr} = \frac{v^2}{gl}$ is selected as qualitative criterion. Geometric similarity ratio is 1:2.5. According to $Fr_m = Fr_s$,

$$\text{Aspect ratio}: \ l_m = \lambda \cdot l_p \tag{1}$$

$$\text{Velocity ratio}: \ v_m = \lambda^{0.5} \cdot v_p \tag{2}$$

$$\text{Flow ratio}: \ Q_m = \lambda^{2.5} \cdot Q_p \tag{3}$$

where $\lambda = 0.4$; v is characteristic velocity; Q is flow of molten steel in tundish; m means model; p means prototype.

Method

The main equipment contains a model of asymmetric four-strand tundish, tracer adding device and data acquisition system (DJ800 system, developed by China Institute of Water Science and Technology), as shown in Fig. 2.

Sample-electrolysing method was conducted to analysis the content of large inclusions in the casting billet, and the size and composition of large inclusions were observed by electron microscope.

The content of oxygen and nitrogen in the casting billet before and after optimization was analyzed.

Experimental Scheme

Water model simulation was adopted to analyze the flow field of the original tundish and optimized tundish, in which mainly changes the flow direction of the diversion hole on the retaining wall. The four-strand tundish is asymmetric, and its structure and the structure of retaining wall are shown in Fig. 3.

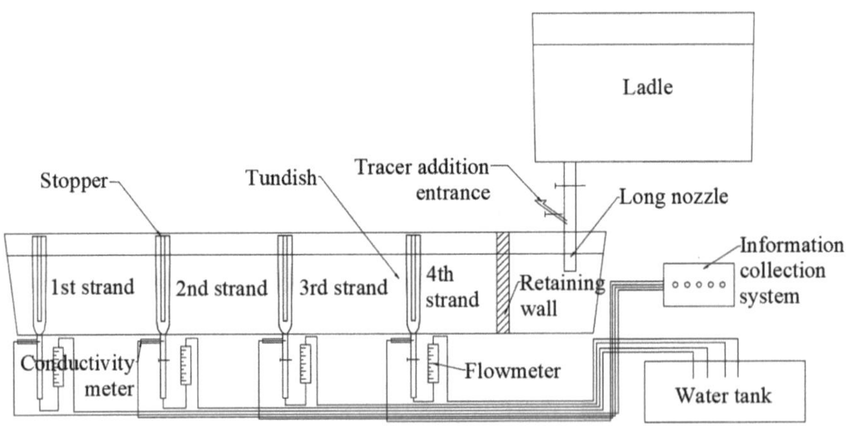

Fig. 2 Sketch of water model experimental setup

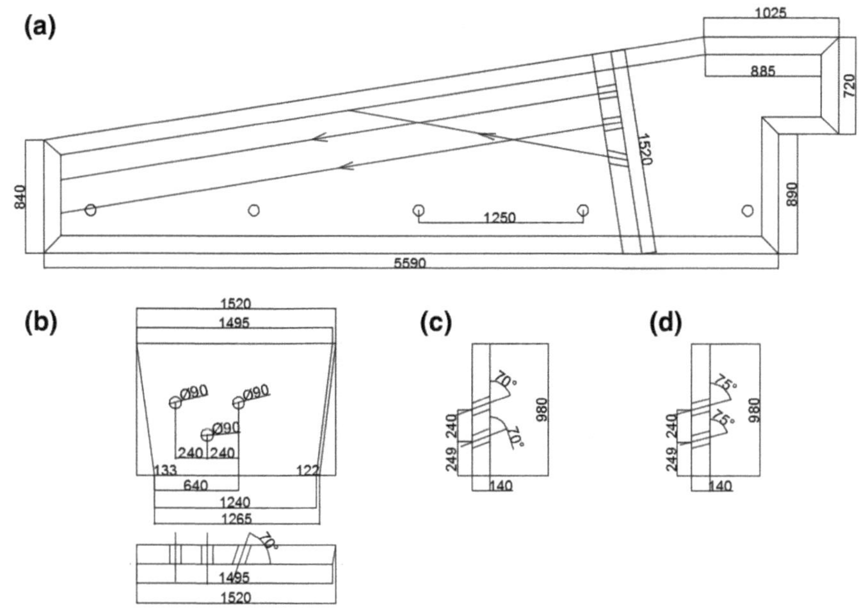

Fig. 3 Vertical view of tundish and the position of diversion holes, **a** is vertical view of tundish, **b** is front view of retaining wall, **c** is left view of original retaining wall, **d** is left view of optimized retaining wall

Results and Discussion

Analysis of the Flow Field of the Original Tundish

The experimental results of using different flow control devices are shown in Table 1. It can be seen that the average residence time is 676.94 s when using the prototype tundish, which is far less than the theoretical average residence time 1029 s. The average residence time increases to 747.46 s after using the optimization scheme. The ratio of the dead zone of the prototype tundish is 17.8%. The homogeneity between each stream is poor. The longest minimum residence time is 116 s of the 1st flow, the shortest minimum residence time is 77 s of the 3rd flow, and the range is 39 in the prototype tundish.

Figure 5 shows the flow field of the prototype tundish. The flow stock flows from the diversion hole and rushes into the surface of tundish, and the impact point is between 3rd and 4th flow and then run to the 2nd and 3rd flow along the liquid surface. Part of the flow fall and come out in the 2nd and 3rd outlets because the kinetic energy is reduced and the flow is affected by the tractive effort of outlets. The other part of the flow is affected by the tractive effort of 1st outlet and moves continuously to the 1st outlet and come out. Because there is no flow strand above the 1st outlet, the liquid steel is relatively inactive and easy to form dead zone.

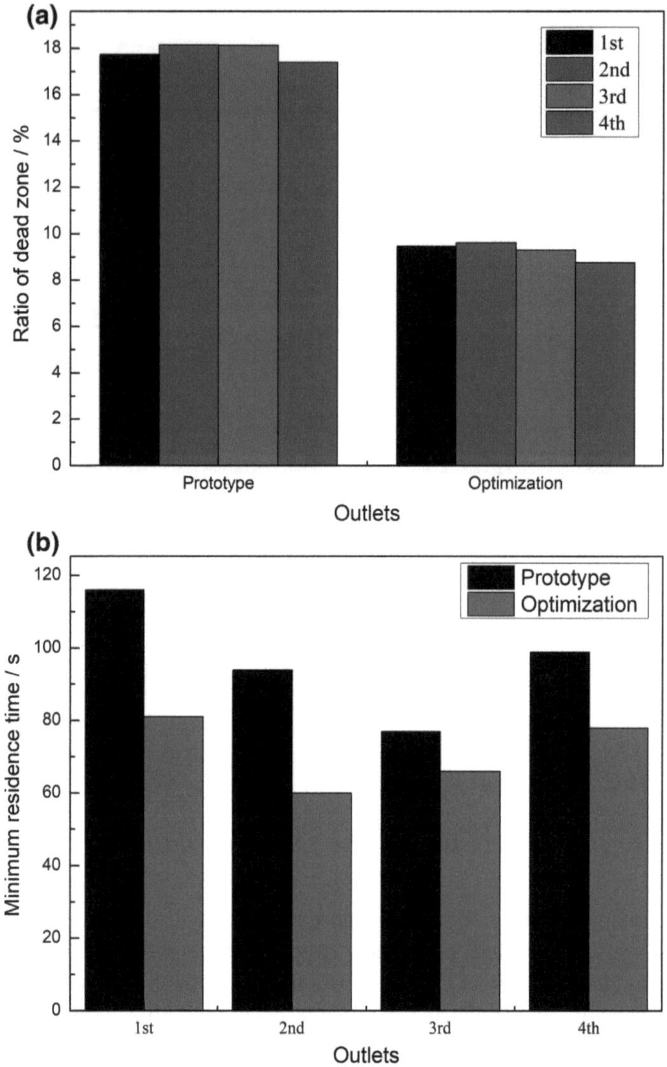

Fig. 4 Comparison of parameters of each strand before and after the optimization of the tundish

Analysis of the Flow Field of the Optimized Tundish

In the optimization scheme, the opening angle of the diversion hole has been changed, and the detailed structure is shown in Fig. 2.

After using the optimization scheme, the dead zone in the tundish improved significantly, and dead zone ratio decreased from 17.8 to 9.3% with a decrease of 47.8%, as shown in Fig. 4a. Due to the decrease of the angle of the diversion hole,

Table 1 Comparison of parameters before and after the optimization of the tundish

		Plug flow ratio/%	Average residence time	The average of average residence time	Range of minimum residence time
The prototype tundish	1st strand	17.41	677.88	676.94	97
	2nd strand	14.44	674.54		
	3rd strand	18.69	674.68		
	4th strand	26.70	680.67		
The optimized tundish	1st strand	14.14	745.91	747.46	71
	2nd strand	15.17	744.73		
	3rd strand	28.76	747.33		
	4th strand	31.55	751.89		

Fig. 5 Flow field of prototype tundish

the impact point that the liquid flow impacts to the liquid surface moved forward, and the kinetic energy of the flow is increased, which leads to the flow came out from the 2nd outlet at first. It improved the update speed of the molten steel above the 1st flow, and the minimum residence time decreased, which reduced the dead zone ratio in this area. It can be seen in Fig. 4b that the minimum residence time of the 1st flow in the optimization scheme was reduced to 81 s, the shortest minimum residence time was 60 s of the 2nd flow. Compared with prototype scheme, the range of minimum residence time was reduced to 21, decreased 46.2%, and the uniformity of each flow is obviously improved.

Optimization Effect Verification

In order to compare the metallurgical effects before and after the optimization of the tundish, the samples of using the prototype scheme and the optimization scheme were taken respectively. The billet in the stable casting was taken to analyze the T. [O] content and the large inclusions content by the total oxygen measurement and large sample electrolysis.

After optimization, the cleanliness of molten steel in tundish was improved, and the content of T. [O] in tundish decreased by 10.6% (Fig. 6).

The large inclusions in tundish obtained by large sample electrolysis were observed by SEM, and the size and composition of large inclusions were obtained. It mainly includes the following types of large inclusions: (1) large inclusions mainly composed of Al_2O_3-SiO_2, approximately spherical; (2) large inclusions mainly based on Al_2O_3-SiO_2-MgO; (3) large inclusions mainly composed of SiO_2, irregular polygons; (4) large inclusions mainly composed of Al_2O_3, irregular polygons; (5) Al_2O_3-SiO_2-based large inclusions with K, Na, Zn, and other elements; and (6) large inclusions mainly composed of MgO-CaO.

The statistical results of the large sample electrolysis in the tundish are shown in Table 2. It can be seen that the number of large inclusions in the tundish is relatively much, and the size distribution of each flow is mainly concentrated on >300 μm; the total amount of large inclusions in the weight steel per unit weight in the optimization tundish is less than the amount of large inclusions in the original tundish, and 17.94 mg * $(10\ kg)^{-1}$ inclusions has been reduced.

Conclusion

(1) The optimization scheme was conducted by changing the direction of diversion holes. After optimization, the average residence time increased by 70.52 s, which provided more sufficient time for the removal of inclusions.

Fig. 6 Typical morphology of large inclusions in cast billets. **a** Al_2O_3–SiO_2; **b** Al_2O_3–SiO_2–MgO; **c** SiO_2; **d** Al_2O_3; **e** Al_2O_3–SiO_2 (containing K, Na, Zn); **f** MgO–CaO

Table 2 Content and size distribution of large inclusions in tundish

Sample	Total amount of inclusions/mg* $(10\ kg)^{-1}$	Size distribution of inclusions/%		
		<140 μm	140–300 μm	>300 μm
The prototype tundish	76.83	0.155	10.578	89.267
The optimized tundish	58.89	0.006	7.330	92.664

The proportion of dead zone dropped by 47.8%, the range of minimum residence time between each flow decreased, and the consistency of each flow was improved.

(2) The amount of large inclusions in the weight steel per unit of the optimization tundish was decreased significantly after the optimization and decreased by 17.94 mg*$(10\ kg)^{-1}$ in total. The T. [O] content decreased by 10.6%. The content of inclusions was reduced, and the metallurgical function of the tundish had been enhanced.

Acknowledgements This work was financially supported by the Open Project of State Key Laboratory of Advanced Special Steel, Shanghai University (SKLASS 2017-13).

References

1. Sahai Y, Emi T (1996) Criteria for water modeling of melt flow and inclusion removal in continuous casting tundishes. ISIJ Int 36:1166–1173
2. Xie WX, Bao YP, Wang M et al (2014) Study on improvement in uniformity of multi-strand tundish. J Univ Sci Technol Beijing 36:213–217
3. Eastman C, Glaws P, Cao D (2016) Liquid steel flow and interactions with nonmetallic phases in the continuous casting tundish using CFD & physical modeling. In: CFD modeling and simulation in materials processing. Cham: Springer
4. He F, Zhang LY, Xu QY (2016) Optimization of flow control devices for a T-type five-strand billet caster tundish: water modeling and numerical simulation. China Foundry 13:166–175
5. Wang JY, Bao YP, Qu Y (2001) Tundish metallurgy. Metallurgical Industry Press, Beijing
6. Zheng SG, Zhu MY (2005) Analysis model for flow characteristics in multi-strand continuous casting tundish. Acta Metallurgica Sin 41:67–70
7. Solorio-Diaz G, Palafax-Ramos J, García-Demedices L et al (2004) Analysis of fluid flow turbulence in tundishes fed by a swirling ladle shroud. ISIJ Int 44:1024–1032
8. Li YH, Zhao LH, Bao YP et al (2014) Effect of diversion hole on flow trajectory of steel in multi-strand tundish. J Univ Sci Technol Beijing 36:21–28
9. Ramos-Banderas A, García-Demedices L, Díaz-Cruz M (2003) Mathematical simulation and modeling of steel flow with gas bubbling in trough type tundishes. ISIJ Int 43:653–662
10. Kim HB, Isac M, Guthrie RI et al (2005) Physical modelling of transport phenomena in a delta shaped, four strand tundish with flow modifiers. Steel Res Int 76:53–58
11. Sheng DY, Jonsson L (2000) Two-fluid simulation on the mixed convection flow pattern in a nonisothermal water model of continuous casting tundish. Metall Mater Trans B 31:867–875

12. Odenthal HJ, Böiling R, Pfeifer H (2003) Numerical and physical simulation of tundish fluid flow phenomena. Steel Res Int 74:44–55
13. Chen C, Jonsson LTI, Tilliander A et al (2015) A mathematical modeling study of the influence of small amounts of KCl solution tracers on mixing in water and residence time distribution of tracers in a continuous flow reactor-metallurgical tundish. Chem Eng Sci 137:914–937
14. Huang J, Zhang Y, Zhang Y et al (2016) Study of flow characteristics of tundish based on digital image velocimetry technique. Metall Mater Trans B 47:3144–3157
15. Cwudziński A (2017) Physical and mathematical simulation of liquid steel mixing zone in one strand continuous casting tundish. Int J Cast Metals Res 30:50–60
16. Li D, Cui H, Liu Y et al (2016) A new method based on the F-curve for characterizing fluid flow in continuous casting tundishes. Metall Mater Trans B 47:1237–1242

The Validation of Feeder Modeling for Ductile Iron Castings

Fu-Yuan Hsu and Yu-Hung Chen

Abstract A precise feeder design for a ductile iron casting is difficult, due to the complexities of its liquid–solid states transformation. During solidification, a negative pressure in castings is developed from the volumetric shrinkage as liquid iron solidified, while a positive pressure is added from the volumetric expansion as spheroidal graphite crystallized from the liquid iron. A unique crystalline structure of Austenite shell surrounds the spherical graphite and also increases the viscosity of feeding fluid flow. First principle model was applied for modeling the flow behavior of the residual liquid iron during solidification. The phenomenon of volumetric expansion as spheroidal graphite crystallized was also considered. The model was validated by real casting a cube-shaped cast. The predictions of the shrinkage area and porosities were confirmed by the hardness contours in the cross-sectional area of the castings. An optimized solidification modulus ratio of the feeder to this cube cast was suggested.

Keywords Ductile cast iron · Feeder design · Hot spot · Modeling validation · Spheroidal graphite

Introduction

In the Campbellology, spherical graphite was precipitated from liquid cast iron through various steps of the nodulation process in ductile cast iron [1]. Using 2.5-mm-alumina refractory tube drawing magnesium-treated ductile cast iron melt, Fredriksson et al. explore various directional solidification experiments and rapidly quenched the tube for metallographic analysis [2]. They found that the volume fraction of austenite in dendrites is about 10 times larger at the early stage of the

F.-Y. Hsu (✉) · Y.-H. Chen
Department of Materials Science and Engineering, National United University, No. 1 Lein-Da, Kung-Ching Li, MiaoLi 36003, Taiwan, ROC
e-mail: willyhsu@nuu.edu.tw

M. Tiryakioğlu et al. (eds.), *Shape Casting*, The Minerals, Metals & Materials Series,
https://doi.org/10.1007/978-3-030-06034-3_22

solidification process (i.e., the solid fraction around 0.1 ~ 0.2) comparing to those of austenite in shell and graphite spheroids.

In this earlier stage of ductile iron solidification, the volume fraction of graphite is less than the critical value of the gray iron solidification given by Schmidt et al. [3]. At this solidification time, volumetric shrinkage occurs because of the liquid transforming to a high density of austenite with a little (or less) low density of graphite. It could be easily compensated by filling in new liquid from a feeder or from parts of casting solidifying later. If no feeders are used, macro-pores will be formed.

In the solid fraction greater than 0.4, the precipitation of graphite nodule inside austenite shell then increases. Austenite shell will thus be plastically deformed dramatically, and these solid austenite grains coherently interact with each other to form networks. At this time, it is difficult to transport the melt through networks of austenite grains (e.g., mass feeding in five feeding mechanisms by Campbell [1]). Also, the melt scattered with the nodules becomes a highly viscous non-Newtonian fluid. The feeding efficiency of a feeder in a ductile iron casting is low, comparing to that in gray iron. The macro-pores will then remain, although some other parts of casting may be expanded overall due to the growing of graphite nodules at solid fraction greater than 0.4 ~ 0.5. In later solidification (e.g., solid fraction above 0.67), the micro-pores are formed from the residual melt encapsulated within networks of austenite shells.

Therefore, the best method to locate macro-pores in a ductile iron casting is to track the level of its liquid region during solidification process. An accurate modeling package is required to predict exact locations of macro-pores in ductile iron castings.

Method

The freezing time of any solidifying body is approximately controlled by its own geometric modulus *(M)*, which is the ratio of its volume *(V)* over its cooling surface area *(A)*. The modulus of a casting geometry has a unit of distance (mm or m), and it is expressed as:

$$M = \frac{V}{A} \tag{1}$$

There are two main parts in a casting: **m**odulus of **c**ast **Mc** and **m**odulus of **f**eeder **Mf**. A dimensionless ratio of feeder modulus (Mf) overcasts modulus (Mc) which named as **Mfc**.

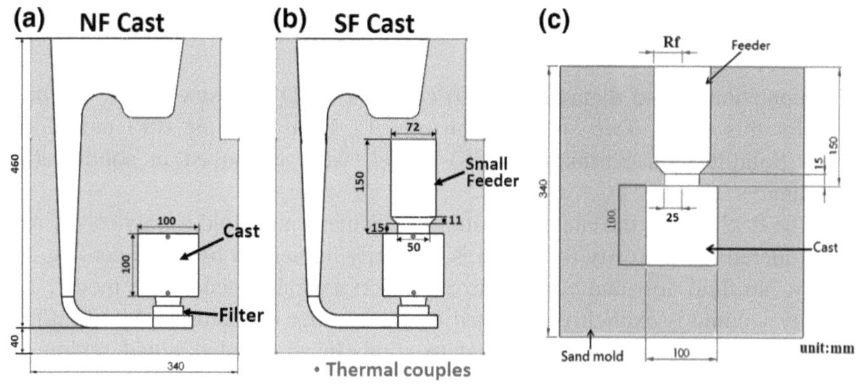

Fig. 1 Two feeding systems with the cube cast: **a** no-feeder (NF) cast, **b** small feeder (SF) cast, and **c** optimized feeder dimensions. (Noted the thermal couples located at the red points.)

Feeding Systems

In this study, a cube cast with dimensions of 100 mm × 100 mm × 100 mm (Mc = 16.67 mm) was used for this experiment. With this cast, two feeding systems and optimized feeder modeling were designed in Fig. 1. Two feeding systems are named as a **n**o-**f**eeder (NF) cast and a **s**mall **f**eeder (SF) cast (Mf = 16.97 mm). In these two runner systems, the total head height of 460 mm was designed. A pouring basin and a ceramic foam filter with dimensions of 50 mm × 50 mm 15 mm and pore size of 10 ppi were applied, in order to provide a quiescent filling and to reduce the bottom gated velocity under the critical value of 0.5 m/s [4]. To probe temperature and filling time of liquid metal into the cast, two thermal couples were attached in the gate and the top of the cast as shown in Fig. 1a, b.

Casting Experiment

To provide a rigid mold for withstanding volume expansion of the cast iron, phenolic urethane cold box (PUCB) binder was used as sand mold material. Ductile cast iron with the composition of carbon 3.5 wt%, silicon 2.25%, and sulfur 0.03% (FCD-450) was melted at 1350 ± 50°C, and it was treated with magnesium nodulant in a tundish ladle. Casting was removed from runner system. Shrinkage holes on the top of casting were sealed with paraffin wax for later Archimedes' floatation measurements. The **b**ulk **d**ensity (D_B) of the casting then can be calculated, and this measurement is correlated with the density of liquid metal as filling the casting cavity. After the D_B calculation, the castings were sectioned in halves for hardness test. Using 100 kg force and 1.588 mm hard steel sphere, Rockwell hardness was measured in every 5 mm distance within this cross section in order to visualize microstructures distributed within the center plane.

The Solidification Modeling

A **c**omputational **f**luid **d**ynamics (CFD) code, Flow-3D™ Cast v4.2 [5], has been used for this study. Two solidification models, **F**irst **P**rinciple (FP) model and **R**apid **S**olidification **S**hrinkage (RSS) model, were employed in solidification modeling.

In the RSS model, the energy equation of the metal and mold is only solved as it is considered that porosity formation is primarily influenced by metal cooling and gravity. No fluid flow and volumetric expansion are calculated in this model. The porosity volume is explicitly evaluated by the cooling rate and the density difference of the transforming from liquid to solid. In an isolated liquid region, its shrinkage volume at each time step is supplied from the liquid on its upper cells along gravity direction. Porosity is then formed on the top portions of this region as these cells begin deficit of liquid.

In the FP model, the continuity and energy equations are modified to include the volumetric source terms. In the solidification of a fluid region, a negative pressure was developed by the density difference of the liquid–solid transformation. Derived from volume loss in shrinkage, this negative pressure can be replaced by pulling the surrounding fluid. If the fluid is confined within a region, it is then built up until a critical pressure is reached. At this critical point, a macro-pore is allowed to open. The critical pressure is relative to one negative pressure to expel gases out of the liquid solution. At each time step, the complete solution of momentum and energy equations is involved in this model. The size of time step is controlled by the stability criteria associated with fluid flow.

The physical and thermal properties of ductile cast iron with carbon 3.5 wt% and silicon 2.25 wt% used in the modeling are listed in elsewhere [5]. In the solidification modeling, the bulk density (D_B) of the cast is also related to the density of liquid as filling the cavities of the cast and feeder. Therefore,

$$D_B = \frac{W_c}{V_o} = \frac{(V_o - V_s) \times \rho_M}{V_o} \tag{2}$$

where W_c is the total weight in the cast and feeder, V_o is the volume of the cast (V_c) plus the volume of feeder (V_f) if available. In the above equation, V_s is the shrinkage volume in the cast, and ρ_M is the averaged density of the solid metal predicted by the modeling at the end of the solidification. If a feeder is used on the cast, the feeder efficiency ($\boldsymbol{e_f}$) is defined as the shrinkage volume in the feeder over the feeder volume (V_f).

$$e_f = \frac{V_s}{V_f} \times 100\% \tag{3}$$

To track the lowest position of a shrinkage macro-pore, the distance of shrinkage (Ds) is defined from the datum line at the connection point between feeder and the cast. Along gravity direction, positive distance is far from the cast while negative distance is inside of the cast.

The Feeder Modeling

Once the solidification models are approved, the validated model can be applied for feeder optimization modeling of the cube cast. In this modeling, initial liquid temperature of 1450 °C was set up at the beginning of solidification modeling. This is because approximately only 1 percent of heat loss of liquid during a very short filling time was expected [6]. To optimize a feeder, dimension of feeder radius (Rf) (i.e., Fig. 1c) is a variable in the modeling while other dimensions are fixed.

Result

Casting Experiment

Using Archimedes' method, the bulk densities (D_B) of NF and SF casts were measured and they are 6887 and 6883 (kg/m^3), respectively. Holes on top of the NF cast (Fig. 2a) and on the feeder of SF cast (Fig. 3a) are clearly observed. On the cross section, porosities (macro-pores) (e.g., Figs. 2c and 3d) are found by naked eyes. The shrinkage distances (D_S) of the NF and SF casts are, respectively, −45 and −35 (±1 mm).

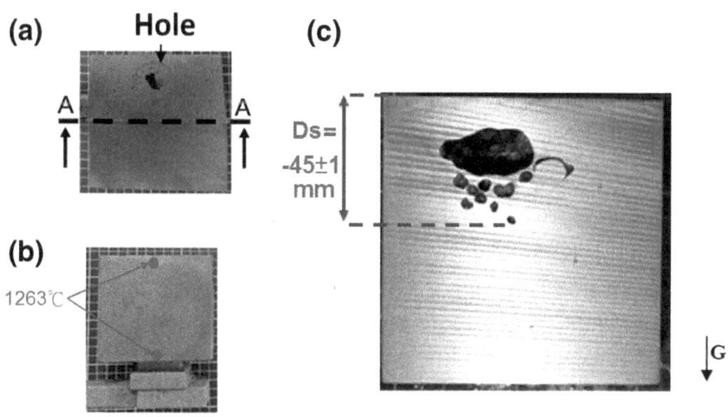

Fig. 2 Casting of the no-feeder (NF) cast in various views: **a** top, **b** front, **c** the A-A cross-sectional views

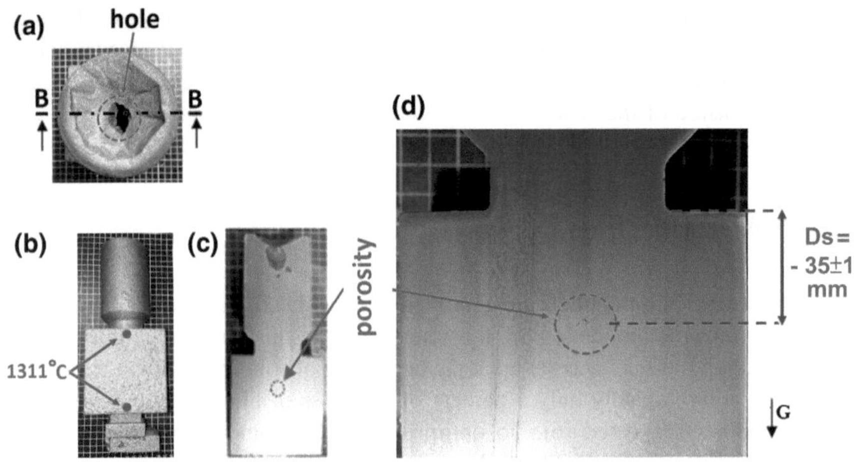

Fig. 3 Casting of the small feeder (SF) cast in various views: **a** top, **b** front, **c** the B-B cross-sectional, and **d** the magnified cross-sectional views

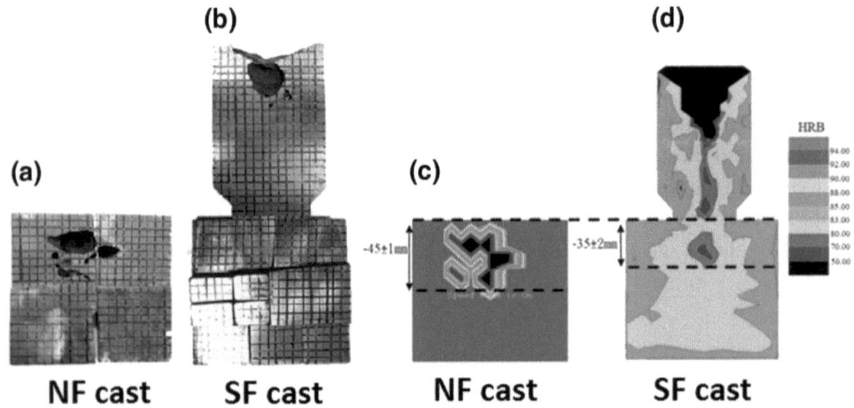

Fig. 4 Hardness measurements on the cross-sectional surfaces in the center of the **a** NF and **b** SF casts at the temperatures of 1263 °C and 1311 °C, respectively, as the filling completion, and their hardness contours plotted against to their coordinate on each cross sections of **c** NF and **d** SF casts

In hardness tests, every 5 mm distances in webs on the cross section in the center of the castings have been measured as shown in Fig. 4a, b. As these hardness measurements were plotted along the coordinates of the cross section of the casts, their hardness contours within these surfaces are illustrated in Fig. 4c, d. In these figures, the color in black is the lowest hardness (lower than 50 HRB) in the scale while the red color is the hardness greater than 92 HRB. In the porosity regions, the dark blue color is clearly displayed where the Ds of −45(±1 mm), −35(±2 mm), and −31 (±2 mm) are marked.

The Solidification Modeling

The modeling results of the casts with two different solidification models at various time frames are shown in Fig. 5. In this figure, the color in red represents regions of

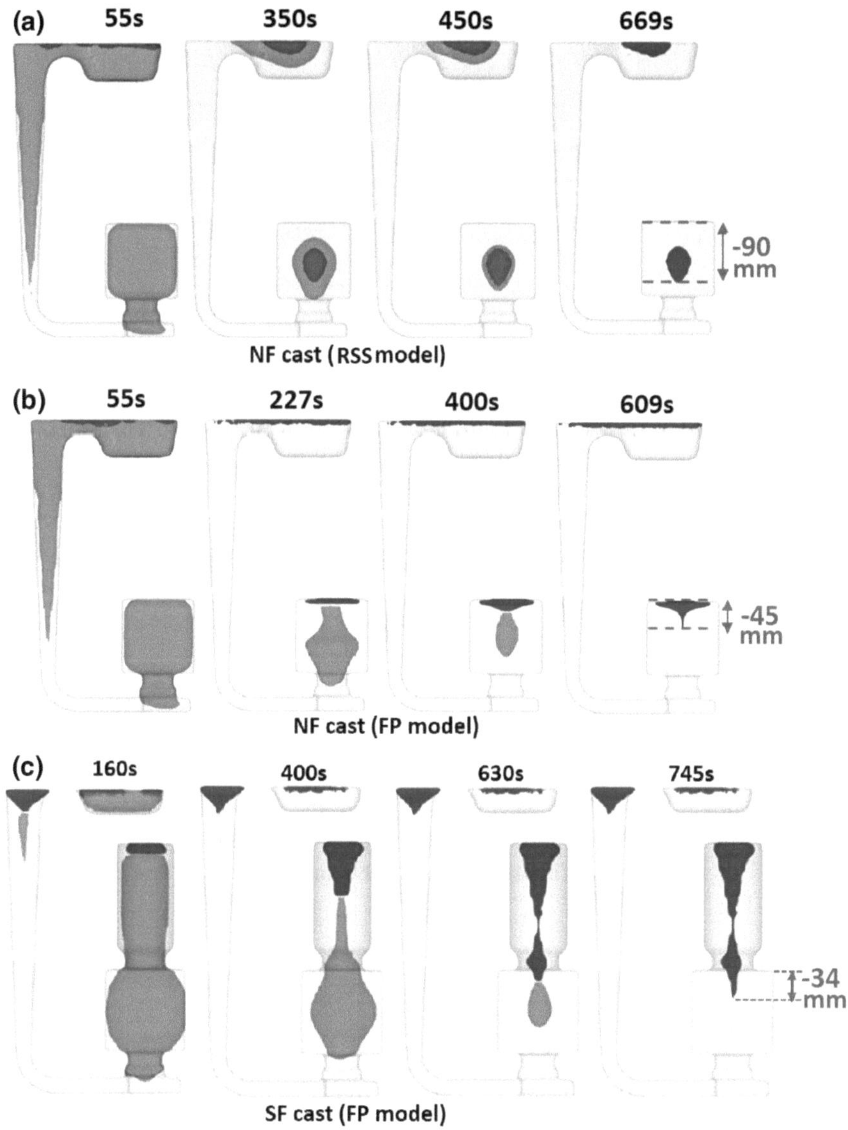

Fig. 5 Modeling results of the casts with two different solidification models: **a** NF cast (RSS model), **b** NF cast (FP model), and **c** SF cast (FP model). Noted the red religion is liquid as the blue is porosity

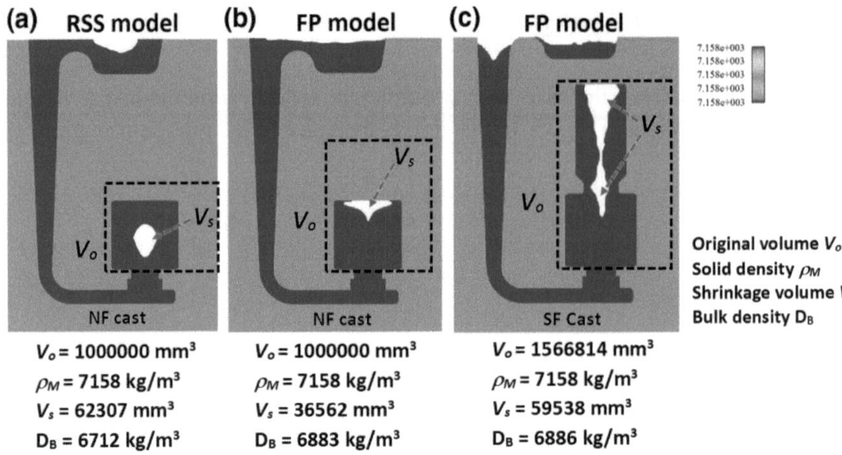

Fig. 6 Bulk density calculations by the solid density and the shrinkage volume of various casts with two different models using Eq. 2. (Noted only blue color that is 7158 kg/m^3)

liquid, while the blue color is the area of porosity (i.e., macro-pore). Also, the Ds distances of −90, −45, and −34 (mm) are measured, respectively, for the NF cast (RSS model), NF cast (FP model), and SF cast (FP model).

Using Eq. 2, the bulk densities D_B for the casts with two different solidification models are calculated as shown in Fig. 6. The uniform solid density ρ_M of 7158 kg/m^3, which is only the blue color in the contours, is presented in this figure.

The Feeder Modeling

To optimize the geometry of feeder, various feeder sizes were used for solidification modeling. Some modeling results at various modulus ratios of feeder overcast (Mfc) are shown in Fig. 7. In each modulus ratio, the shrinkage distance (Ds) and the feeder efficiency (e_f) are measured at the end of solidification modeling. Against the values of Mfc, the two relationships, the ***Ds*** and the e_f, are respectively presented in Fig. 8a, b.

Discussion

In the NF cast, the Ds' distances derived from the real casting (Fig. 2c), the hardness contour test (Fig. 4c) and the FP model (Fig. 5b), are all very much closed to −45 mm. The D_B measurements, obtained from the real casting and the FP model, are almost the same, and they are 6887 and 6883 kg/m^3 (Fig. 6b),

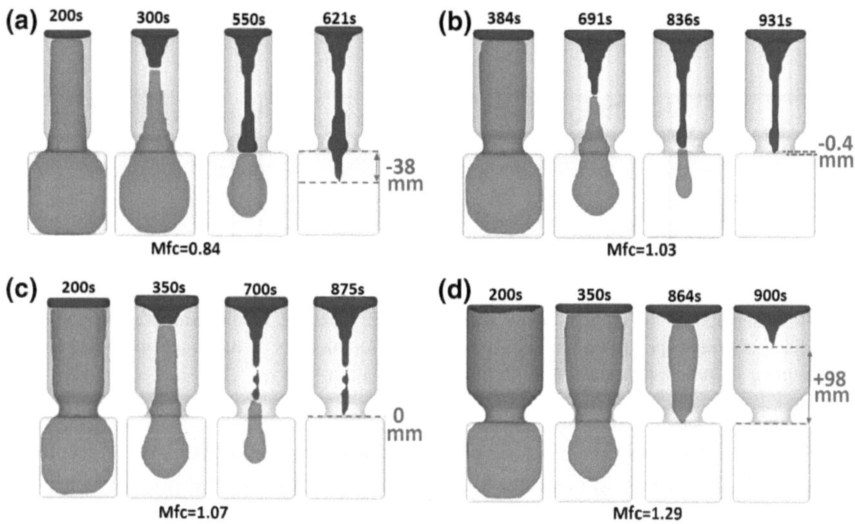

Fig. 7 Modeling results (used the FP model) at various modulus ratios of feeder overcast (Mfc): **a** Mfc = 0.84, **b** Mfc = 1.03, **c** Mfc = 1.07, **d** Mfc = 1.29. (Noted the red region is liquid as the blue is porosity.)

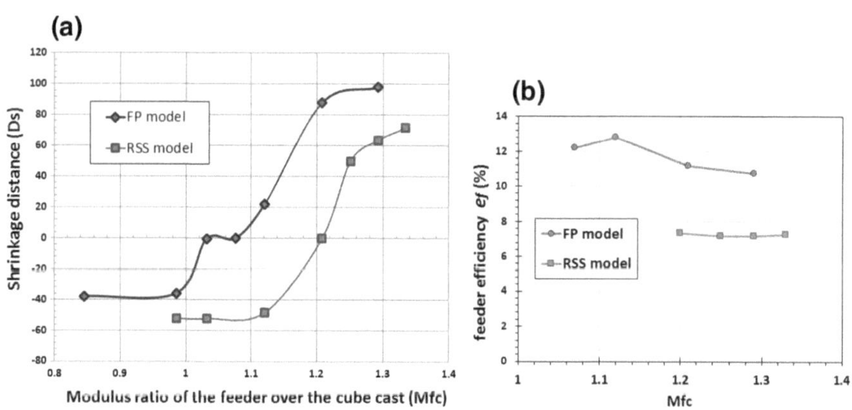

Fig. 8 Shrinkage distance and feeder efficiency plotted against to the modulus ratio Mfc in two solidification models

respectively. In contrast, the Ds (−90 mm) (i.e., Fig. 5a) and D_B (6712 kg/m^3) (i.e., Fig. 6a) values predicted by the RSS model are largely different from those in the casting. Although some shapes of the porosity predicted by the FP model are not similar to that shown in the cross section of the casting, the reverse triangle profile with its hardness lower than HRB 88 (i.e., the green color in the scale of Fig. 4c) is much correlated to a reverse conical shape of the porosity, predicted by the FP model, as sectioning in half. There is a metal skin on the top surface of the casting.

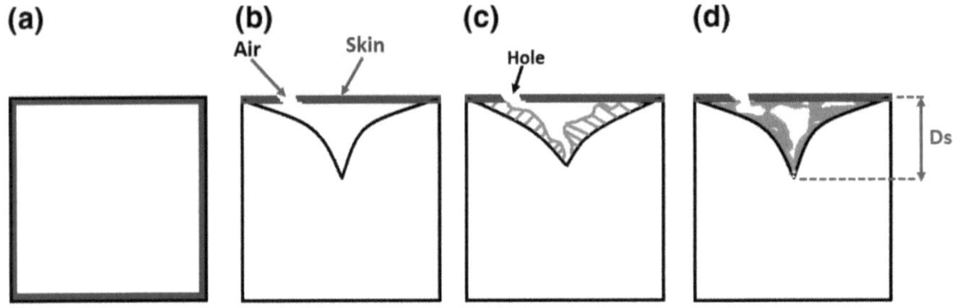

Fig. 9 Schematic of the conical cavity region, derived from the volume shrinkage at the earlier solidification process and filled with the less dense microstructures by the squeezing process due to the graphite volumetric expansion at the later solidification time

In Fig. 2a, a hole on this top skin is clearly observed and porosity is connected underneath. This top skin does not exist on the top of the conical porosity predicted by the FP modeling. The formation of this top skin is explained in Fig. 9.

Firstly, at the end of the filling process, the metal is completely full in the cavity of the NF cast (Fig. 9a). As contacting the wall of mold, the skull of liquid metal on the surface is quickly solidified since the liquid metal on surfaces losing heat into surrounding mold.

Secondly, a negative pressure is then developed due to the volume shrinkage at the liquid–solid transformation in cooling. The negative pressure is accumulated until a critical point where the atmosphere punched a hole on the top skin of the cast to release it (Fig. 9b). At this same time, a high original hydraulic pressure derived from total head height of the pouring basin is then shut by the liquid metal prematurely solidified in the spruce or in small runners. The rest of the liquid in the cast cavity becomes an isolated region without the help of this hydraulic pressure. Thus, the level of the liquid in this region starts to descend as air entering through the hole.

Thirdly, at the beginning of the solidification of ductile cast iron, the fraction of the dendrite austenite is normally 10 times larger than those of shell austenite and nodular graphite [2]. Before the solid fraction of 0.4, low density of the liquid cast iron is transformed to the high density of the austenite and the negative pressure is thus developed (Fig. 9b). Not until the solid fraction reached 0.5 or greater, the low density of nodule graphed encapsulated by austenite can grow. At this time, the remaining liquid is losing its flow fluidity within networks of solid grains. In the FP modeling, at the critical solid fraction of 0.67, above which the liquid has no fluidity, the volume expansion resulted from the nodular graphite formation is ignored.

The conical shape of the porosity in the FP model is related to the region at the solid fraction of 0.67. After this solid fraction, growing (or expanding) nodular graphite then can push the residual liquid through the inter-grains networks into the conical cavity, early developed from the volume shrinkage. This event just likes squeezing water out a porous sponge (Fig. 9c).

Finally, at each time step of growing and pushing event during the solidification proceeded, low dense microstructures with weak strength are accumulated in the conical region at the end of solidification (Fig. 9d). These weak microstructures are normally referred as micro-porosities, and they easily can be picked up by the hardness test.

The Ds distance in a casting also refers to the lowest liquid level (e.g., solid fraction of 0.65 in FP model), below which a sound microstructure can be produced, when liquid metal still has fluidity. Although the exact shape of macro-porosity cannot be predicted in the FP model, its calculation in bulk density D_B is closed to that measured by Archimedes' method. It implies that the modeling density of liquid in the cast cavity before the start of the solidification is the same as that in real casting experiment. In the FP model, the volumetric expansion at the later solidification process is ignored but the liquid metal level can be precisely located at the earlier solidification.

This FP model is efficiently enough to locate the region where sound casting structure can be produced. Also, it helps us to validate the volumetric expansion of graphite growing at the later solidification (at the solid fraction greater 0.5) correlated to the works of Fredriksson et al. [4].

In the SF casting (e.g., Fig. 3d), the Ds (i.e., −35 mm) and D_B (i.e., 6883 kg/m^3) values are closed to those (i.e., −34 mm and 6886 kg/m^3) predicted by FP modeling (Fig. 5c). The hardness contour test (Fig. 4d) is a sensitive way to reflect the variation of microstructures resulting from cooling rate. The middle blue color region (i.e., the range of HRB 70 ~ 80) in the hardness scale in Fig. 4d is a suitable hardness range to pinpoint the macro-porosity formed before the metal has fluidity (e.g., the porosity predicted by the FP modeling, such the blue region in Fig. 5c). Near the end of solidification, such few amounts of the squeezed liquid cannot free fall by gravity or, in the other word, it cannot fill the lowest location of the porosity because it has no fluidity at this time.

Figure 8 shows the feeder optimization modeling results. In Fig. 8a, as the Ds distance is zero, the modulus ratios of Mfc are 1.07 and 1.21 from interpolating the curves predicted by the FP and RSS models, respectively. Since the FP model is more accurate to predict the Ds distance, its curve can be used to optimize feeder design. Against to the same ratio Mfc (i.e., 1.07) found in zero Ds distance, the feeder efficiency *ef* (i.e., 12.2%) can be derived from the FP modeling curve in Fig. 8b. This feeder with a high feeder efficiency and a minimum size is recommended for future usage in feeding ductile cast iron.

In the FP modeling curve in Fig. 8a, a plateau in the range between 1.03 and 1.07 of Mfc ratio is closed to the Ds of zero. This abrupt transforming curve is an evidence of that the FP model considers the volumetric expansion of spheroidal graphite growing while the RSS model does not. In the FP modeling, as Mfc is 1.07, the residual liquid with volume of 2089.0 mm^3 is remained within the cast (e.g., at 700 s in Fig. 7c). The last residual liquid is eventually transformed to eutectic structure of nodular graphite and shell austenite without creating any porosity inside of the cast. That is Ds of zero.

Conclusion

1. The shrinkage distance Ds distance predicted by the first principle modeling is an accurate measuring method to locate the lowest liquid level where macro-porosity is formed.
2. A large volumetric expansion by growing spheroidal graphite at solid fraction greater than 0.5 has been validated by the different results found between the FP modeling and real casting. Conical shape of the porosity in the FP model is related to the region at the solid fraction of 0.67.
3. For a cube cast with Mc of 16.67 mm, the optimized feeder, which is a minimum size (i.e., Mf = 17.84 mm or Mfc = 1.07) and a large feeder efficiency (12.2%), can be proposed.
4. In the diagram of the Ds distance plotted against to the Mfc ratio, the plateau in the FP modeling curve is an evidence of that the FP model considers the volumetric expansion of spheroidal graphite growing.

Acknowledgements The authors personally acknowledge the help of the colleagues in Shou Chen Industry Co., Ltd No. 168, Deli Rd., Taiping District, Taichung City 411, Taiwan (R.O.C.), for casting experiments.

References

1. Campbell J (2011) Complete casting handbook, metal casting process, metallurgy, techniques and design, first edition, pp 347–352 [1a], p 406 [1b], p 663 [1c], p 666 [1d]
2. Fredriksson H, Stjerndahl J, Tinoco J (2005) On the solidification of nodular cast iron and its relation to the expansion and contraction. Mater Sci Eng, A 413–414:363–372
3. Schmidt WA, Sullivan E, Taylor HF (1954) AFS Trans 9:70
4. Hsu F-Y, Lin H-J (2011) Foam filters used in gravity casting. Metall Mater Trans B 42 (6):1110–1117
5. Flow 3D Casts, Flow Science. https://www.flowsciences.com/2018.06
6. Richins DS, Wetmore WO (1952) Hydraulics applied to molten aluminum. Trans ASME, 725–732

The Contactless Electromagnetic Sonotrode

Koulis A. Pericleous, Valdis Bojarevics, Georgi Djambazov, Agnieszka Dybalska, William Griffiths and Catherine Tonry

Abstract Ultrasonic pressure waves generated using a tuned electromagnetic induction coil promote cavitation in alloy melts as an alternative to the immersed sonotrode technique. The method targets the same benefits offered by traditional UST (degassing, microstructure refinement, dispersion of particles), but without some of its drawbacks. The method is contactless, meaning it can be applied equally to high temperature/reactive melts, avoiding contamination due to probe erosion, and consequently, it is maintenance free. Due to induction stirring, larger volumes of melt can be treated (a major limitation of the traditional method), as the liquid is forced to pass repeatedly through zones of cavitation activity. The coil configuration used will depend on application. In the installation shown, a top conical coil immersed in aluminium melt (contactless due to EM repulsion) was used. Simulations of sound, flow and EM fields are given, compared with experiments and indicating strong stirring, evidence of cavitation through emitted sound signals and, most importantly, grain refinement.

Keywords Ultrasonic melt treatment · Electromagnetic vibration · Gas cavitation

Introduction

Intense vibration of the liquid metal during solidification is known to refine microstructure, remove trapped gas and reduce porosity [1–3]. Vibration can be achieved through various means: shaking the crucible, inserting a mechanical stirrer in the melt or, more commonly, using an immersed ultrasonic horn, e.g. in the DC casting of aluminium ingots [3]. Given the range of methods used, J. Campbell used

K. A. Pericleous (✉) · V. Bojarevics · G. Djambazov · C. Tonry
Centre for Numerical Modelling and Process Analysis, University of Greenwich, London, UK
e-mail: k.pericleous@gre.ac.uk

A. Dybalska
School of Metallurgy and Materials, University of Birmingham, Birmingham, UK

W. Griffiths
University of Birmingham, Birmingham, UK

M. Tiryakioğlu et al. (eds.), *Shape Casting*, The Minerals, Metals & Materials Series,
https://doi.org/10.1007/978-3-030-06034-3_23

an extensive literature review on the effects of vibration on solidification [1] to present a unifying frequency–amplitude *f-a* map concept, further stating '… the *f-a* map shows a close correlation between the theoretical threshold for cavitation and the onset of grain refinement'. In ultrasonic processing, frequencies around 20 kHz were found to be the most effective [3] for aluminium treatment, a fact attributed to the presence of cavitation. Gas micro-bubbles exist in the liquid due to either dissolved gases coming out of solution as the melt cools (e.g. H_2 in Al), air trapped with oxides during mould filling or gas attached to the surface of immersed particles. Bubbles, subjected to an imposed pressure sine wave, first expand to many times their original size and then collapse violently generating high-speed micro-jets and shock waves. These events fragment emerging dendrites or facilitate the breakup and dispersion of particle clusters [4] to generate nucleation sites. Evidence of cavitation has been seen in situ [5], found in post-solidification analyses, and shown to depend on a critical pressure threshold [3]. More recently, ultrasonic processing has been adopted as a promising candidate technique for the production of MMNCs in the EU Exomet project [6].

There are, however, several disadvantages in immersive stirring/vibration techniques: contact with the liquid metal leads to contamination of the melt and conversely erosion of the immersed probe surface, requiring then frequent replacement. These problems are limiting when the technology is to be applied to high-temperature alloys (ODS steel, nickel superalloys) or to highly reactive metals (Ti, Zr). Another limiting factor of the immersed sonotrode approach is the localised effect of the cavitation region, which leads to long processing times, high cost/energy usage and consequently small treatment volumes. EM induction is an attractive alternative. The effects of the Lorentz force in stirring the melt and the Joule heating are well known and widely used in practice. What is often overlooked is the ability of the AC component of the induced Lorentz force to generate a strong sound field within the melt, equivalent to that produced by an immersed sonotrode, but without the contact. Vives [7] was the first to investigate experimentally such a non-contact technique, using a combination of static and AC magnetic fields. Other investigators followed this lead, using different configurations; for AlSi hypereutectic alloys [3, 8], for grey iron melts aluminium alloys [9] and for titanium aluminides [10]. In each case, different thresholds of pressure amplitude were found to be necessary and a range of frequencies was used from 50 Hz to 50 kHz. There has been no systematic study of the sound field generated in these situations, which led us in a previous publication [11], to examine the sound field generated by the cylindrical induction coil in an induction crucible, coupling the compressible sound wave equations with the pressure source generated by the electromagnetic field and with the Rayleigh–Plesset equations for cavitating bubbles. This simulation demonstrated that pressure amplitudes likely to cause cavitation of dissolved gas can be generated, provided the AC frequency is tuned to approach wave resonant conditions—a similar conclusion was reached by Vives, who considered the design of a resonant EM cavity for this purpose [7]. In our interpretation, we extended this idea via the introduction of a *tuned top coil* [12], which is gradually inserted into the liquid metal volume, avoiding contact via EM repulsion of the melt-free surface.

At this very close proximity, intense vibrations and melt stirring can be produced which can be controlled via the depth of immersion, current magnitude and frequency. The concept developed theoretically through a coupled MHD-acoustic numerical simulation has now been implemented in practice with encouraging results as shown in this paper.

The remainder of the paper is arranged as follows: A description of the experimental setup used is given, followed by the mathematical model. Illustrative results of simulations using aluminium melts are then given, together with experimental findings, finishing with discussion and conclusions.

The 'Top Coil' Sonotrode

The 'top coil' sonotrode consists of a conical induction coil, which is inserted into the melt in a crucible as shown in Fig. 1. The coil is water-cooled with sufficient current running through it to prevent contact between the liquid metal and the coil surface. A protective ceramic coating is used to eliminate spark erosion.

The Lorentz force due to the interaction between the coil current and the opposing current induced in the melt repels the free surface, but also drives bulk stirring as shown in Fig. 1a. The AC component of the force operates at *twice* the supply frequency (see Eq. 12) to vibrate the melt. The applied force is concentrated in a thin '*skin layer*' on the free surface, but its effect is transmitted into the bulk through sound waves, as pressure fluctuations. Provided these fluctuations are strong enough, gas bubbles oscillate and then cavitate. To reach the necessary pressure threshold, the coil frequency has to be tuned to produce sound resonance in the treatment vessel, a condition that needs to account for the vessel geometry, volume of melt and crucible sound absorption characteristics [13]. In contrast to the immersed sonotrode where the cavitation energy is concentrated around the probe, active cavitation zones are expected to lie in resonant nodes deep in the melt.

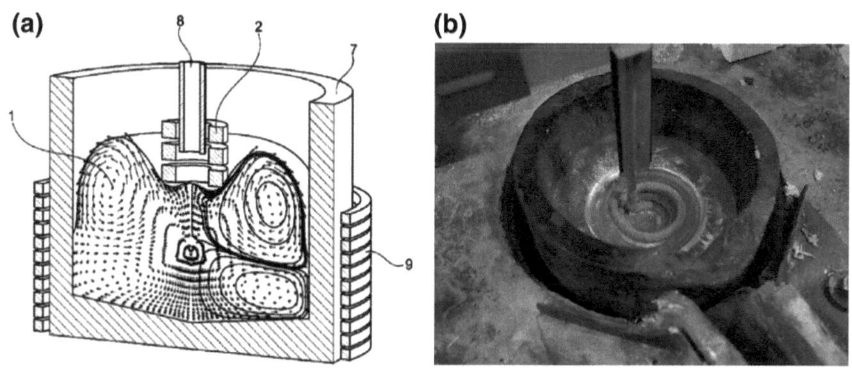

Fig. 1 **a** The sonotrode concept (VibroEM patent application #PCT/EP2013/067896) [12] and **b** shown in prototype form

Induced flow then ensures gas bubbles can have multiple passes through these nodes, improving cavitation efficiency and reducing process time.

An external Ultramic 200 K digital ultrasonic microphone was used to capture sound emitted from the crucible to detect cavitation activity and the tracks of radioactive particles were used to determine the extent of induced stirring (see [14]).

Mathematical Basis

As can be inferred from the process description, this is an application covering a combination of traditional engineering fields (flow, heat transfer, sound and electromagnetic). Due to space limitation, only a brief summary of the mathematical/numerical model used is given here and the reader is referred to the accompanying references for detailed mathematical formulations.

MHD Flow and Heat Transfer

The time-dependent conservation equations for mass, momentum and energy are solved, accounting for the effects of electromagnetic interaction and change of phase. An in-house software code 'SPHINX' solves the equations numerically using a spectral collocation method on a continuously deformable mesh coincident with the metal volume as described in Pericleous and Bojarevics [15]. The method allows real-time deformation of the liquid-free surface based on the balance of forces acting. A description of the main equations follows:

Momentum

$$\partial_t \mathbf{v} + (\mathbf{v} \cdot \nabla)\mathbf{v} = -\rho^{-1}\nabla p + \nabla \cdot (\nu_e(\nabla \mathbf{v} + \nabla \mathbf{v}^T)) + \rho^{-1}\mathbf{f}_m + S_D \tag{1}$$

where $\mathbf{f}_m$ is the volumetric electromagnetic force and S_D is a 'Darcy' resistance term introduced to account for the phase change 'mushy' zones and to stop flow in solidifying regions.

Continuity for an incompressible fluid

$$\nabla \cdot \mathbf{v} = 0 \tag{2}$$

The boundary conditions are the no-slip condition at the solid walls ($\mathbf{v} = 0$) or the free surface dynamic and kinematic conditions when the liquid metal is detached from the solid wall.

Energy conservation

$$\frac{\partial(\rho C_p T)}{\partial t} + \nabla(\rho C_p \mathbf{v} T) = \nabla(k_{eff} \nabla(T)) + q_e + q_L \tag{3}$$

The last two terms in Eq. (3) represent Joule heating, $q_{e,}$ and latent heat release, q_L.

$$q_e = \frac{J^2}{\sigma}; q_L = -\frac{\partial}{\partial t}(\rho f_L L) - \nabla.(\rho \mathbf{v} f_L L) \tag{4}$$

The liquid fraction is defined as a function of T:

$$f_L = \begin{cases} 1 & T > T_L \text{ liquid} \\ \left(\frac{T-T_S}{T_L-T_S}\right) & T_S \le T \le T_L \text{ mushy zone} \\ 0 & T < T_S \text{ solid} \end{cases} \tag{5}$$

Then,

$$S_D = \frac{\mu}{K}\mathbf{v};\ K = \frac{f_L^3}{\zeta(1-f_L)^2} \tag{6}$$

Heat transfer boundary conditions are stated for free surface radiation and wall loss, described in [16]. The effective heat transfer at solid walls is then given by $-\rho C_p \alpha_e \partial_n T = h(T - T_w)$, where *h(T)* is the heat transfer coefficient, with free surface radiation given by: $-\rho C_p \alpha_e \partial_n T = \varepsilon \sigma_b (T^4 - T_w^4)$, with α_e being the turbulent thermal diffusion coefficient.

Turbulence is modelled using the *k*-ω model of Wilcox [17]:

$$\left.\begin{array}{c} \partial_t k + \mathbf{v}\cdot\nabla k = \nabla\cdot[(v + \sigma_k v_T)\nabla k] + G - \beta^* \omega k \\ \partial_t k + \mathbf{v}\cdot\nabla k = \nabla\cdot[(v + \sigma_k v_T)\nabla k] + G - \beta^* \omega k \\ v_T = \alpha^* \frac{k}{\omega} \end{array}\right\} \tag{7}$$

where ω is the frequency of vorticity fluctuations and k the turbulence kinetic energy per unit mass, and the various constants are functions of R_T, the local turbulent Reynolds number (see Wilcox [17]).

Figure 2 shows the velocity and temperature field created by the top coil, acting together with the normal furnace coil surrounding the crucible. Two alternative coil arrangements are shown to demonstrate the flexibility of the system. AC current of 1700 A RMS at 10 kHz is supplied to the top coil and the furnace coil operates at 2000 A, 2.5 kHz. There is evident deformation of the free surface due to the top coil, accompanied by a 'hot spot' due to Joule heating. The Lorentz force is responsible for strong induction stirring, leading to a strong toroidal vortex below the free surface, countered by a reverse vortex deeper in the crucible. In the cases chosen, the vortex rotation changes direction from counterclockwise (2a) to clockwise (2b).

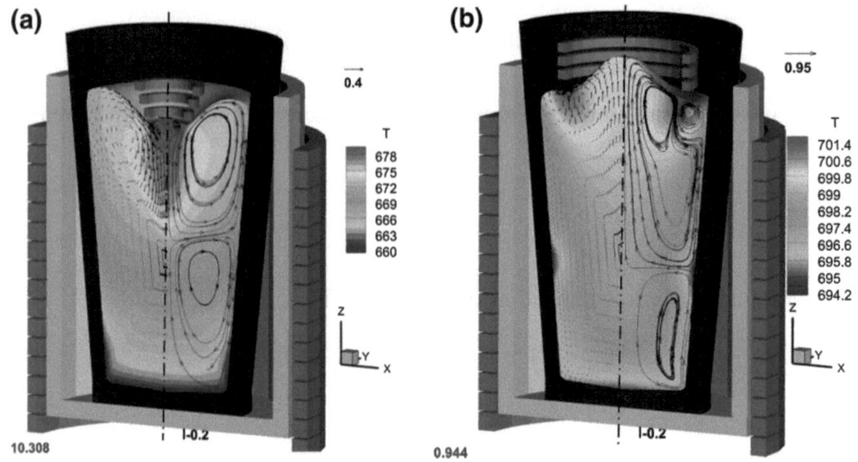

Fig. 2 **a** Simulated flow field in Al crucible, with immersed conical and free surface shape deformation. The temperature map is accompanied with flow streamlines (right) and velocity vectors (left). **b** Alternative top coil suggestion producing a very different flow field (note the furnace coil surrounding the crucible is also active in both cases)

Electromagnetic Sound Generation

The electromagnetic force exerted by the induction coil(s) acts in a thin skin layer δ (see Eq. 11 below); its time-averaged influence is responsible for the induction stirring appearing in Fig. 2. The time-dependent part acts as a vibration source, exciting sound waves in the melt. Sound propagation is then modelled by solving the wave equations in the time domain, using a fourth-order staggered variable scheme on a regular Cartesian grid [18].

As an illustration, the 1D harmonic sound propagation caused by the coil AC magnetic field **B** and induced current, **J,** can be split into real and imaginary components:

$$\boldsymbol{B} = B_R \cos \omega t + B_I \sin \omega t \tag{8}$$

$$\boldsymbol{J} = J_R \cos \omega t + J_I \sin \omega t \tag{9}$$

where

$$J_R = \sigma \frac{\omega}{2} \delta (B_R + B_I); J_I = \sigma \frac{\omega}{2} \delta (-B_R + B_I) \tag{10}$$

The skin depth indicating the penetration of induced current is given by

$$\delta = \sqrt{(\frac{2}{\mu\omega\sigma})} \tag{11}$$

The Lorentz force, being the cross product of magnetic field and current, can be split into mean and sinusoidal components

$$\begin{aligned} \boldsymbol{F} &= \boldsymbol{J} \times \boldsymbol{B} = \bar{F} + \tilde{F}, \\ \bar{F} &= \frac{1}{2}(J_R B_R + J_I B_I) = \frac{1}{2\mu\delta} B_o^2 e^{-2\frac{x}{\delta}} \\ \tilde{F} &= \frac{1}{2\mu\delta} B_o^2 e^{-2\frac{x}{\delta}} \sqrt{2}\cos(2\omega t - 2\frac{x}{\delta} + \frac{\pi}{4}) \end{aligned} \tag{12}$$

The mean component $\bar{F}$ is then responsible for bulk stirring. The pressure waves generated by the time-dependent component of the Lorentz force are computed by solving the Euler form of the momentum equations, generating a perturbation velocity field, $\bar{v}$. The source S then represents pressure contributions due to cavitating bubbles:

$$\frac{\partial p}{\partial t} + \rho c^2 \frac{\partial \bar{v}}{\partial x} = S; \; \rho \frac{\partial \bar{v}}{\partial t} + \frac{\partial p}{\partial x} = \tilde{F} \tag{13}$$

Boundary conditions take into account transmission and reflection of sound through the crucible walls dependent on the acoustic impedance of the materials, coupled with constant pressure at the liquid-free surface. Details of the approach used are given in [13].

The instantaneous sound field corresponding to the case in Fig. 2a is given in Fig. 3, where deep pressure nodes/anti-nodes are seen to penetrate deep into the melt, likely to spread the active cavitation zone. Cavitation is handled using a multi-scale approach (in both space and time) based on an extension of the Rayleigh–Plesset equations [19].

Gas Cavitation

In aluminium, bubbles are assumed to remain spherical as they oscillate radially in a pressure wave, a reasonable assumption, since the surface tension between aluminium and hydrogen (the common gas phase in aluminium melts) is large, compared to water say. The Rayleigh–Plesset equation can then be used to represent the bubble dynamics:

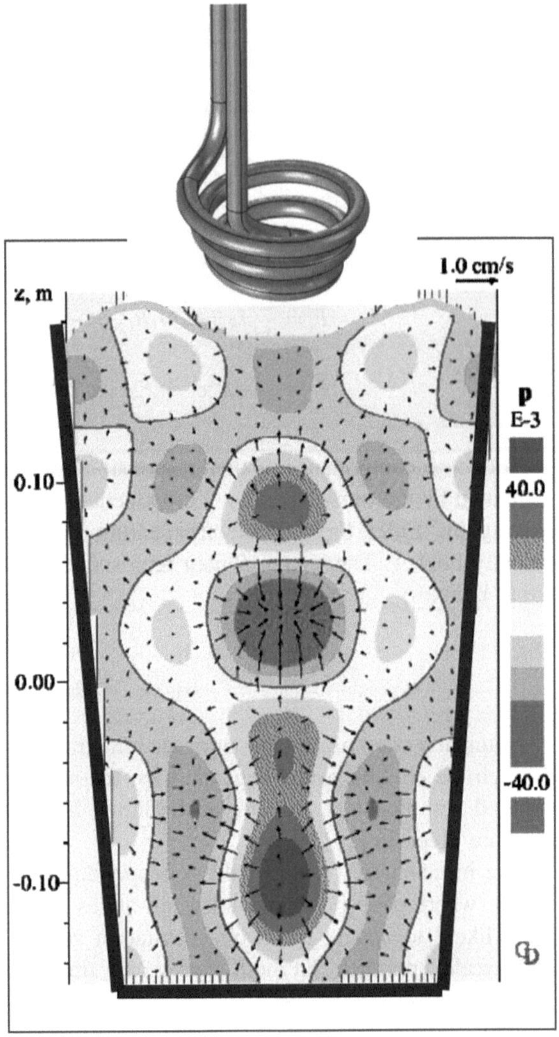

Fig. 3 Computed sound field in a crucible with the top coil acting as the sound source

$$R\ddot{R} + \frac{3}{2}\dot{R}^2 = \frac{p_s}{\rho} \tag{14}$$

with

$$p_s = p_b + p_v - \frac{2\sigma}{R} - \frac{4\mu\dot{R}}{R} - p_0 - p_\infty - p \tag{15}$$

R is the bubble radius, p_0 is the atmospheric pressure, p_b is the pressure inside the bubble, p_v is the liquid vapour pressure, p_∞ is the pressure from the ultrasonic source, σ is the surface tension between the liquid and the bubble gas, and μ is the dynamic viscosity of the liquid.

The bubble pressure p_b is given by

$$p_b = p_{g,0}(R_0/R)^{3\kappa} \tag{16}$$

where $p_{g,0}$ is the initial bubble pressure and κ is the polytropic exponent. See Lebon et al. [19] for a full description of an advanced cavitation model used in this work and Manoylov et al. [4] for an extension of this model to the dispersion of particle clusters in metallic alloy melts.

Selected Results

In this section, sample results obtained in pioneering experiments are shown, in part confirming the usefulness of the contactless sonotrode process method in the treatment of aluminium and in part validating the theoretical analysis. These experiments were carried out in Birmingham University using the conical coil depicted in Figs. 1b and 2a for different crucible sizes. In the example given, a cylindrical crucible was used 300 mm in depth, internal diameter 140 mm and external diameter 168 mm. The coil was lowered into the liquid metal deforming the free surface and generating strong radial surface flow towards the crucible axis. The sound emitted from the crucible was recorded over 60 s using an Ultramic 200 K digital ultrasonic microphone and the frequency spectrum analysed using MATLAB, Adobe Audition and Audacity software. Figure 4 shows the FFT spectrum captured for the case with AC current in the top coil set at 9.5 kHz.

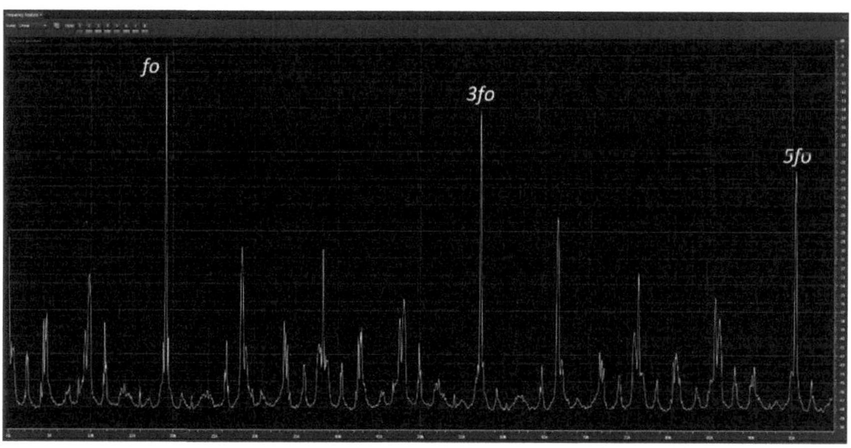

Fig. 4 FFT sound spectrum for top coil driving at fo ~20 kHz; dominant peaks at fo, 3fo, 5fo denote radial modes

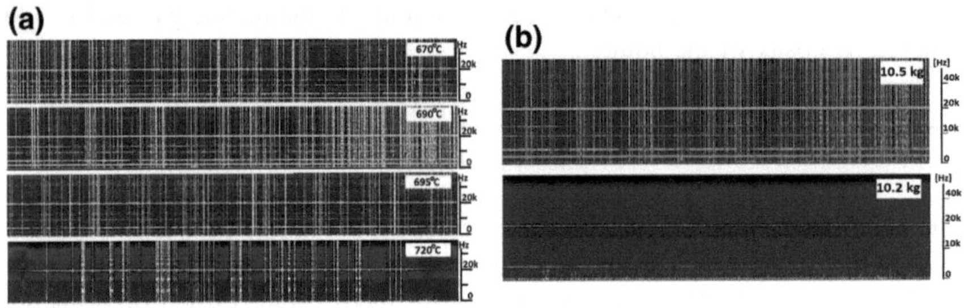

Fig. 5 Spectrograms showing the appearance of wideband noise bursts (vertical lines), indicating the onset of cavitation. The continuous horizontal lines correspond to resonant peaks caused by the driving frequency in the top coil (20 kHz) and the melting coil (5 kHz). **a** Varying temperature between 670 °C and 720 °C shows maximum cavitation activity at 690 °C, while by 720 °C most cavitation activity disappears. **b** Reducing the volume of metal from 10.5 kg to 10.2 kg also has a drastic effect on cavitation with indications of cavitation disappearing

As expected (see Eq. 12), the first dominant sound peak appears at fo = 19 kHz, that is at twice the electrical frequency. There are several other modes present, but the dominant ones are odd multiples of fo, at 57 kHz and 95 kHz, typical of a cylindrical container with an open top.

The intensity of sound radiated from the metal surface gives an indirect indication of the pressure intensity generated inside the crucible by the Lorentz force, a measure critical in reaching the threshold necessary for cavitation. To detect cavitation, we are looking for broadband noise generated by oscillating or collapsing bubbles. Figure 5a shows spectrograms (frequency vs. time) measured at different melt temperatures. It is clear that temperature plays an important role in cavitation activity as noted in previous publications [20]. For this particular configuration, with a top coil AC driving frequency of 9.57 kHz acting in a cylindrical crucible containing 10.5 kg of CP-Al, the optimum temperature appears to be 690 °C. This may be related to the release of dissolved hydrogen at this temperature. Figure 5b shows the sensitivity of the process to small changes in geometry, since even a 5% reduction in the volume of the metal treated (in this case removing a sample for the AA Test Procedure-1 (TP-1)) can alter the conditions for resonance, and as a result, cavitation activity ceases.

Characterisation of the effect of operating the top coil on the Al alloy and, specifically on its grain size, was carried out using the Aluminium Association Test Procedure-1 (TP-1), with a constant cooling rate obtained using a constant water flow rate of 3.8 l/min. The TP-1 samples were cut along their vertical axis and prepared by grinding, polishing and etching with Poulton's Solution to reveal the microstructure. The mean grain size was determined using the mean line intercept method.

The effect of ultrasonic vibration on grain refinement has been widely discussed and explained by cavitation-enhanced heterogeneous nucleation when the liquid metal is processed above the liquidus temperature [1, 21]. Cavitation bubbles grow, pulse and finally collapse, causing local pressure changes leading to several

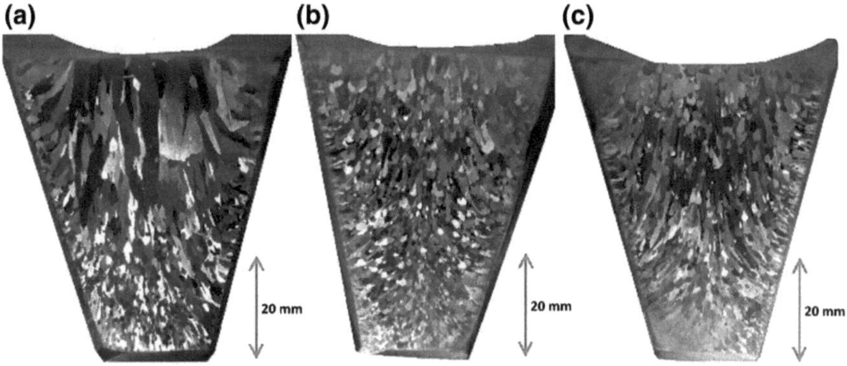

Fig. 6 Treatment of CP-Al for 2 min. **a** Untreated sample taken at 690 °C, mean grain size (A) 6.76 ± 3.76 mm; **b** treated with a top coil AC frequency f = 9.43 kHz at 700 °C, A = 0.96 ± 0.11 mm; **c** treated with a top coil AC frequency f = 9.41 kHz at 690 °C, A = 0.97 ± 0.41 mm

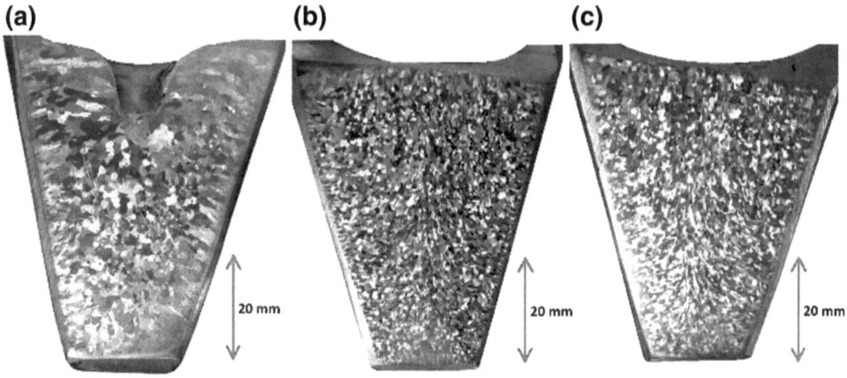

Fig. 7 **a** Untreated sample taken at 683 °C, A = 2.09 ± 0.58 mm; **b** treated for 3–4 min with a top coil AC frequency f = 9.57 kHz at 690 °C, A = 0.59 ± 0.05 mm, **c** treated with a top coil AC frequency f = 9.57 kHz at 680 °C, A = 0.55 ± 0.07 mm

possible effects, such as dispersion and deagglomeration of non-metallic particles already present in the melt, or breakup of evanescent dendrites. Figures 6 and 7 show how grain sizes from TP-1 test samples taken from the experiments with CP-Al and, without any prior grain refining treatment, were decreased in size following treatment by the top coil.

The TP-1 samples presented were taken in conditions thought to be optimal by the microphone check. Observed cavitation noise was reflected in the microstructure of the samples, as one can compare Fig. 7b with the spectrogram recorded at 690 °C. The second, most potent, frequency in the experiments was at 9.43 kHz (axial mode), which was associated with a grain size reduction by a factor of about seven, (see Fig. 6). The maximum reduction in grain size was observed when the samples were treated with an AC frequency of 9.57 kHz at a temperature range between 680 and 690 °C, where a grain size reduction of about 70% was obtained (Fig. 7). Processing time ranged from 2 to 4 min and some of the observed differences may be attributed to that. Further research is needed to that effect.

It is argued that convection can distribute the effects of cavitation, and in the traditional immersed sonotrode technique, convection is due to acoustic streaming, which is strongest in a cone below the immersed probe. In contrast, the top coil technique combines ultrasonic vibrations to strong induction stirring, which easily involves the full volume of the melt. This was indicated numerically in Figs. 1 and 2 and detected experimentally by observing strong flow streaming on the surface of the melt. The ability of the top coil to disperse particles in the melt is also demonstrated in Fig. 8, which compares particle tracks obtained numerically in a small crucible using the Lagrangian technique (accounting for the effects of turbulence and electromagnetophoretic Kolin–Leenov forces) with tracks of 200-micron radioactive particles obtained experimentally using the positron emission particle tracking (PEPT) technique [14]. The crucible is quickly filled with particles in both cases.

Scaling up of the contactless ultrasonic process is straightforward, since the coil dimensions and frequency can be designed to suit the application. Similarly, there are plans within the current project to extend the technique to DC casting, where the acoustic volume is determined by the size of the sump. Being contactless, the technique can be used for high temperature or reactive materials. This is demonstrated numerically in Fig. 9, where nickel has replaced aluminium in the crucible. The difference now is that the separation between the coil and melt surface is reduced due to the increased metal density.

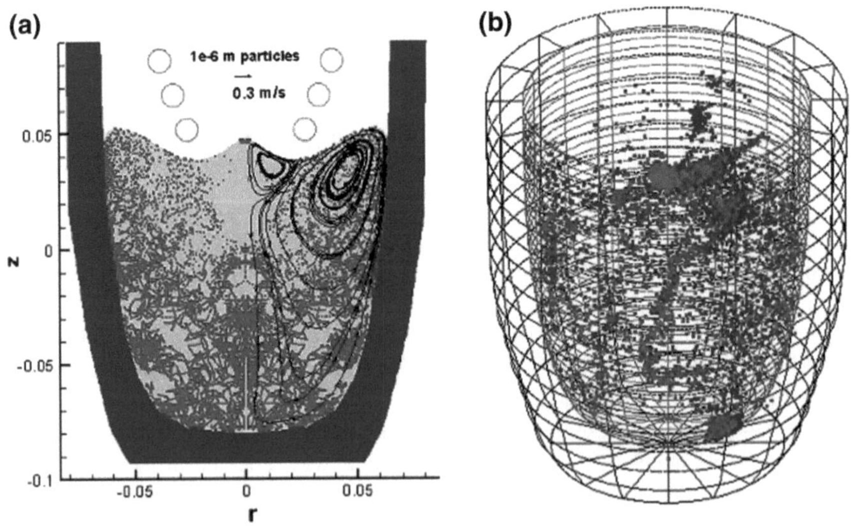

Fig. 8 Comparison of **a** numerical result showing dispersion of 1 mm particles due to induction stirring, against **b** experimental PEPT traces, in a 13 cm crucible, containing A357 alloy

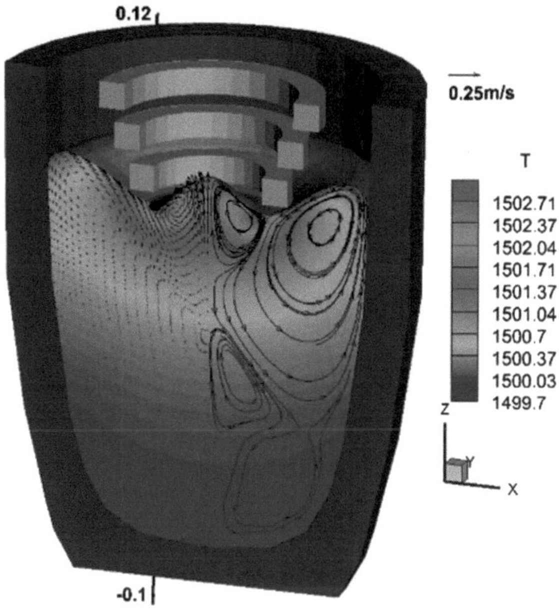

Fig. 9 Treatment of nickel using the top coil sonotrode (1.8 kA, 8.8 kHz)

Concluding Remarks

It has been shown that ultrasonic waves strong enough to produce cavitation in a crucible containing liquid metal can be generated using a contactless electromagnetic technique. This technique, originally based on a theoretical concept, was proven experimentally in the treatment of liquid aluminium. The experiments provide validation for the accompanying computational model, which couples fluid flow, heat transfer, electromagnetic interaction and sound field simulations. It was found that resonant conditions are necessary to produce sound waves of sufficient strength for cavitation, which means process geometrical details and the sound absorbing properties of the crucible become important for a successful implementation. Although the experiments are so far conducted in aluminium, since the technique is contactless it can be equally applied to investigate its use in steels, nickel alloys and titanium.

Acknowledgements The authors acknowledge financial support from the ExoMet Project (co-funded by the European Commission (contract FP7-NMP3-LA-2012-280421), by the European Space Agency and by the individual partner organizations) and the UK Engineering and Physical Sciences Research Council (EPSRC) through grants EP/P034411/1, EP/R000239/1 and EP/R002037/1.

References

1. Campbell J (1981) Effects of vibration during solidification. Int Metals Rev 26(1):71–108
2. Eskin GI, Eskin DG (2014) Ultrasonic treatment of light alloy melts, 2nd edn. CRC Press
3. Meek T, Jian X, Xu H, Han Q (2006) Ultrasonic processing of materials, ORNL/TM-2005/125
4. Manoylov A, Lebon GSB, Djambazov G, Pericleous K (2017) Coupling of acoustic cavitation with DEM-based particle solvers for modeling de-agglomeration of particle clusters in liquid metals. MMTA 48(11):5616–5627
5. Tzanakis I, Xu WW, Eskin DG, Lee PD, Kotsovinos N In situ observation and analysis of ultrasonic capillary effect in molten aluminium, Ultrason. Sonochem. 27:72–80
6. Sillekens W, Jarvis D, Vorozhtsov A, Bojarevics V, Badini C, Pavese M, Terzi S, Salvo L, Katsarou L, Djeringa H (2014) The ExoMet Project: EU/ESA Research on High-Performance Light-Metal Alloys and Nanocomposites, Metall Mater Trans A, 2014 45A(8):3349–3361. 13
7. Vives C (1996) J Crystal Growth 158(1–2):118
8. Grants I, et al (2015) J Applied Phys 117:204901
9. Tzanakis I, Xu WW, Lebon G, Eskin DG, Pericleous K, Lee PD (2015), Phys Proced 70:841
10. Ruirun C, Deshuang Z, Tengfei M, Hongsheng D, Yanqing S, Jingjie G, Hengzhi F (2017) Effects of ultrasonic vibration on the microstructure and mechanical properties of high alloying TiAl. Sci Rep 7:41463. https://doi.org/10.1038/srep41463
11. Bojarevics V, Djambazov GS, Pericleous KA (2015) Metall Mat Trans A 46(7):2884–2892
12. Bojarevics V, Pericleous K, et al (2016) Eur Patent 13756442:3–1373
13. Tonry CEH, Djambazov G, Dybalska A, Bojarevics V, Griffiths WD, Pericleous KA Resonance from contactless ultrasound in alloy melts, In: Light Metals 2019
14. Griffiths WD et al (2012) Metal Mater Trans B 43B:370–378
15. Pericleous KA, Bojarevics V (2007) Progress Comput Fluid Dyn 7, Nos. 2/3/4
16. Bojarevics V, Harding RA, Pericleous KA, Wickins M (2004) The development and experimental validation of a numerical model of an induction skull melting furnace. Metall Mat Trans B, 35(4):785–803
17. Wilcox DC (1998) Turbulence modelling for CFD. 2nd ed., DCW Industries, California
18. Djambazov GS et al (2000) AIAA J 38(1):16
19. Lebon GSB, Tzanakis I, Djambazov G, Pericleous K, Eskin DG (2017) Numerical modelling of ultrasonic waves in a bubbly Newtonian liquid using a high-order acoustic cavitation model. Ultrasonics Sonochemistry 37:660–668
20. Tzanakis I, Lebon GSB, Eskin DG, Pericleous K (2016) Investigation of the factors influencing cavitation intensity during the ultrasonic treatment of molten aluminium. Mater Des 90:979–983
21. Eskin GI (1998) Ultrasonic Treatment of Light Alloy Melts, 1st edn. Gordon and Breach Science Publishers, Amsterdam

Simulation Analysis Techniques for Investment Casting Process of Ni-Base Superalloy Components

Kosuke Fujiwara, Hidetaka Oguma, Masaki Taneike, Ikuo Okada, Kyoko Kawagishi, Tadaharu Yokokawa and Hiroshi Harada

Abstract Development of the Ni-base single crystal (SC) superalloy, named MGA1700 (TMS-1700), has been carried out in a collaborative research between National Institute for Materials Science (NIMS) and Mitsubishi Heavy Industries (MHI). Besides the excellent creep strength and dwell thermo-mechanical fatigue strength, the newly developed alloy for high-efficiency gas turbine blade has no expensive elements such as Rhenium. On the other hand, the mechanical properties or manufacturing yield of SC superalloy blades is quite sensitive to casting process conditions. Therefore, in this study, simulation analysis techniques for various steps of investment casting process, including the solidification process and the mold manufacturing process, have been established to find adequate casting conditions. These analysis techniques can be also applied to other components, such as directionally solidified (DS) rotating blade or conventional casting (CC) rotating blade/stationary vane.

Keywords Investment casting · Casting simulation · Single-crystal superalloy

Introduction

Gas turbine combined cycle (GTCC) power plants have higher-power generation efficiency than the other kinds of power plants operated by fossil fuel. However, there is a strong urge to improve the thermal efficiency further to satisfy the strong demand for a reduction of CO_2 emissions. One of the effective ways to increase the thermal efficiency of GTCC is increasing the gas turbine inlet temperature. Therefore, the gas turbine with turbine inlet temperature of 1700 °C is being

K. Fujiwara (✉) · H. Oguma · M. Taneike · I. Okada
Mitsubishi Heavy Industries, Ltd., 2-1-1 Shinhama, Arai-cho, Takasago
Hyogo 676-8686, Japan
e-mail: kosuke1_fujiwara@mhi.co.jp

K. Kawagishi · T. Yokokawa · H. Harada
National Institute for Materials Science, 1-2-1 Sengen, Tsukuba, Ibaraki 305-0047, Japan

M. Tiryakioğlu et al. (eds.), *Shape Casting*, The Minerals, Metals & Materials Series,
https://doi.org/10.1007/978-3-030-06034-3_24

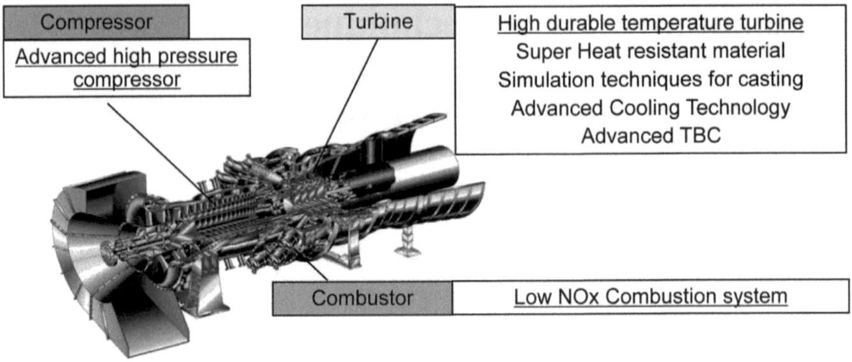

Fig. 1 Technical items for 1700 °C class gas turbine

Fig. 2 Turbine inlet gas temperature of MHI gas turbine and creep durable temperature of superalloys (Creep durable temperature at 138 MPa x 50000 h) [8, 9]

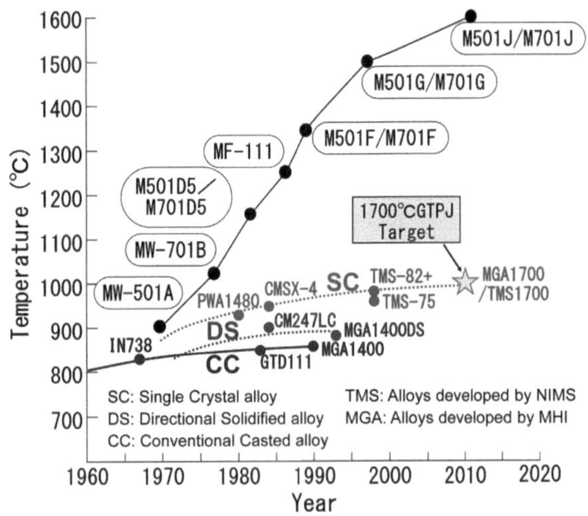

developed in the Japanese national project (Fig. 1) [1, 2]. In this project, some key technologies in the advanced high-pressure compressor, low NOx combustion system, and high durable temperature turbine have been developed. The technologies for high durable temperature turbine include the development of heat resistant materials and the establishment of simulation analysis techniques for investment casting process.

Heat-resistant material is one of the important technologies to realize 1700 °C class gas turbine project. Figure 2 shows the turbine inlet gas temperature of MHI gas turbine and the development trend of superalloys for gas turbine blade. The high-temperature durability of the superalloy has been constantly increased with increasing turbine inlet gas temperature, where SC superalloys have the highest strength due to their high γ’ phase volume fraction without grain boundary.

In general, it is difficult to make a SC superalloy having both excellent creep strength and thermo-mechanical fatigue strength at the same time. However, the new SC superalloy MGA 1700 (TMS-1700), developed in a collaborative research between NIMS and MHI, has both of these properties [3, 4]. In addition, MGA1700 has no expensive elements such as Rhenium since they were developed based on the second-generation SC superalloys [5–7].

Simulation analysis for investment casting process is important for SC superalloy blades. Since the mechanical properties or the manufacturing yield are strongly affected by the casting defects and the crystal orientation, simulation analysis is used to find a suitable casting process condition to suppress the casting defects and the unfavorable crystal orientations. The casting defects mainly occur during the solidification process of molten metal, while it is also expected that some defect formations and low-yield rate are depending on the condition of mold manufacturing process. For example, if the injection molding condition of wax pattern model is inadequate, core cracking or short shot may occur. Therefore, in this study, simulation analysis for various steps of investment casting process, not only for solidification process but also for mold manufacturing process, has been established.

Design of Alloy Composition

The aim of development of new SC superalloy is to obtain good high-temperature properties (creep rupture strength, fatigue strength, and oxidation resistance), low manufacturing cost (alloy cost, casting cost, heat treatment cost) and good castabilities. The SC alloys, after the second generation, contain expensive elements, such as Re and Ru, in order to improve the alloy strength remarkably [10]. However, in this project, the total amount of Re and Ru was designed to 3% or less to reduce the alloy cost.

Generally, it is considered that the high-temperature strength of Ni-base superalloy is due to the solid solution strengthening and γ' precipitation strengthening. Therefore, in the superalloy, Al and Ti, etc. are added to promote the γ' phase formation, while W and Mo are added as solid solution strengthening elements. However, the γ' phase forming elements and the solid solution strengthening elements amount should be properly balanced; otherwise, an excess of either of those elements causes phase instability and precipitation of brittle TCP phase such as σ phase.

The composition of the experimental alloys was determined according to the above concept using the materials design flow chart, as shown in Fig. 3. NIMS has been involved in the development of the superalloy for a long time and developed the alloy design program based on the data base of material properties. The alloy design program can estimate metallurgical factors such as γ/γ' phase composition, γ' phase volume fraction, phase stability factor and γ/γ' phase misfit parameter. By applying multiple regression analysis with these factors, the creep strength and

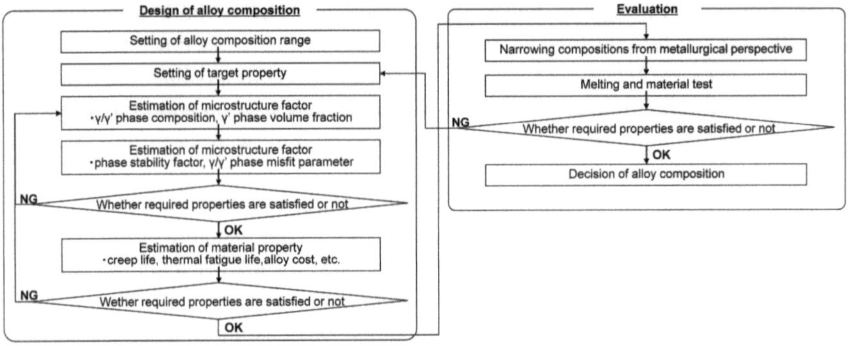

Fig. 3 Alloy design flow of a new superalloy

thermo-mechanical fatigue strength can be estimated. Especially, the creep strength was estimated with good accuracy because of availabilities of a large amount of data of NIMS, but it was not possible for thermo-mechanical fatigue strength due to lack of adequate amount of data. Therefore, the estimation program of thermo-mechanical fatigue strength has been improved by reflecting the tested results in this study.

Mechanical Tests

Microstructures

Based on the results of alloy design program, dozens of superalloys were designed and size of 200 × 200 × 20 mm cast plates was manufactured using ceramic molds. After the heat treatment, round bar and plate shape specimens were machined for mechanical properties measurements. Figure 4 shows the microstructures of a plate specimen of the developed alloy (MGA1700/TMS-1700), which contains no Rhenium. The microstructure was homogenized and no eutectic γ/γ' phase was observed.

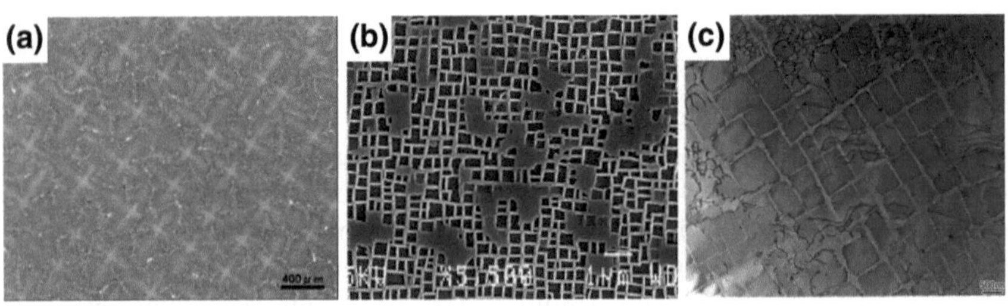

Fig. 4 **a** Optical micrograph, **b** SEM image, and **c** TEM image of MGA1700 (TMS-1700)

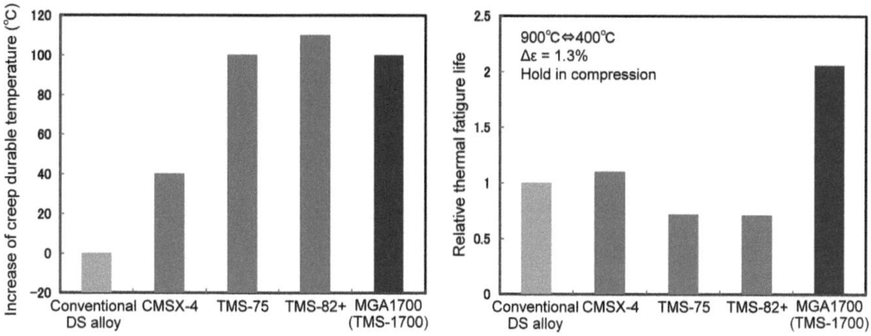

Fig. 5 Increase of creep durable temperature and relative thermal fatigue life

Mechanical Properties

The creep test was carried out at temperature from 900 to 1100 °C. The summary of the creep tests are shown in Fig. 5 in comparison with that of conventional DS superalloy. The creep strength of MGA1700 (TMS-1700) was superior to that of CMSX-4 and identical with the TMS-82 + and TMS-75. The result confirmed that MGA1700 had excellent creep strength.

As considering the actual operating conditions of blades in gas turbine, the thermo-mechanical fatigue test was carried out. The temperature and strain were controlled under out-of-phase condition. The temperature was oscillated in the range of 400–900 °C, and test pieces were held for an hour at 900 °C in compression stress. Since results of fatigue tests tended to vary easily, multiple tests were carried out at the same conditions. The fatigue test results are shown in Fig. 5. As it can be seen, the MGA1700 had excellent fatigue strength, and thus, it satisfied our development objectives.

Moreover, the absence of expensive elements makes the MGA1700 alloy as a potential material for practical application.

Simulation Analysis

Castability of SC Superalloy

For SC superalloy, simulation analysis techniques to evaluate castability are quite important because castability is directly linked to mechanical properties and manufacturing yield. Typical defects in SC alloys are shown in Fig. 6. Since the size of industrial gas turbine blade is quite larger than that of air craft engine, it is very prone to casting defect formation. Therefore, the castability of the developed SC superalloy was evaluated by simulation technique and the best casting conditions were selected to inhibit the casting defects and control the crystal orientation.

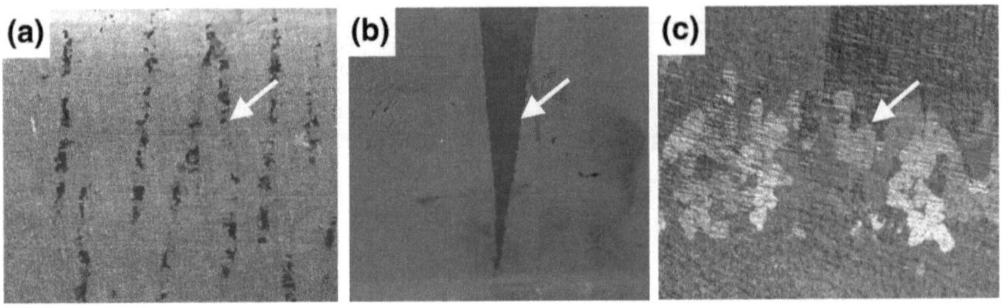

Fig. 6 Typical casting defects in the SC superalloy **a** freckle, **b** stray grain, and **c** re-crystallization

Simulation of Casting Process

Solidification simulation results are used to predict the casting defect generation and the crystal orientation. In order to conduct the simulation more precisely, some casting experiments were carried out. Figure 7 shows the appearance and the schematic diagram of the test furnace and a casting mold. Specimens were cast under various conditions of withdrawing rate and mold heater temperature. After the casting, the casting defects on the cast specimens were checked. The simulation parameters were adjusted based on the comparison of experimental results and previously conducted simulation results.

Figure 8 shows the basic mechanism of freckle generation. Driving force for freckle generation is segregation of compositional elements. In typical Ni-base superalloys, low-density elements (such as Ti or Al) tend to segregate to the liquid and high-density elements (such as W, Re) tend to segregate the solid and concentrate in the dendrite core. These segregations make the liquid lighter in the vicinity of liquid/solid interface and promote convection flow of molten metal. The convection flow, promoted by segregation, causes the freckle generation. On the other hand, the friction force between the liquid and dendrite arms acts as a resistance force of the freckle formation.

To evaluate the freckle formation, the Rayleigh number has been used. The Rayleigh number is the ratio between driving force (liquid density variation) and

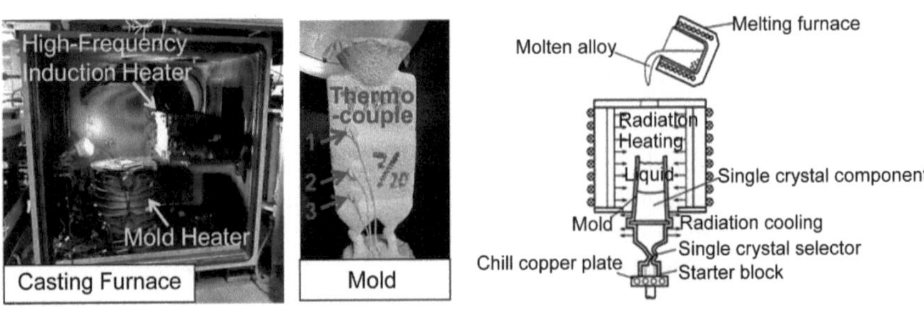

Fig. 7 Appearance and schematic diagram of test furnace and casting mold

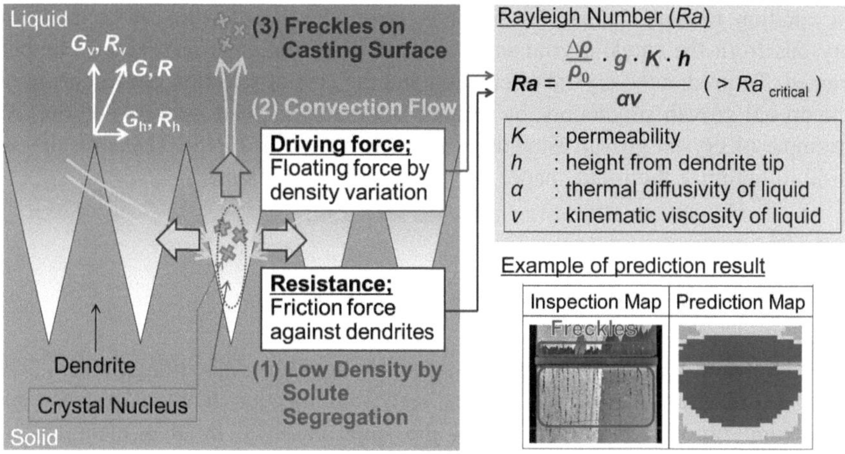

Fig. 8 Basic mechanism and an example of prediction result of freckle generation

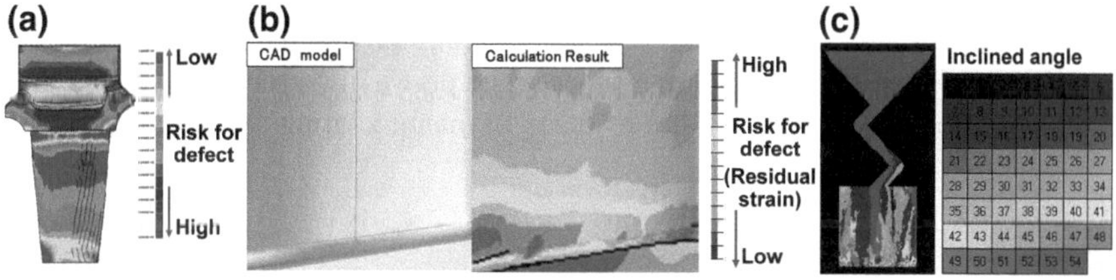

Fig. 9 Predicted results of **a** stray grain, **b** re-crystallization, and **c** crystal orientation

resistance force (permeability of dendrites or liquid properties) [11, 12]. In Fig. 8, the result of experiment and simulation was compared and showed good match between the results.

Figure 9 shows the predicted result of the stray grain and the re-crystallization generation. The stray grain is a crystal grain grows irregularly on the way of continuous crystal growth from the bottom side of the mold to top. The stray grain generation is mainly caused by an imbalance between two solidification parameters, called solidification speed and thermal gradient. Therefore, the stray grain can be predicted using these parameters obtained from the solidification simulation results. Similarly, the re-crystallization occurs in the heat treatment process because of the residual strain generated during the casting process. To predict the re-crystallization, the temperature variation results of solidification simulation were transferred to FEM code, and then the residual strain generated during the solidification and the post-solidification was calculated.

The predicted place of casting defect generation by simulation was in good agreement with the place of casting defect generation on the real blade.

The angle difference between crystal orientation and withdrawing direction in casting process should be minimized to obtain sufficient material strength.

In addition to that, the crystal selection error, developed when more than two crystals from the single-crystal selector enter the component part, should be prevented. To predict the crystal orientation and the crystal selection error generation, the crystal growth simulations by cellular automaton method were carried out. An example of crystal growth simulation result is shown in Fig. 9. These results are used to optimize the mold shapes and the casting conditions.

Simulation of Mold Manufacturing Process

The simulation technique was also used for the mold manufacturing process. The mold manufacturing process includes core/wax injection molding process and de-waxing process. Figure 10 shows the steps involved in investment casting process and the simulation results of filling time in core/wax injection molding process for stationary vanes.

The flow behavior of core/wax and the stress distribution, mainly caused by the thermal expansion, are calculated by the simulations. These results are used for short-shot prediction in core/wax injection molding process or core/mold cracking prediction in wax injection molding process and de-waxing process. By using these simulation techniques, the conditions of mold manufacturing are optimized.

Trial Cast of Real Blade

These analysis techniques were applied to real blade shape and a trial casting of SC blade was carried out. As a result, the large-size IGT blade (over 300 mm) with no crystal defects was obtained. An appearance of trial cast blade is shown in Fig. 11.

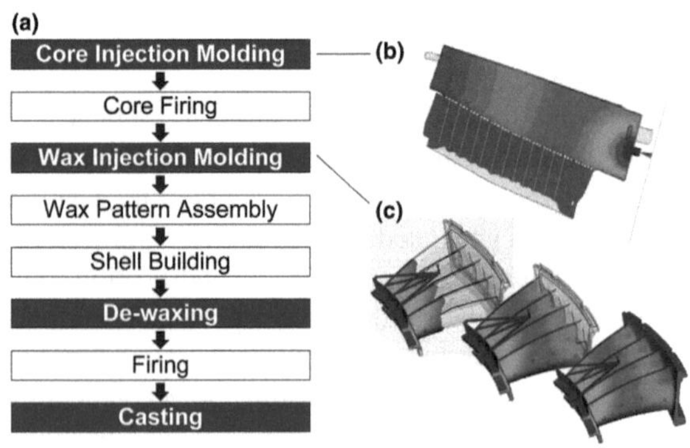

Fig. 10 **a** Various steps of investment casting and examples of simulation results of filling time in injection molding processes of **b** core and **c** wax for stationary vanes

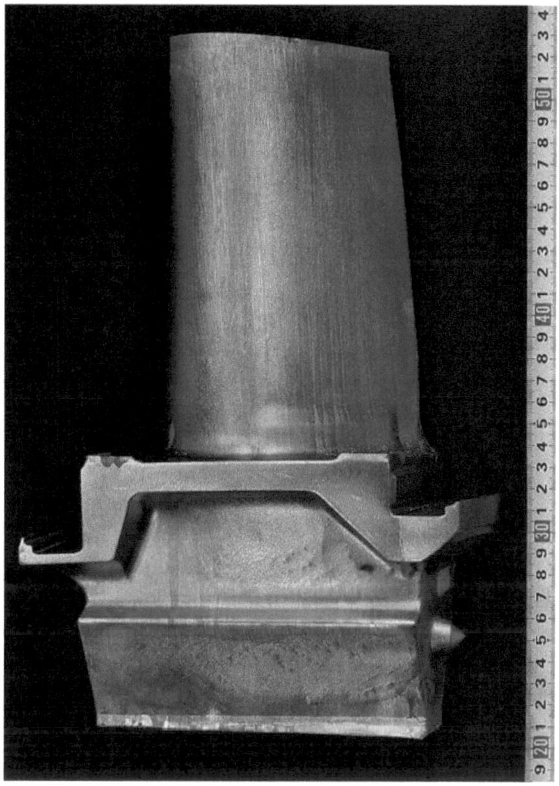

Fig. 11 Appearance of trial cast blade

Conclusions

The development of Ni-base single crystal superalloy was carried out in collaboration research between National Institute for Materials Science (NIMS) and Mitsubishi Heavy Industries (MHI). The developed alloy (MGA1700/TMS-1700) had the excellent creep strength properties and the excellent dwell thermo-mechanical fatigue properties; besides, it also had no expensive elements such as rhenium.

The simulation analysis technique for various steps of investment casting process, including the solidification process, core/wax injection molding process, and de-waxing process, was established to find adequate casting conditions. The established technique was very helpful to predict the casting defects, the crystal orientation, the short shot of core/wax and the core/mold cracking.

A trial casting test was also carried out according to the results obtained from the experiment and simulation and found that the large-size IGT blade was defect-free.

Acknowledgements These research results were obtained from a project subsidized by Ministry of Economy, Trade and Industry (METI) and New Energy and Industrial Technology Development Organization (NEDO).

References

1. Tsukagoshi K et al (2007) Mitsubishi Techn Rev 44(4)
2. Ito E et al (2007) Development of key technologies for the next generation gas turbine. ASME Paper No. GT2007- 28211
3. Kawagishi K et al (2016, August) Superalloys 2016: Proceedings of the 13th international symposium of superalloys. John Wiley & Sons, Inc. Hoboken, NJ, USA, pp 115–122
4. Oguma H et al (2015) In Proceedings of international gas turbine congress, pp 173–178
5. Hino T et al (1999, November) International gas turbine congress 1999 kobe, 169–174
6. Kobayashi T et al. (1997) Adv Turbine Mater Des Manuf, 766–773
7. Koizumi Y et al (2001) High Temp Mater, May 31–June 2, 2001
8. Saeki H et al (1996) J Gas Turbine Soc Jpn 24(93):43–47
9. Harada H (2000) J Gas Turbine Soc Jpn 28(4):278–284
10. Zhang JX et al (2002) Metall Mater Trans A 33(12):3741–3746
11. Beckermann C et al (2000) Metall Mater Trans A, 31(10):2545–2557
12. Rad MT et al (2013) Metall Mater Trans A, 44(9):4266–4281

Improvement in Metallurgical Properties of Gravity Die Cast 2024-T6 Aluminum Alloy via Cryogenic Process

Engin Tan, Sinan Aksöz, Yavuz Kaplan, Hilal Can and Derya Dışpınar

Abstract The aim of this study is to investigate the effects of cryogenic process (CP) on the properties of 2024-T6 aluminum alloy. So, samples were held at −40 °C medium for 24 h in CP after solution heat treatment. Finally, samples were subjected to aging process at 120 °C for 8 h after CP. Furthermore, aging process was carried out after 0, 1, and 4 h later after CP. Examination of microstructure was executed by SEM, EDS, and optical microscope (OM) observations. Hardness tests were also conducted through HV. Improvement in metallurgical properties of 2024-T6 aluminum alloy was tested with these parameters, in detail.

Keywords Cryogenic · Die cast · 2024-T6 Al alloy · SEM and EDS analyses · Hardness

Introduction

2024-T6 high-strength aluminum alloy has a wide usage area in industry like aviation, automobile, and also several engineering applications [1–3]. One of the improved methods in the properties of this alloy is cryogenic process (CP) [4–6]. CP is a simple, low-cost, and environmentally-friendly process that has an effective role to stabilize the sizes and to advance the rigidity [7–9]. However, it is necessary to determine the basic mechanisms between CP parameters and materials'

E. Tan (✉) · S. Aksöz · H. Can
Department of Metallurgical and Materials Engineering, Technology Faculty, Pamukkale University, Kınıklı Campus, Denizli, Turkey
e-mail: etan@pau.edu.tr

Y. Kaplan
Department of Mechanical and Manufacturing Engineering, Technology Faculty, Pamukkale University, Kınıklı Campus, Denizli, Turkey

D. Dışpınar
Department of Metallurgical and Materials Engineering, Engineering Faculty, Istanbul University, Istanbul, Turkey

M. Tiryakioğlu et al. (eds.), *Shape Casting*, The Minerals, Metals & Materials Series,
https://doi.org/10.1007/978-3-030-06034-3_25

properties. So in recent years, many researchers study the effect of CP parameters on microstructure and performance of materials [10–14].

In this study, the influence of CP process on the metallurgical properties of gravity die cast 2024-T6 aluminum alloy was investigated. So, microstructure and hardness characteristics of the samples were studied.

Materials and Method

Commercially available 2024 aluminum alloy was used in experimental studies. The chemical composition of alloy was given in Table 1.

The samples were subjected to solution heat treatment to 520 °C, holding for 2 h and then quenched in water for three different temperatures 0, 20, and 80 °C. CP was carried out to solutionize samples at −40 °C for 24 h in dry air medium. Finally, CP samples were aged in resistance furnace from 2 h to 8 h at 120 °C to optimize microstructure and hardness characteristics. Furthermore, CP samples were held at room temperature for three different times before aging treatment 0, 1, and 4 h. The detailed specification along with samples labels is given in Table 2.

Nomenclature definitions: T6: Traditional artificial aging process, SHT: Solution heat treatment, CP: Cryogenic process, QW0: Samples quenched in 0 °C water, QW20: Samples quenched in 20 °C water, QW80: Samples quenched in 80 °C water, CP0: Samples held for 0 h at room temperature after CP, before aging, CP1: Samples held for 1 h at room temperature after CP, before aging, CP4: Samples held for 4 h at room temperature after CP, before aging. Example: QW0T6 means samples quenched in 0 °C water, aged in 120 °C for 2–4–6–8 h (no CP), and QW80CP4 means samples quenched in 80 °C water, held for 4 h at room temperature after CP, before aging.

After processes given in Table 2, the measurement of the HV hardness was carried out in accordance with ASTM E92-17. For each sample, at least five points were measured to obtain an average value. Microstructure analysis was conducted using optical microscope (Nikon Eclipse LV150) and field-emission scanning electron microscope (Zeiss-Supra40 V FESEM). OM and SEM specimens were prepared by mechanical grinding and polishing. Keller's reagent (ingredient: 5 ml of nitric acid, 3 ml of hydrochloric acid, 2 ml of hydrofluoric acid and 190 ml distilled water) was used for etching.

Results and Discussion

Figure 1 shows the measurement results of hardness tests. Firstly, as seen in Fig. 1, the highest hardness values were obtained for 8 h aging time in all samples. However, the CP process significantly influenced the hardness of the basic alloy.

Table 1 Chemical compositions of 2024 aluminum alloy (wt%)

Element	Cu	Fe	Si	Zn	Mn	Mg	Ti	Cr	Al
In this study	4,23	0,562	0,307	0,0375	0,571	1,48	0,017	0,152	Balance
Nominal [15]	3,8-4,9	≤0,50	≤0,50	≤0,25	0,3-0,9	1,2-1,8	≤0,15	≤0,10	Balance

Table 2 T6 heat treatment and CP parameters

Nomenclature	SHT	Quench temperature (in Water Medium) (°C)	CP	Aging
QW0T6	520 °C	0	–	120 °C
QW20T6	2 h	20	–	2–4–6–
QW80T6		80	–	8 h
QW0CP0		0	−40 °C	
QW0CP1			24 h	
QW0CP4				
QW20CP0		20		
QW20CP1				
QW20CP4				
QW80CP0		80		
QW80CP1				
QW80CP4				

Higher hardness values were obtained in samples treated with CP, compared to samples treated traditionally artificial aging, T6. In general, in order to analyze the effect of quenching temperatures in the T6 process, it is seen that the slope of the hardness change curve is quite steep in the quenching process at 0 °C degree and the samples are affected more than the hardness change. This can be explained by the fact that the low quenching temperatures in the aging process delay the decomposition of the secondary phases by the abrupt cooling effect, thus obtaining a microstructure with a more frequent and homogeneous distribution. Furthermore, the formation of grain refinement occurred more easily during CP due to the well-distributed second phases [16]. At the 80 °C quenching temperature, the slope of the hardness change curve for all samples appears to be low. That is, the samples harden with increasing aging time, but the hardness increase is lower than the lower quenching temperatures. This can be explained by the fact that the vapor film formed on the surface of the samples at high temperatures slows down the cooling. In the quenching process, the (atomic) void concentration increases as the rate of quenching increases [17, 18]. The highest hardness values of all samples were obtained for 20 °C quenching temperatures. In these conditions, it can be said that the optimum dew pointing temperature is room temperature. At 80 °C degree water, internal stresses were expected to be less, and as a result, hardness values were decreased [19–21]. At high water temperatures, the cooling rate decreases, and coarse and localized secondary phases form which dilutes the alpha phase (depleting the Cu from the solid solution). Therefore, the hardness decreases.

Figure 2 compares the grain characteristic the samples with T6 and CP treatments. In Fig. 2 a, it is seen that the grains of the T6 sample are relatively large and the size is not uniform. After CP, the grains were refined, as shown in Fig. 2b, c, and d. It is reflected that CP can supply finer and more homogeneously distributed grains compared with the single T6 aging process. The effects of CP on the grain characteristic of 2024-T6 aluminum alloy were also investigated with the SEM and EDS tests shown in Figs. 3, 4, and 5. It was found that the compound of the second

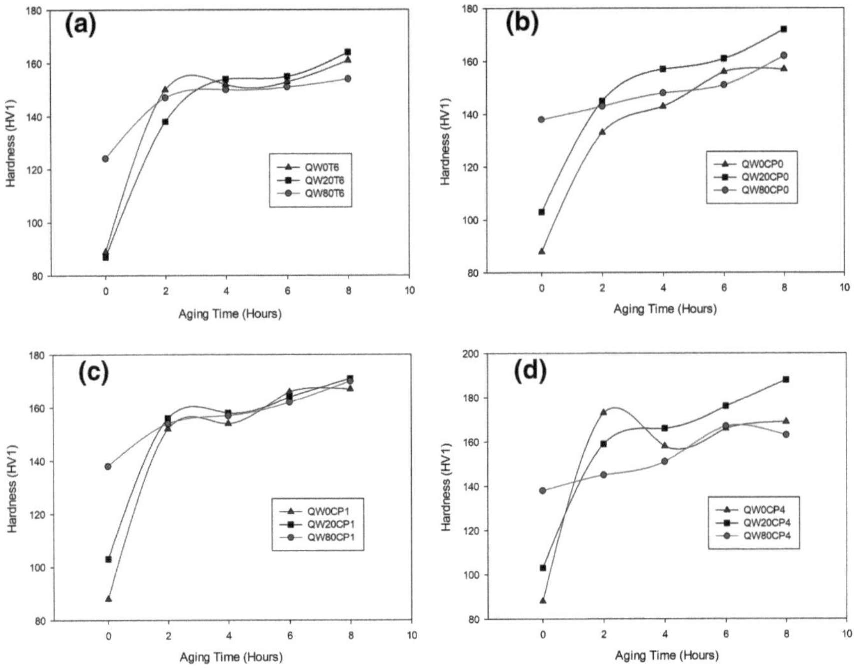

Fig. 1 Hardness results of samples **a** traditionally artificial aged, **b** held for 0 h at room temperature after CP, before aging, **c** held for 1 h at room temperature after CP, before aging, **d** held for 4 h at room temperature after CP, before aging

phase mainly consists of Al, Cu, and Mg in SEM and EDS analyses. Hereby, Al_2CuMg and $CuAl_2$ precipitate probably formed as second phase. The morphology of these phases showed that after the CP, the size was small and the distribution was highly uniform, as shown in Fig. 2b, c, d. CP can provide finer grains and more homogeneously distributed secondary phases. So secondary phases in grain boundaries make the grains more uniform [16].

According to the literature, it is expected that there will be secondary phases in grain boundaries [16, 22]. Secondary phases in grain boundaries seen in OM images were also observed as a result of SEM analyses (Fig. 3). The weight of the Cu atom is greater than the weight of the Al atom. Therefore, when the backscatter process is applied to the SEM device, it is expected that the Cu atoms will appear in white due to the different contrast [22, 23]. Thus, white-colored Cu phases at the grain boundaries were seen (Fig. 3). The results of EDS analysis also support this situation (Fig. 4 and 5) [22, 23] so that Al_2CuMg and $CuAl_2$ precipitates are probably formed as the second phase. Meanwhile, these precipitates were well distributed in the microstructure and their average size was reduced [6].

Aluminum has a high affinity for oxygen, so it oxidizes in nanoseconds to form an oxide layer on its surface. This oxide is not a problem as long as it is on the

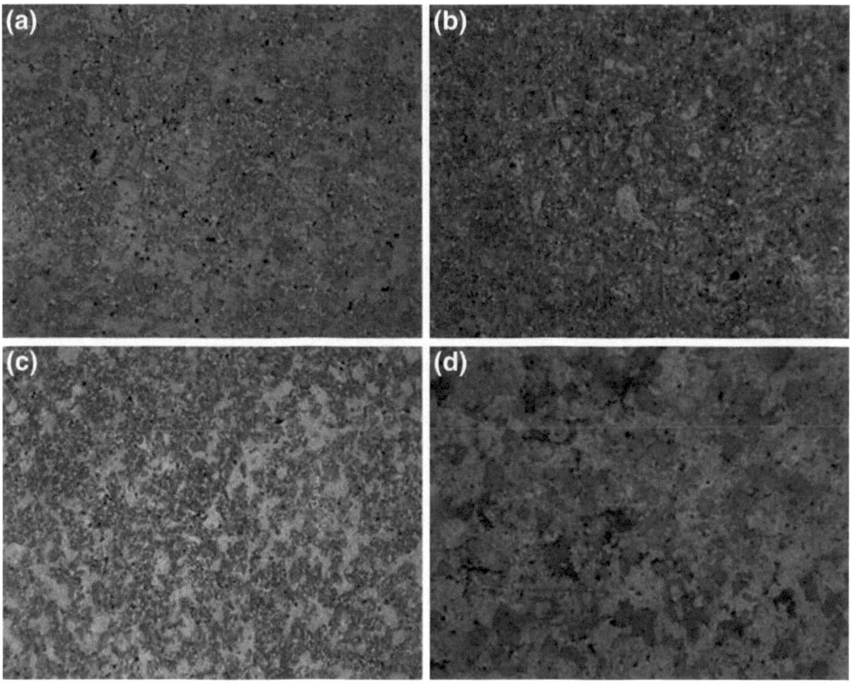

Fig. 2 OM images of **a** QW20T6, **b** QW0CP4, **c** QW20CP4, **d** QW80CP4 samples (100x)

surface. However, this solid oxide layer can be miscible in the liquid, especially in liquid state, during various turbulence and similar surface defects during production [22]. If the oxide layer is likened to a sandpaper, then there will be no bond between the two oxide surfaces after this folding and will normally include a gap. These defects are called bifilm [24], and bifilms, in addition to causing porosity, cause crack-like defects that will cause mechanical properties to deteriorate [22].

From Fig. 4, the amount of Mg is observed to be around 1.9%. Mg is combined with aluminum oxide to form oxides of MgO • Al_2O_3 called spinel oxide. The presence of spinel oxides and bifilms in the SEM and EDS analyses has clearly demonstrated this situation (Figs. 4 and 5).

Figure 5 also shows some cracked areas in the microstructure. These cracks, seen in the samples quenched in water at 0 °C, are thought to occur due to (i) internal stresses, (ii) void concentration, and (iii) formation of precipitation phases [22]. Therefore, it is important to determine the optimum quenching temperatures in aging processes.

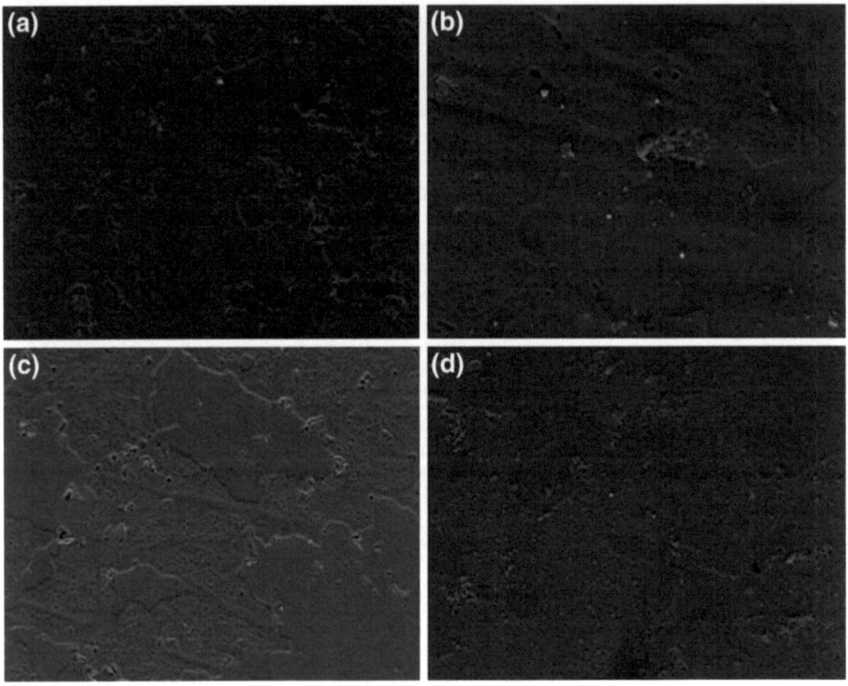

Fig. 3 SEM images of **a** QW20T6, **b** QW0CP4, **c** QW20CP4, **d** QW80CP4 samples (2000x)

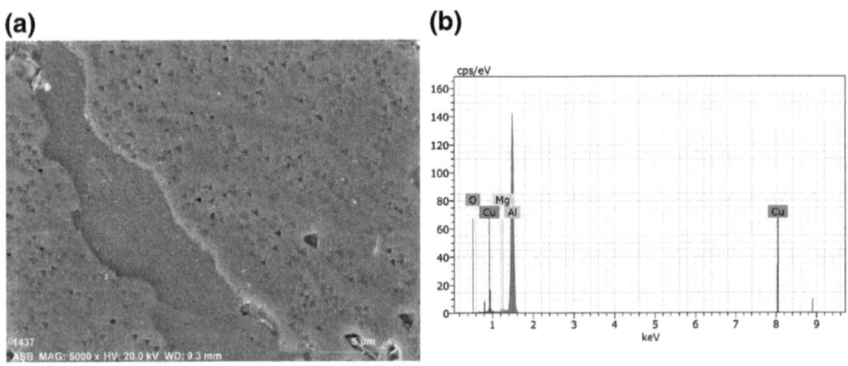

Fig. 4 QW20CP4 sample, **a** SEM image, **b** elemental peaks distribution taken on the SEM image

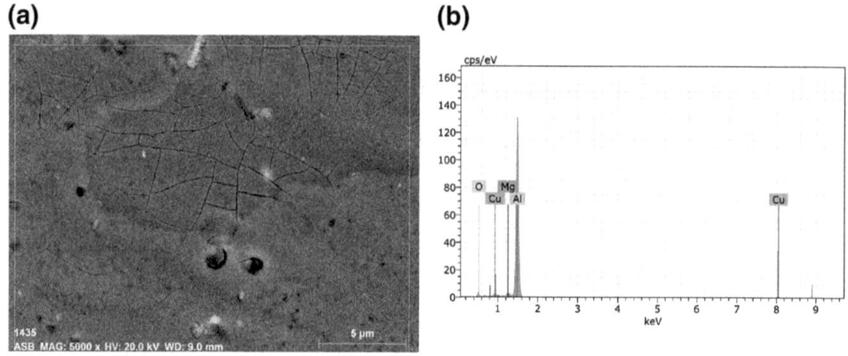

Fig. 5 QW0CP4 sample, **a** SEM image, **b** elemental peaks distribution taken on the SEM image

Conclusions

In this study, the effects of cryogenic process (CP) on properties of 2024-T6 aluminum alloy were investigated. The CP effect on the microstructure, hardness, SEM, and EDS can be summarized as follows:

- The CP process significantly influenced the hardness of the basic alloy. Higher hardness values were obtained in samples treated with CP, compared to samples treated traditionally artificial aging, T6.
- The slope of the hardness change curve is quite steep in the quenching process at 0 °C degree, and the samples are affected more than the hardness change. This can be explained by the fact that the low quenching temperatures in the aging process delay the decomposition of the secondary phases by the abrupt cooling effect, thus obtaining a microstructure with a more frequent and homogeneous distribution.
- The formation of grain refinement occurred more easily during CP due to the well-distributed second phases.
- The highest hardness values of all samples were obtained for 20 °C quenching temperatures. At high water temperatures, the cooling rate decreases, and coarse and localized secondary phases form which dilutes the alpha phase (depleting the Cu from the solid solution). Therefore, the hardness decreases.
- The hardness values were increased with the CP. The best hardness value is determined in the QW20CP4 (quenched in 20 °C water, held for 4 h at room temperature after CP, before aging) sample.
- CP can provide finer grains and more homogeneously distributed secondary phases. So secondary phases in grain boundaries make the grains more uniform.
- Bifilms can cause porosity formation and affect the metal quality in aluminum alloys.

References

1. Tan E, Tarakcilar AR, Dispinar D (2015) The effect of melt quality and quenching temperature on the Weibull distribution of tensile properties in aluminium alloys. Materialwiss Werkstofftech 46(10):1005–1013
2. Zhou J, Xu S, Huang S, Meng X, Sheng J, Zhang H, Li J, Sun Y, Boateng EA (2016) Tensile properties and microstructures of a 2024-T351 aluminum alloy subjected to cryogenic treatment. Metals 6(279):1–10
3. Shahsavaria A, Karimzadeha F, Rezaeiana A, Heydaria H (2015) Significant increase in tensile strength and hardness in 2024 aluminum alloy by cryogenic rolling. Proced Mater Sci 11:84–88
4. Aksoz S, Bostan B (2018) Effects of ageing and cryo-ageing treatments on microstructure and hardness properties of AA2014–SiC MMCs. Trans Indian Inst Met 71(8):2035–2042
5. Singh AK, Ghosh S, Mula S (2016) Simultaneous improvement of strength, ductility and corrosion resistance of Al2024 alloy processed by cryoforging followed by ageing. Mater Sci Eng, A 651:774–785
6. Araghchia M, Mansouria H, Vafaeia R, Guo Y (2017) A novel cryogenic treatment for reduction of residual stresses in 2024 aluminum alloy. Mater Sci Eng, A 689:48–52
7. Li GR, Cheng JF, Wang HM, Li CQ (2017) The influence of cryogenic-aging circular treatment on the microstructure and properties of aluminum matrix composites. J Alloy Compd 695:1930–1945
8. Nayan N, Narayana Murty SVS, Jha AK, Pant B, Sharma SC, George KM, Sastry GVS (2014) Mechanical properties of aluminium–copper–lithium alloy AA2195 at cryogenic temperatures. Mater Des 58:445–450
9. Taskesen A, Aksoz S, Ozdemir AT (2017) The effect of cryogenic treatment on ageing behaviour of B_4C reinforced 7075 aluminium composites. Kovove Mater 55:57–67
10. Park DH, Choi SW, Kim JH, Lee JM (2015) Cryogenic mechanical behavior of 5000- and 6000-series aluminum alloys: Issues on application to offshore plants. Cryogenics 68:44–58
11. Lee WS, Lin CR (2016) Deformation behavior and microstructural evolution of 7075-T6 aluminum alloy at cryogenic temperatures. Cryogenics 79:26–34
12. Khedekar D, Gogte CL (2018) Development of the cryogenic processing cycle for age hardenable AA7075 aluminium alloy and optimization of the process for surface quality using gray relational analysis. Mater Today Proc 5:4995–5003
13. Wang J, Xue X, Yang Z, Zhang H, Zhou Y (2011) Effect of cryogenic treatments on mechanical properties of 2A11 aluminum alloy. Adv Mater Res 146–147:1646–1650
14. Ko DH, Ko DC, Lim HJ, Lee JM, Kim BM (2013) Prediction and measurement of relieved residual stress by the cryogenic heat treatment for Al6061 alloy: mechanical properties and microstructure. J Mech Sci Technol 27(7):1949–1955
15. International Alloy Designations and Chemical Composition Limits for Wrought Aluminum and Wrought Aluminum Alloys (Revised 2001), The Aluminum Association, Inc.
16. Zhou J, Li J, Xu S, Huanga S, Menga X, Shenga J, Zhanga H, Suna Y, Feng A (2018) Improvement in fatigue properties of 2024-T351 aluminum alloy subjected to cryogenic treatment and laser peening. Surf Coat Technol 345:31–39
17. Porter D (1981) Phase transformations in metals and alloys. Van Nostrand Reinhold Publish
18. Dieter G (1986) Mechanical metallurgy. McGraw–Hill Science/Engineering/Math, 3rd Edn
19. Tan E, Tarakcilar AR, Dispinar D (2012) Quench sensitivity of 2024, 6063 and 7075. Paper presented at the 141st TMS Annual Meeting, Orlando, FL, USA, 11–15 March 2012
20. Tan E, Tarakcilar AR, Dispinar D (2012) Correlation between melt quality and fatigue properties of 2024,6063 and 7075. Paper presented at the 141st TMS Annual Meeting, Orlando, FL, USA, 11–15 March 2012

21. Tan E, Tarakcilar AR, Dispinar D (2012) Effect of melt quality and quenching temperature on the mechanical properties of SIMA 2024 and 7075. Paper presented at the 14th international conference on advances in materials and processing technologies (AMPT), Istanbul, 13–16 July 2011
22. Tan E (2011) Blister formation in aluminium alloys. PhD thesis, Pamukkale University
23. Tan E, Tarakcilar AR, Dispinar D (2011) Blistering problems observed in strain induced melt activated aluminium alloys. Trans Indian Inst Metal 64(6):555–563
24. Campbell J (2003) Casting. Butterworth, 2nd edn. UK

Melt Cleaning Efficiency of Various Fluxes for A356 Alloy

Çağlar Yüksel, Uğur Aybarc, Eray Erzi, Derya Dispinar and Mustafa Cigdem

Abstract Degassing operations are common methods to remove inclusions and oxides from the melt. In many occasions, fluxes are used through the melt in order to increase efficiency of oxide removal. In this work, a dirty melt was prepared by means of adding 50 wt% swarf to primary A356 alloy. Fluxes with various ratios of NaCl–KCl and $MgCl_2$–KCl salts were prepared. Different weight ratio of Na_3AlF_6 and AlF_3 was added. Melt cleanliness was measured by means of change in bifilm index (ΔBI) before and after degassing. It was concluded that most of the fluxes had cleaning efficiency below 40% while $MgCl_2$ based fluxes had over 70% cleaning effect.

Keywords A356 · Flux · Bifilm (BI) · Cleanliness

Introduction

Metal cleanliness is vital importance to obtain the desired mechanical properties. To determine the metal cleanliness as numerically, researchers have used the many different techniques. bifilm analysis is one of these techniques. Especially, the last years bifilm analysis has been seen great interest and begun to use to determine the melt quality and cleanliness.

Lots of works were done to identify and also determine these inclusions [1–5]. Dispinar [5] succeeded to quantify melt quality of aluminium alloys via Bifilm Index (BI). This index is useful to understand if the liquid metal is ready to cast or not. Uludag

Ç. Yüksel
Metallurgical and Materials Engineering, Ataturk University, Erzurum, Turkey

U. Aybarc
CMS, Izmir, Turkey

E. Erzi · D. Dispinar (✉)
Metallurgical and Materials Engineering, Istanbul University-Cerrahpasa, Istanbul, Turkey
e-mail: deryad@istanbul.edu.tr

M. Cigdem
Metallurgical and Materials Engineering, Yildiz Technical University, Istanbul, Turkey

M. Tiryakioğlu et al. (eds.), *Shape Casting*, The Minerals, Metals & Materials Series,
https://doi.org/10.1007/978-3-030-06034-3_26

et al. [6] offered that if the BI is higher than 50 mm, the melt must be processed, such as degassing and fluxing operations. Flux addition is known as ordinary melt treatment to purify liquid metal [7]. As well as no quantitative data were reported about melt quality, most of researchers have attempted to remove inclusion by adding fluxes and try solely to determine melt quality with mechanical properties [8–10].

Liu et al. treated Al-7Si-0.5 Mg aluminium alloy with flux injection method and concluded that by using flux injection method both higher strengths and elongation values can be obtained for aluminium alloys [9]. Duan, et al. [10] tried to purify aluminium can by using NaCl-KCl based flux with minor addition of MgCl2, CaCl2, Na3AlF6, Na2SiF6, NaF and CaF2 with different ratio in 3 different formula. And also they used also different temperature (at 720, 740 and 760 °C) to optimize the process. They declared that appropriate melting temperature range is 720–740 °C and flux addition is 4%. And also they reported that if the composition of flux consist of excess CaF2, it can be obtained inclusion removal efficiency as 88,6%. Liu, et al. [11] used the flux which consist of 40wt% NaCl, 30wt% KCl, 10wt% NaF and 20wt % Na_3AlF_6 to remove inclusion from melt. They applied 3 cycle of purifying during 6 min. Then the amount of inclusions was decreased from 2.1% to 0.35% (with a decreasing rate of 83.3%). And also, hydrogen concentration dropped was decreased from 0.37 ml/100gr Al to 0.12 ml/100gr Al (with a decreasing rate of 68%). Mashhadi, et al. [12] pressed aluminium alloy chips by using cold pressing method at different pressures, then melted different melt medias at 750 °C. They were confirmed no big difference within melts that were melted in KCl, NaCl and KF flux mixture and without flux. Also, Yuksel et al. [13] studied on mechanical properties, i.e. toughness and strength, among three different charge of aluminium alloy. 100% secondary alloy exhibits large differences on mechanical properties compared to 100% primary and %50 primary +50% secondary alloy melts.

Experimental Work

In present study both primary A356 alloy and shavings from machining were used as melting material. 10 kg of charge material which consist of equal amount of primary and shavings was put into SiC crucible. The chemical composition of both swarf and ingot is given in Table 1. When the temperature of melt reached at 750 °C, degassing operation was applied for 10 min with Ar gas. Fluxes were added after 5 min. The list of fluxes and wt% that they are added are given in Table 2.

In order to quantify the cleanliness of the melt reduced pressure test (RPT) was used as 100 gr of sample was solidified at 100 mbar. The cross section of the samples was subjected to image analysis and BI was measured. At least three RPT

Table 1 Chemical compositions of ingot and swarf

Si	Mg	Fe	Ti	B	Ni	Sr	Al	
7,31	0,32	0,07	0,11	0,00	0,01	0,02	92,16	**Shavings**
7,28	0,227	0,07	0,11	0,003	0,004	0,013	92,29	**Ingot**

Table 2 Flux and ratios used in this study

Flux	wt%
Na_3AlF_6 (Commercial Cryolite)	10, 20, 30, 40
Na_2SiF_6(NaCl)	10, 15, 20, 25, 30, 35, 40
NaF/AlF_3 (CM) = 1,85	10, 15, 20, 25, 30, 35, 40
NaF	10, 20, 30, 40
NaF/AlF_3 (CM) = 2,7	10, 20, 30, 40
NaF/AlF_3 (CM) = 3	10, 20, 30, 40
$Na_2SiF_6(MgCl_2)$	10, 20, 30, 40

samples were collected before and after degassing operation. The change in BI was calculated as a percentage to indicate efficiency of cleanliness.

Results

Seven different fluxes were used in this study as given in Table 2. For each flux, different weight ratios according to the weight of charge material were investigated. The change in the BI was quantified and the results are given in Figs. 1, 2, 3, 4, 5, 6, 7 and 8.

Commercial cryolite is the most common fluxing agent in aluminium industry to remove inclusions from the melt [7]. Therefore, cryolite was used for the first set of tests. The change in BI can be seen in Fig. 1. The negative change in BI indicates that the melt had become worse. On the other hand, positive values show increase in the melt cleanliness as a percentage. In this case, up to 25 wt% flux addition has almost no effect on the cleaning process. At 25%, there is a 75% cleaning

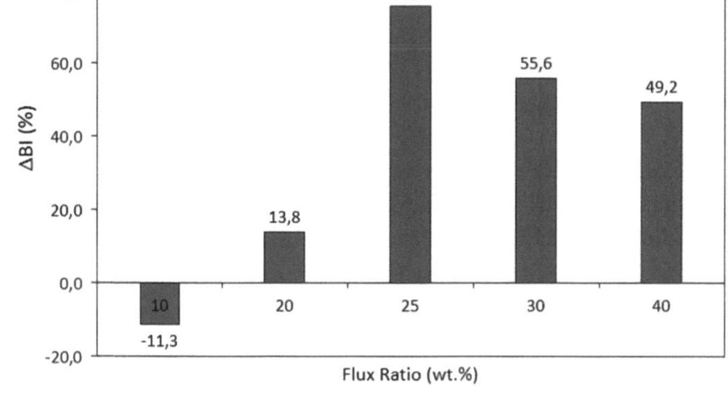

Fig. 1 Variation of BI versus various wt% cryolite

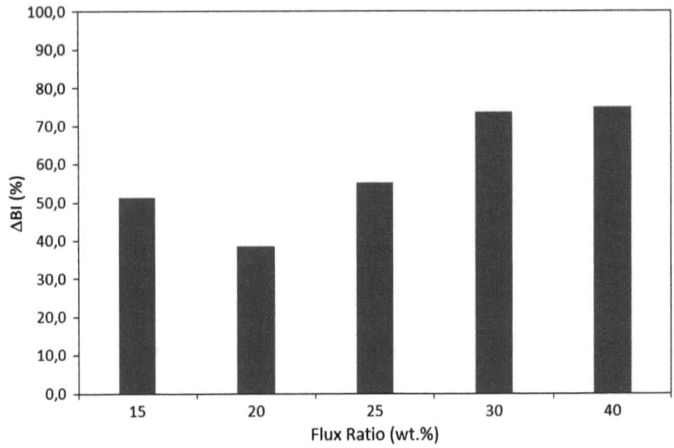

Fig. 2 Variation of BI versus various wt% Na2SiF6 (nacl)

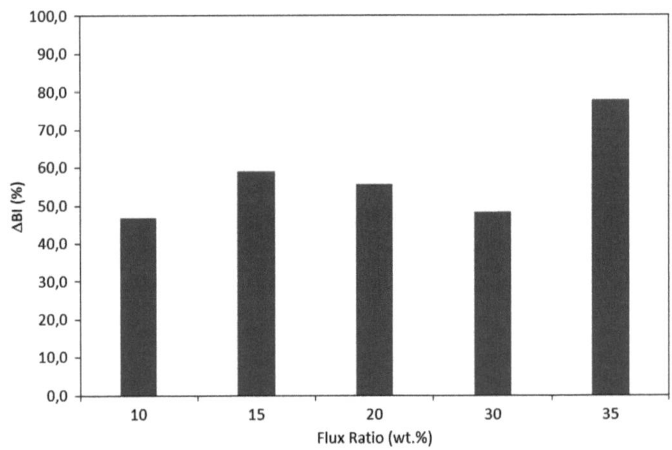

Fig. 3 Variation of BI versus various wt% NaF/AlF3 1.85

efficiency. At higher flux additions the cleanliness decreases but remains at approximately 50%.

When Na2SiF6 (NaCl based) was used, the change in BI before and after degassing remains around 50% regardless of the change in wt% flux, as seen in Fig. 2. According to the Fig. 2, it can be seen that there is an increasing trend in melt cleanliness as flux content is increased.

When NaF/AlF3 = 1,85 was used, there appears to be a stable cleanliness effect, where BI change is 50%. The results show that (Fig. 3) there is no effect of changing the flux content. Thus, the capacity of removing inclusions of this flux seems to be limited to 50% cleanliness.

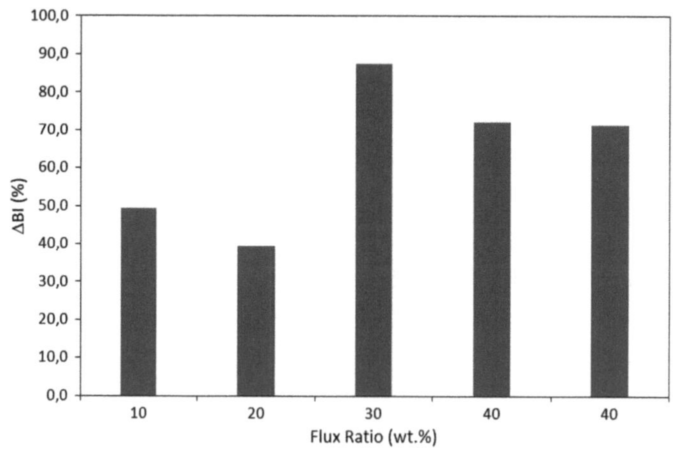

Fig. 4 Variation of BI versus various wt% NaF

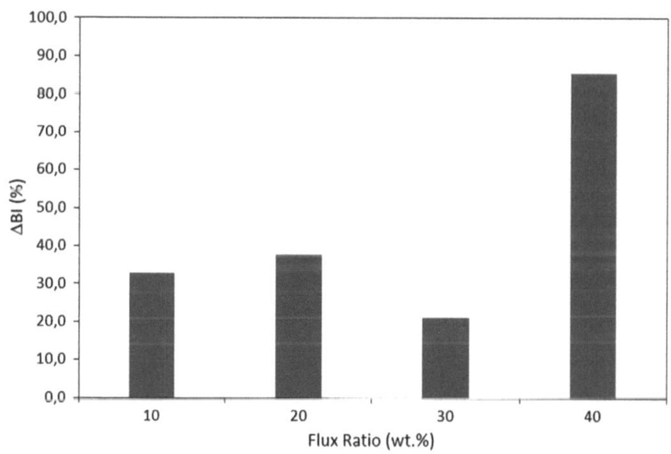

Fig. 5 Variation of BI versus various wt% NaF/AlF3 = 2.8

Amongst the fluxes studied in this work, NaF gives the highest BI changes. As seen in Fig. 4 minimum efficiency is 45% for low wt% addition. As quantity of NaF is increased, efficiency is increases to 70%. This shows that even at low wt% NaF usage at least 50% of the melt can be cleaned.

Figure 5 shows the effect of NaF/AlF3 = 2,8 addition on the cleaning efficiency. It can be seen that this flux has very low effect on the melt cleanliness. Even with the increase of quantity of the flux efficiency is less than 30%. Surprisingly, at 40% addition of the flux 90% of the melt was cleaned. However, this value corresponds to low initial BI. Thus, this shows that this flux can only be used when the starting metal quality is quite high.

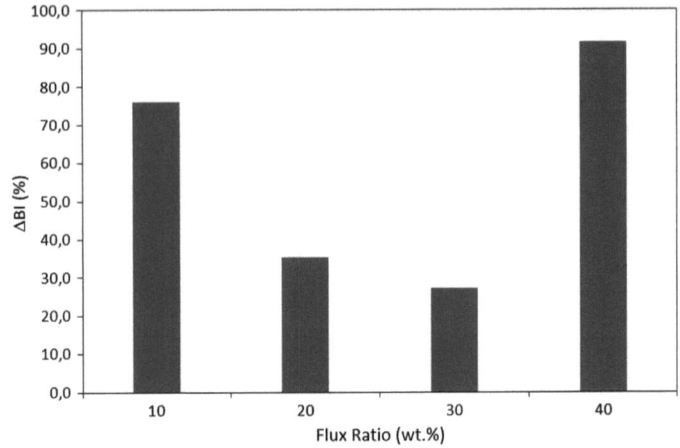

Fig. 6 Variation of BI versus various wt% NaF/AlF3 = 3

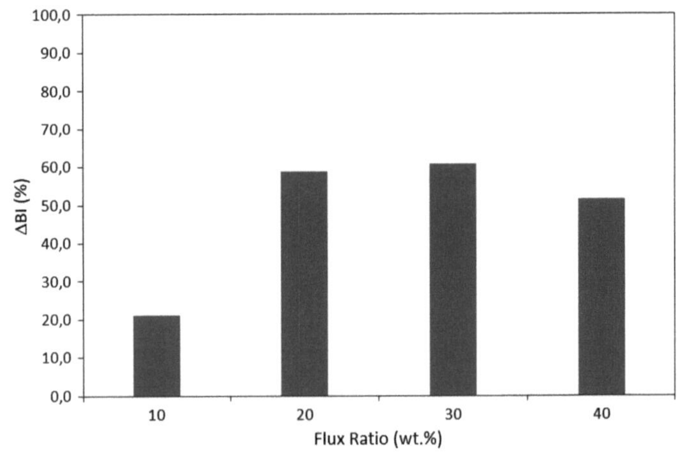

Fig. 7 Variation of BI versus various wt% Na2SiF6 (MgCl2)

The studies with NaF/AlF3 = 3 has quite interesting results (Fig. 6). At low wt% flux addition, there is a high cleaning efficiency around 80%. This value decreases with increased flux content to 30%. However, at 40% flux the efficiency increases to 90%, which corresponds to almost perfect cleaning.

When MgCl2 based flux is used instead of NaCl, the cleaning efficiency remains constant around 60% for flux ratios higher than 20%. This is 10% less than NaCl based flux. However, the stability of MgCl2 fluxes is higher NaCl.

Figure 8 summarizes the effect of use of different fluxes on the cleaning efficiency of A356. Figure 8a shows the average values, whereas Fig. 8b shows the best scenario. When the best situation is considered it can be seen that depending on

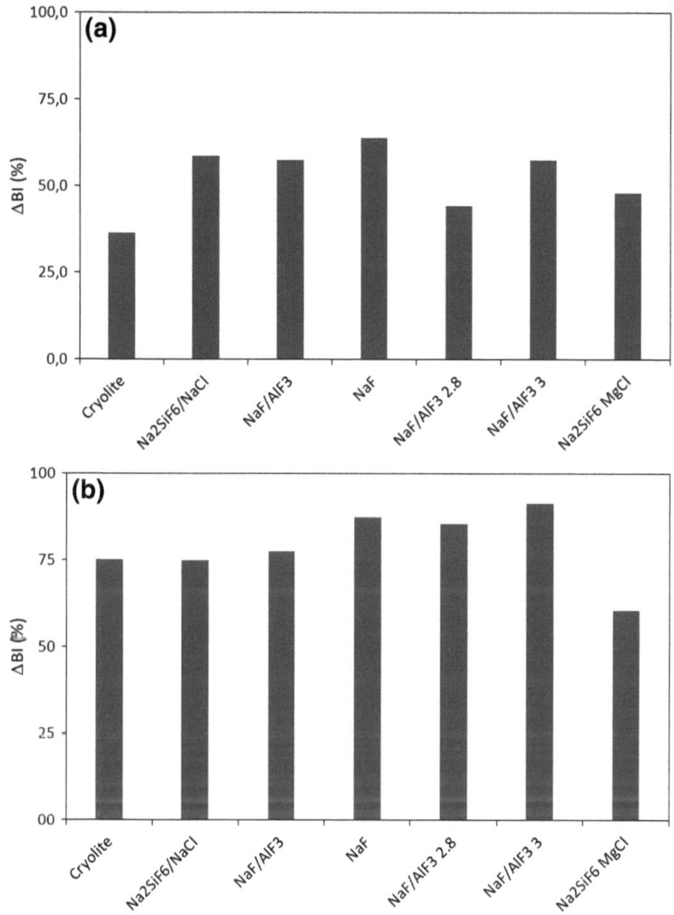

Fig. 8 Variation of BI for different fluxes. **a** Average values, **b** Best scenario

the melt quality level 75% of bifilms can be removed from the melt by the use of fluxes. Considering the average cleaning efficiency which actually depends on various levels of bifilm quantity, only 50% of the inclusions can be cleaned from the melt.

Conclusions

In this work different novel fluxes were prepared from NaCl–KCl and MgCl2–KCl based fluxes with mixtures of commercial cryolite, Na2SiF6, NaF/AlF3 (1.8, 2.8 and 3) and NaF. The addition of the flux to the melt was purging through the rotary impeller at different weight percentages of the melt; as 10, 20, 30 and 40. The cleaning efficiency was measured by the change in the BI before and after degassing. It was found that commercial cryolite had the lowest cleaning efficiency of 30%. On the other hand, the best cleaning efficiency was obtained by NaF with a value of about 65%. The average efficiency values of other fluxes were found to be around 50%. From the experiments with various parameters, it was concluded that the use of fluxes to clean melt which consist of 50% A356 and 50% swarf, could only achieve 50% of cleaning rate. It is necessary to note that bifilm content of the initial melt plays an important role for the degassing processes. The higher the bifilm content much easier and faster the cleaning process would be.

References

1. Dispinar D, Campbell J (2006) Use of bifilm index as an assessment of liquid metal quality. Int J Cast Metal Res 19(1):5–17
2. Dispinar D, Campbell J (2014) Reduced pressure test (RPT) For Bifilm assessment. In: Shape casting: 5th international symposium 2014, pp 243–251. Springer, Cham
3. Dispinar D, Campbell J (2004) Metal quality studies in secondary remelting of aluminium. Foundry Trade J 178(3612):78–81
4. Dispinar D, Campbell J (2007) A comparison of methods used to assess aluminium melt quality. In: Shape casting: 2nd international symposium, The Minerals, Metals & Materials Society, pp. 11–18
5. Dispinar D (2006) Determination of metal quality of aluminium and its alloys (Doctoral dissertation, University of Birmingham)
6. Uludağ M, Çetin R, Dişpinar D, Tiryakioğlu M (2018) On the interpretation of melt quality assessment of A356 Aluminum alloy by the reduced pressure test: the Bifilm index and its physical meaning. Int J Metalcast, 1–8
7. Yüksel Ç (2016) Improving Liquid metal quality of Aluminum and its alloys. Yıldız Technical University, Turkiye
8. Caceres CH, Selling BI (1996) Casting defects and the tensile properties of an Al-Si-Mg alloy. Mater Sci Eng, A 220(1–2):109–116
9. Liu PY, Jia J, Guo JJ (2000) Melt treatment of A357 alloy by flux injection. Mater Sci Forum 331:283–288. Trans Tech Publications
10. Duan RB, Bai PK, Yang J, Zhang WD, Ding H (2013) Study on Waste Aluminum cans remelting and purification technology. Appl Mech Mater 372:360–363. Trans Tech Publications
11. Liu C, Hu ZL, Zeng JM (2012) Removal of impurities in Aluminum by uses of fluxes. Adv Mater Res 509:152–155. Trans Tech Publications
12. Mashhadi HA, Moloodi A, Golestanipour M, Karimi EZV (2009) Recycling of aluminium alloy turning scrap via cold pressing and melting with salt flux. J Mater Process Technol 209 (7):3138–3142
13. Yuksel C, Tamer O, Erzi E, Aybarc U, Cubuklusu E, Topcuoglu O, Cigdem M, Dispinar D (2016) Quality evaluation of remelted A356 scraps. Arch Foundry Eng 16(3):151–156

Part IV
Properties of Castings

Characterisation of Lead Sheet Manufactured Using Traditional Sand-Casting Technique

Arun Prabhakar, Konstantinos Salonitis and Mark Jolly

Abstract Lead sheet is widely used for roofing and flashing applications. Continuous casting and rolling are the most common processes used for manufacture of such sheet. Sand casting of lead sheet is a traditional process used since the Romans. Sand cast lead sheet is presently used for restoration of heritage buildings, churches and cathedrals. It possesses a distinct texture due to the uniqueness of the manufacturing process and is considered a premium product. In the present study, sand cast lead sheet samples were collected and characterised. The observations are then compared with characteristics of continuously cast lead sheet obtained from the literature and are reported in this paper.

Keywords Sand casting · Lead sheet · Casting · Microstructure

Introduction

Lead has been used since 5000 BC along with metals like gold, silver, copper, iron and tin. The Chinese used lead to make coins around 2000 BC. The Romans made water pipes using lead sheet by rolling them and welding them together at the joint using molten lead [1]. Lead is ideal for roofing applications and architectural cladding as it is highly malleable and resistant to corrosion. Around 85% of the demand for lead sheet is for these kinds of applications [2]. Lead has a high density (11340 kgm^{-3}) and possesses a high attenuation coefficient (5.549 cm^2g^{-1} at 100 keV), especially for high-energy X-rays which makes it an ideal material for

A. Prabhakar · K. Salonitis · M. Jolly (✉)
Manufacturing Department, School of Aerospace Transport and Manufacturing, Sustainable Manufacturing Systems Centre, Cranfield University, Bedford, UK
e-mail: m.r.jolly@cranfield.ac.uk

A. Prabhakar
e-mail: a.prabhakar@cranfield.ac.uk

K. Salonitis
e-mail: k.salonitis@cranfield.ac.uk

M. Tiryakioğlu et al. (eds.), *Shape Casting*, The Minerals, Metals & Materials Series,
https://doi.org/10.1007/978-3-030-06034-3_27

radiation protection [3, 4]. Moreover lead sheet is durable, long lasting, needs low maintenance and is fully recyclable at the end of its life [5].

Lead sheet used for construction purposes is mostly produced from recycled lead by two techniques. The majority of sheet is produced by rolling (also called milling), which involves casting lead slabs and then rolling them in a mill until the desired thickness (code) is obtained. Another technique is a direct method (DM) in which sheet of a specific thickness is semi-continuous cast. In this method, a water-cooled drum is immersed in a bath of molten lead and a semi-continuous sheet is cast onto the surface of the drum. There is a third method of manufacture of lead sheet by the traditional sand casting method [6]. The main market for this sheet is the heritage industry and is generally for renovation of older churches, cathedrals, stately homes and castles. Sand casting is one of the earliest methods of manufacturing sheet lead and has been used by the Romans since 300 BC for making water pipes and coffins for example. Blaskett and Boxall [7] thus giving the product an historical importance. The sheet produced using the sand casting method looks very different from rolled and machine cast sheet. The upper side, or the side that is exposed to the atmosphere, has a shiny appearance with the presence of numerous white patches (Fig. 1a). On the other hand, the sand side (surface that is in contact with the sand bed) as shown in Fig. 1b retains the texture of the sand bed and has a dull appearance similar to the case of DM sheets where one of the surfaces retains the texture of the casting drum [8].

Usually the lead used in sand cast sheet contains a higher level of impurities whereas the lead used in DM sheets are 99.8–99.9% pure [9]. Use of pure lead (refined) in sand casting is observed to result in formation of interference colours on the top surface and the pattern (as in Fig. 1c) is not observed. The interference colours are due to oxidation is related to the thickness of the oxide film [10]. The pattern (as observed in Fig. 1a) is observed to form when the melt consists of a mix of refined lead and scrap lead.

Since unrefined scrap is mixed with pure lead, every cast has a different chemical composition.

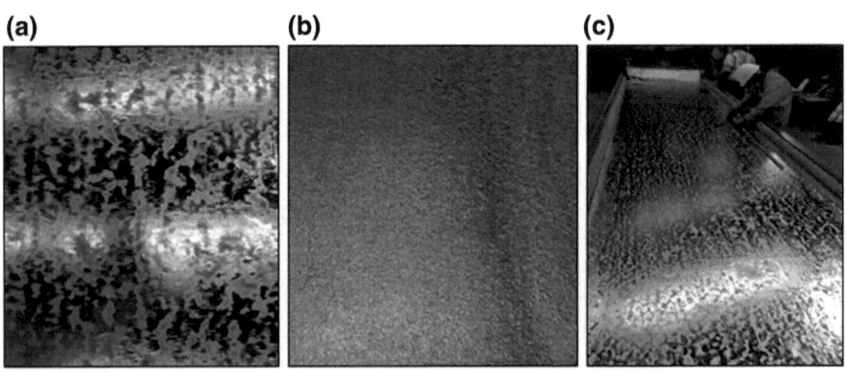

Fig. 1 **a** Top surface of sand cast lead sheet (90 cm × 50 cm); **b** Sand side of sand cast lead sheet (90 cm × 50 cm); **c** Sand side of sand cast lead sheet (90cm × 50cm)

Lead sheet used for roofing purposes should not only be visually attractive, but also should possess good mechanical properties. Understanding the microstructure is therefore very important. Previously published work on continuously cast lead sheet and milled sheet (Whillock [7]) has contained extensive research on the microstructure and mechanical properties of continuous cast lead sheet and rolled lead sheet. Whillock investigated relationships between mechanical properties such as tensile, creep and thermal fatigue behaviours, and microstructure for rolled and continuous cast lead sheet at different concentrations of copper [8]. However, not a lot of research has been conducted in understanding properties and metallography of sand cast lead sheet. Sand cast lead sheet contains more impurities when compared to DM or rolled sheet and Lead is highly sensitive to changes in properties with the presence of even trace amounts of alloying elements. Variations in grain size and hardness values of lead have been reported when only traces of other elements present [9]. For example, the addition of Sb is found to increase the tensile strength and hardness of lead progressively [7]. This paper investigates the microstructure of sand cast lead sheet with different impurity contents and compares it with DM sheet.

Materials and Methods

a. *Experimental design*

Lead sheet used for this study was manufactured using Sand Casting and direct method at a foundry based in Derbyshire, UK.

Direct Method is a continuous casting process in which a rotating drum which is water cooled, internally, dips into a pool of molten lead, picking the liquid metal and solidifies it into a sheet with the thickness depending on the depth of dip, speed of rotation, temperature of melt, etc. The process uses 99.98% pure refined lead [11].

The process of manufacturing sandcast lead involves pouring of molten lead at around 350C on a rectangular sand bed and subsequent flattening of the surface using a screed by smearing the top surface to wipe off the excess liquid metal. Temperature of pour, clearance between the screed and the sand bed, speed of motion of the screed and smoothness of the bed surface affects the quality of the cast sheet.

Three different ratios of pure lead to unrefined scrap were used to manufacture sand cast sheet (Table 1) on a sand bed that is 7m $\times$ 1.3m . A 6-cm thick layer of silica sand was used. 99.98% lead ingots and secondary lead scrap which consists

Table 1 Sample names, composition and manufacturing method

Sample name	Material mix	Method of manufacture
A1	Lead ingots	Sand casting
A2	50:50 Lead ingots: Scrap	Sand casting
A3	75:25 Lead ingot: Scrap	Sand casting
A4	Similar material available in the market	Sand casting
A5	Refined lead	Direct method

mainly of lead from batteries, building and construction industry were used for sand casting. Temperature of melt was fixed at 350 C for all casts. 99.98% pure DM sheet of 1.8-mm thickness (Code 4) manufactured using direct method was also used to compare hardness values in this study.

As described in Table 1, samples from A1 to A4 were manufactured using sand casting and A5 was manufactured using direct method. Material mix for A4 is unknown since it is sourced from the market and is used for comparison purposes only.

b. ***Metallographic preparation of samples***

Pb alloys are difficult to polish to a finish suitable for metallographic inspection due to its softness [12]. Grinding of samples smears the metal and deposits abrasive grit from the grinding paper onto the surface of the specimen. Care needs to be taken as any process that leads to heat generation can alter the microstructure as lead recrystallises even at room temperature [13]. Care must also be taken as lead is prone to oxidation upon contact with air. There are different methods available in the literature [8, 11, 12] for metallographic preparation of lead samples; however, most of the techniques proved to be practically difficult when tried to reproduce.

Samples were grinded using 600 and 1200 grit silicon Carbide (Sic) paper. Cutting oil was used to lubricate the grinding process. This also helped prevent lead from sticking on to the paper and forming a glaze that could drag and distort the surface of the sample. After 1200 grit paper, 6 microns and 1-micron diamond paste were used on a chemically resistant cloth to reduce mechanical damage and excessive smearing of the sample. Polishing was performed using 0.04-micron colloidal silica and soap. Finally, samples were chemically polished for 2–3 s with a mixture of 70 ml hydrogen peroxide and 30 ml oxalic acid and immediately rinsed with water. Etching was performed using a solution of 9 g ammonium molybdate, 15 g citric acid and 80 ml distilled water. Samples were cleaned in an ultrasound bath in between steps.

c. ***Characterisation***

The samples in the study were studied using a Nikon Optiphot image acquisition system equipped with Leica Application suite for optical imaging and a Phillips XL30ESEM environmental scanning electron microscope (ESEM) equipped with an energy dispersive spectroscope (EDX). The optical imaging was carried out under 10 $\times$ magnification. Elemental analysis of samples was conducted using Petromax DCM2752 mass spectrometer. Analysis was conducted at ten different points on each sample and the average was taken. To measure hardness values of samples, a Zwick Roel hardness tester was used.

Results

a. ***Elemental analysis***

As mentioned previously, a combination of pure and unrefined scrap lead is used to manufacture sand cast lead sheet in order to achieve the patterned surface. As a

Table 2 Mean and standard deviation (SD) values of common impurities obtained from elemental analysis of 80 different scrap melt batches

	Cu (%)	Zn (%)	Sn (%)	Sb (%)	Bi (%)	Ag (%)
Mean	0.0283	0.0029	0.086	0.0315	0.0179	0.0041
SD	0.0255	0.0101	0.0011	0.079	0.0030	0.0184

result, the impurity content on each cast sheet varies. An elemental analysis of the scrap used for the cast showed presence of mainly copper (Cu), zinc (Zn), tin (Sn), antimony (Sb), bismuth (Bi) and silver (Ag) whose concentrations vary with the scrap supplied. Table 2 shows the average percentage and standard deviation (SD) of each of these elements out of 80 different scrap batches.

Figure 2 shows elemental analysis and different impurity contents in each of the samples used for this study. Analysis was conducted at ten different points on each sample and the average taken. Presence of copper improves creep resistance properties of DM sheets [8]. Hence, DM sheet are manufactured with small amounts of copper as it is evident from Fig. 2e. The sample manufactured using 75% pure ingots and 25%

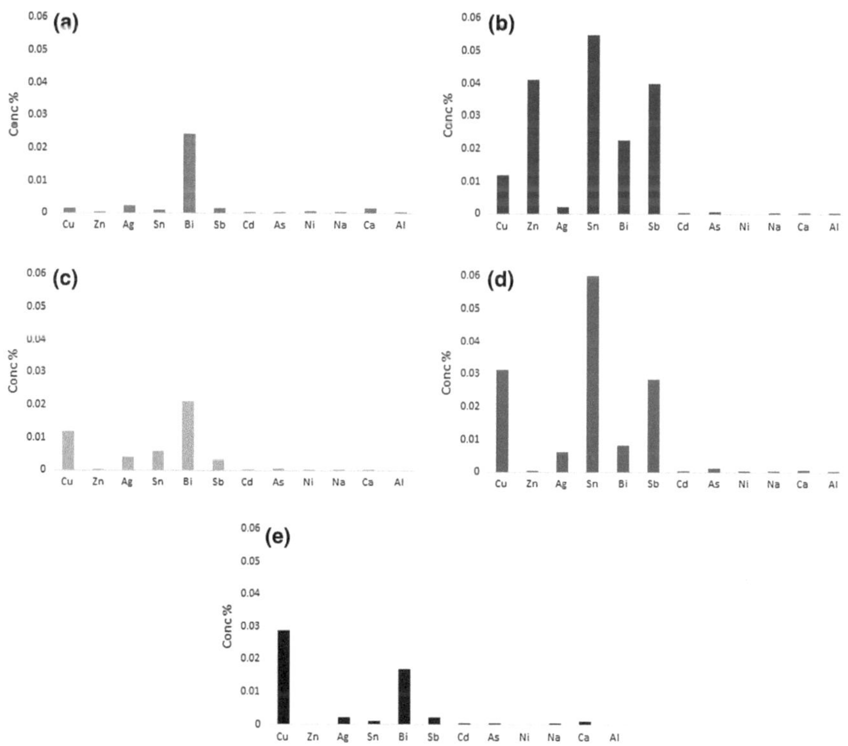

Fig. 2 Elemental composition of different samples: **a** Sample A1, **b** Sample A2, **c** Sample A3, **d** Sample A4, **e** Sample A5

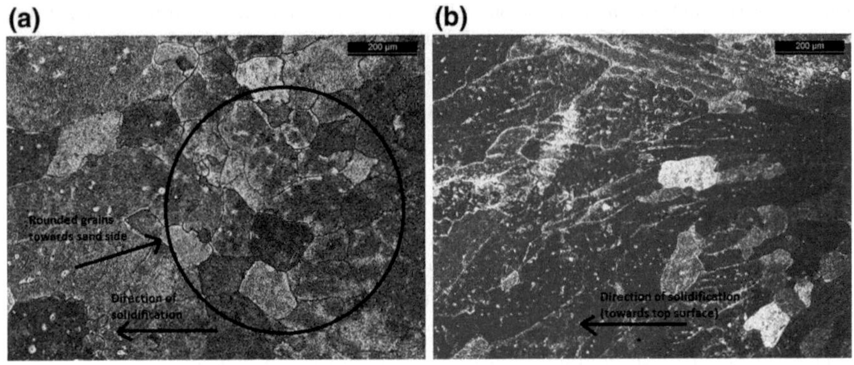

Fig. 3 Micrographs of a cross section of sand cast lead sheet (S2); **a** Grains at the edge corresponding to sand side **b** Long branched grains oriented towards the top side

scrap has almost similar impurity content as pure ingots. Yet, there are differences in the appearance of the final cast product. As mentioned before, use of pure ingots is observed to result in the formation of a colourful patina on the top surface.

b. ***Microstructure***

Sand cast lead sheet samples were grinded, polished, etched and observed under a microscope. The grains were observed to be small and circular in shape near to the sand side of the sheet (Fig. 3a) which branched into large long grains oriented towards the top surface of the sheet (Fig. 3b). This indicates the presence of temperature gradient during solidification and is partly in line with the observations of Rocca et al. who investigated ancient gallon-roman lead sarcophagus which potentially could have been manufactured via sand casting technique [6]. Some areas near the top surface of the sheet were observed to have small circular grains.

Sample A4 also possessed a similar grain structure but the grains were observed to be larger in size (Fig. 4).

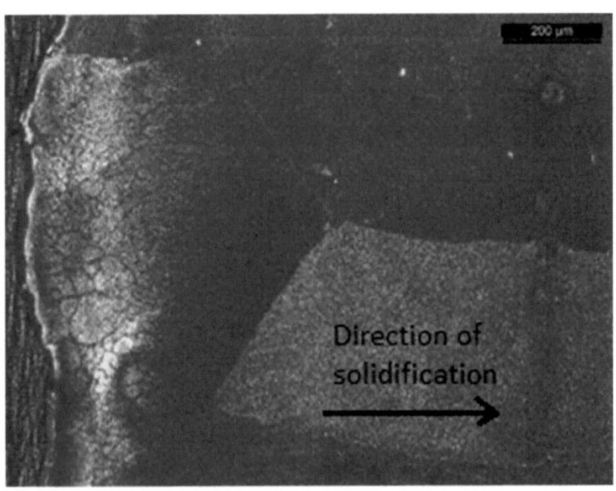

Fig. 4 Micrograph of a cross section of sample A4 showing long and large grains oriented towards the top side of the sheet

c. *SEM analysis*

Top surface of sand cast lead sheet was observed under SEM to understand the differences in morphology of the dull and shiny texture that appear during solidification. And elemental analysis in the area showed presence of mainly lead, carbon and oxygen in the region. As it can be observed from Figs. 5a and 6, the dull area has a branched rounded structure compared to the smooth surface of the shiny/grey area.

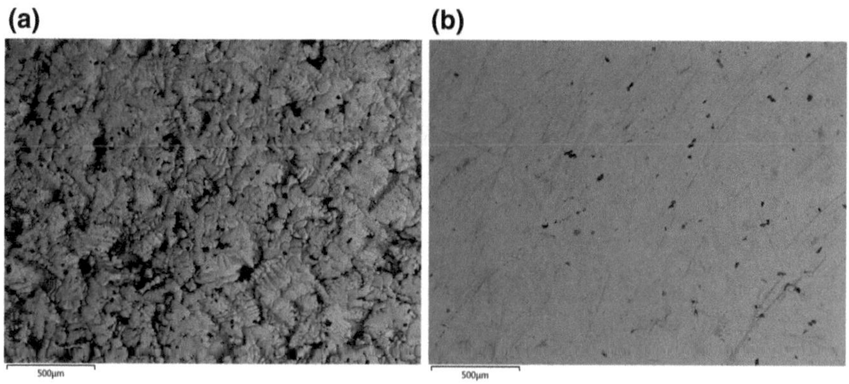

Fig. 5 SEM images of dull and shiny patches on top side of sand cast lead sheet; **a** Dull/Whitish areas **b** Shiny/Grey area

Fig. 6 Secondary electron image of the dull area showing dendrites with oxide concentrated at boundaries

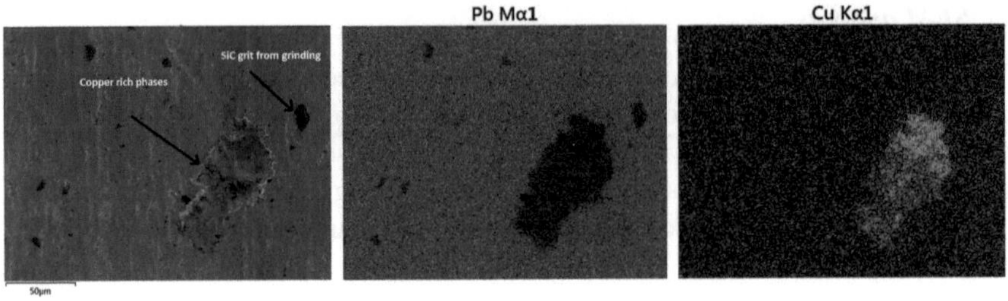

Fig. 7 SEM Images of S3 and elemental maps showing copper-rich areas

Table 3 Hardness values of samples

Sample number	Vickers hardness number	Percentage of lead
A1	4.9	99.97
A2	4.1	99.83
A3	4.7	99.95
A4	4.8	99.70
A5	5.1	99.95

SEM images of A4 revealed a patch like appearance on the surface. The sample consisted of mainly copper and tin as impurity. Elemental mapping was performed in the region and it showed presence of copper-rich areas as shown in Fig. 7.

d. ***Hardness***

The samples were cut and grinded using a 1200 grit Sic paper in water to prepare a flat surface for hardness testing. 25 g load was used for 15 s along the cross section of samples. Ten tests were conducted for each sample and the average value was taken. Table 3 shows the hardness values of different samples.

Discussion

Lead is one of the most recycled materials and as a result scrap consists of mostly lead from batteries, building and construction industries. This is often clean though refining may be required [14]. The melt used in sand cast lead sheet is often a mixture of pure (refined) lead and scrap lead to attain the pattern on the surface. The addition of scrap results in formation of the patterned surface which appears as white and dark (or dull and shiny) patches on the top surface (Fig. 1). From SEM observations, white (or dull) areas are observed to appear so, due to rounded and branched dendrites with oxides and carbonates concentrated in those areas, potentially due to an increased surface area (Fig. 7). The region on the top surface of the melt which is smeared by the screed is flat and solidification occurs due to contact with the screed which could potentially be the reason of occurrence of rounded grains in some areas on the top edge of the cross section as observed.

From literature, it was found that addition of even trace amounts of impurities to lead alters its properties [1]. Addition of 1.25% antimony increases its tensile strength from 17Mpa to 29 MPa [15]. However, such prominent differences were not observed in hardness values of the collected samples probably due to the effect of getting neutralised by other impurities.

Dilute alloys sometimes have unstable growth conditions [8]. The rough sand surface results in a number of nucleation sites for solidification [16]. Solute particles that have low solubility in lead are rejected to the regions ahead of the sand interface. A low concentration of solute closes to the sand surface results in formation of circular grains, potentially due to moderate undercooling. As the concentration of solute particles increases with distance from the sand interface (due to solute particles being pushed away from the sand interface), it is believed that dendritic solidification is forced to happen due to severe undercooling. This is in line with microstructure observations in the previous section. Susan Whillock has observed similar results during solidification studies of continuously cast lead sheet with varying amounts of copper as impurity [17].

The hardness values of all samples were almost similar to that of continuously cast sheet. There are slight variations in hardness values which could be due to the presence of impurities. Further study needs to be conducted to understand effect of these trace impurities on other properties like tensile and thermal fatigue.

Conclusions

The melt used in sand cast lead sheet is often a mixture of pure (refined) lead and scrap lead to attain the striped pattern. The addition of scrap lead results in formation of a patterned surface. Elemental analysis of different batches of scrap was conducted and was found to be relatively clean with trace amounts of impurities. The overall impurity content in melt batches used in manufacture of sand cast lead sheet is very low even after being mixed with scrap lead (to obtain the patterned surface) since scrap lead consists of mostly secondary lead. Hardness values of sandcast samples were found to be almost similar to that of continuously cast lead sheet and the presence of trace amounts of impurities in the samples were not observed to alter the values much. However, higher amounts of these impurities could result in drastic changes in properties of the cast sheet. Solidification starts at the sand-melt interface and proceeds to the top surface. Grains were observed to be rounded close to the sand surface which developed into large long grains in the direction of solidification. Further study needs to be conducted to understand the effect of these impurities on tensile and thermal fatigue properties of sand cast lead sheet as these are vital parameters for roofing sheet.

References

1. Hofmann W (1970) Lead and lead alloys. In: Lead and lead alloys. Springer, Berlin, Heidelberg, pp 25–320
2. Davidson AJ, Binks SP, Gediga J (2016) Lead industry life cycle studies: environmental impact and life cycle assessment of lead battery and architectural sheet production. Int J Life Cycle Assess 21(11):1624–1636
3. Haynes W (2015) CRC handbook of chemistry and physics. CRC Press, Boca Raton, FL
4. Hubbell JH, Seltzer SM Tables of X-Ray mass attenuation coefficients and mass energy-absorption substances of dosimetric interest
5. Construction|European Lead Sheet Industry Association. http://elsia.org.uk/uses-benefits/construction/. Accessed 03 Aug 2017
6. Rocca E, Mirambet F, Steinmetz J (2004) Study of ancient lead materials: a gallo-roman sarcophagus—contribution of the electrolytic treatment to its restoration. J Mater Sci 3 (4):2767–2774
7. Blaskett DR, Boxall D (1990) Lead and its alloys. Ellis Horwood
8. Whillock S (1987) Microstructure and mechanical properties of milled and continuously cast lead sheet. University of Cambridge
9. Scott DA (1996) A Note on the metallographic preparation of ancient lead. Stud Conserv 41 (1):60
10. Weyand TE (1970) The oxidation kinetics of liquid lead and lead alloys. Missouri University of Science and Technology
11. Prabhakar A, Mielnicka J, Jolly M, Salonitis K Improving energy efficiency in direct method for continuous casting of lead sheets
12. Ednie KH (1996) Metallographic preparation materials: lead alloys of soft. ElsezTier Sci Inc
13. Wong SF, Hodgson PD, Thomson PF (2013) Room temperature deformation and recrystallisation behaviour of lead and lead-tin alloys in torsion and plane strain compression
14. Thornton I, Rautiu R, Brush S (2001) Lead the facts. London
15. Bagherian E-R, Fan Y, Cooper M, Frame B, Abdolvand A (2016) Effect of antimony addition relative to microstructure and mechanical properties of continuous cast lead alloy. In: Met. 2016—25th Anniversary international conference metallurgy materials conference proceedings
16. GOFF LI, Semmens RD (1967) Lead products by continuous casting. Lead 65:295–306
17. Whillock S, Charles JA, Smith GC (1989) Microstructures and mechanical properties of milled and continuously cast lead sheet Part 1 Microstructures. Mater Sci Technol 5 (11):1074–1083

On the Intrinsic and Extrinsic Microstructure-Property Effects in Cast Aluminum Alloys

Murat Tiryakioğlu

Abstract The metallurgy of cast aluminum alloys has been thought to be different from that of wrought alloys. It has been widely accepted that mechanical properties are controlled by dendrite arm spacing, the presence of Fe-bearing particles, and the size of Si particles in Al–Si alloys, which fracture and debond during deformation, leading to premature failure. Whether these effects are intrinsic or mere correlations because of the structural quality of the castings is questioned in this study. By comparing with ideal properties, it is shown that the effect of dendrite arm spacing on elongation, precipitation, and subsequent fracture of β-Al_5FeSi platelets and finally Si particle fracture and debonding are all extrinsic.

Keywords Dendrite arm spacing · β-Al_5FeSi platelets · Si particle fracture · Debonding

Introduction

Metallurgy of aluminum alloys has been taught to many practicing engineers based on the type of final product: cast or wrought. Although the metal is the same, the behavior and performance of cast aluminum alloys have been regarded as different from, usually inferior to, those of their wrought siblings. The current thinking in the metallurgy of cast aluminum alloys can be summarized as follows:

- Pores are intrinsic in aluminum castings.
- Important microstructural parameters to control in order to control properties and performance are dendrite arm spacing, and in Al–Si alloys, particle size of Si and intermetallics, especially β-Al_5FeSi platelets.
- β-Al_5FeSi platelets are weak and brittle, causing premature fracture in castings.

M. Tiryakioğlu (✉)
University of North Florida, Jacksonville, FL 32256, USA
e-mail: m.tiryakioglu@unf.edu

M. Tiryakioğlu et al. (eds.), *Shape Casting*, The Minerals, Metals & Materials Series,
https://doi.org/10.1007/978-3-030-06034-3_28

- In Al–Si alloys, Si eutectic particles fracture and debond in early stages of plastic deformation, causing cracks to form, leading to premature fracture.

To the author's knowledge, the points listed above are thought to apply only to cast aluminum alloys. Moreover, many models to estimate mechanical properties from microstructure have been developed, by building the "physics" around the observed behavior. This phenomenological approach filled the literature with papers that can describe the behavior of cast aluminum alloys in some cases, but not in general. Whether the observed correlation between microstructure and property is based on causation has not been asked. This paper is motivated by the above-mentioned duality in metallurgy books, and the need to reevaluate observed correlations listed above on whether there is a cause and effect behind them.

Intrinsic and Extrinsic Effects

Pores

The first statement was first challenged by Campbell [1, 2] and recently by Yousefian and Tiryakioğlu [3]. It is well known that the necessary condition for the nucleation of a pore in a solidifying liquid, i.e., the critical radius above which a pore is stable, r^*, is found by:

$$r^* = \frac{-2\sigma}{\Delta P^*} \tag{1}$$

where σ is the surface tension of the liquid (N.m) and ΔP^* is the pressure differential and is a negative number. Note that while surface tension for liquid metals is known, either ΔP^* or r^* need to be estimated or alternatively assumed. A review of the literature by Yousefian and Tiryakioğlu showed that ΔP^* is generally assumed to be −1 atm. (−0.1 MPa). This is more than three orders of magnitude higher than the intrinsic strength of aluminum (−4 GPa) at its melting temperature, estimated from a combination of experimental and molecular dynamics data [4]. Therefore, liquid aluminum is expected to be intrinsically resistant to nucleation of pores homogeneously or heterogeneously, even with high levels of hydrogen dissolved in it [3].

Dendrite Arm Spacing

Ductility is known to increase with decreasing primary (λ_1) and secondary dendrite arm spacing, λ_2 [5]. The effect of dendrite arm spacing on elongation (e_F) follows a Hall–Petch-type equation [6–9]

$$e_F = e_0 + \frac{\Lambda}{\sqrt{\lambda_2}} \quad (2)$$

A similar correlation between λ_2 and tensile strength has been reported for various cast aluminum alloys. To normalize elongation for different levels of yield strength, quality index, Q_T [10–14] is used:

$$Q_T = \frac{e_F}{\beta_0 - \beta_1 \cdot \sigma_Y} \quad (3)$$

Combining Eqs. 2 and 3, we obtain:

$$Q_T = Q_0 + \frac{\Lambda_0}{\sqrt{\lambda_2}} \quad (4)$$

Data from ten studies using various cast Al–Si alloys have been analyzed to determine the correlation between secondary arm spacing and Q_T. The details of the datasets are provided in Table 1. The correlation between secondary dendrite arm spacing and Q_T for the ten datasets are presented in Fig. 1. Note the slope and intercept values are all different, which are also listed in Table 1.

Let us first focus on datasets 1 and 2, which come from the same study [15], and the main difference between them is the initial hydrogen content. When Q_T is plotted versus volumetric percentage of pores, f_{pore}, data from both high and low hydrogen groups follow the same trend, as presented in Fig. 2. It is noteworthy that the extrapolation of the curve to $f_{Pore} \approx 0$ gives an approximate y-intercept of 0.22. Therefore, the loss in Q_T of 0.78, estimated with almost no pores in the structure, is an indication of the inherent entrainment damage of the experimental (or production) system.

Recently, Tiryakioğlu et al. [20] showed that there are bifilms in A356 aluminum alloy castings that remain inactive, i.e., are not needed for the formation of pores,

Table 1 Details about the datasets used in Fig. 1 and estimated values for coefficients of Eq. 4

Dataset	Alloy	References	Q_0	Λ_0 ($\mu m^{0.5}$)	Comments
1	319	[15]	−0.0062	0.1717	High hydrogen content
2	319	[15]	−0.0445	0.7928	Low hydrogen content
3	A380	[16]	−0.0490	0.4122	High -pressure die cast
4	A380	[17]	−0.0496	0.4765	Permanent mold
5	A357	[18]	−0.2803	2.8440	No modification
6	A357	[18]	−0.1504	2.6931	Sr modified
7	A356	[19]	−0.3759	4.1723	Aluminum association dataset
8	A356	[18]	−0.3935	4.3961	No modification
9	A356	[18]	0.1626	2.4443	Sr modified
10	A356/7	[12]	1.0000	0.0000	Maximum points

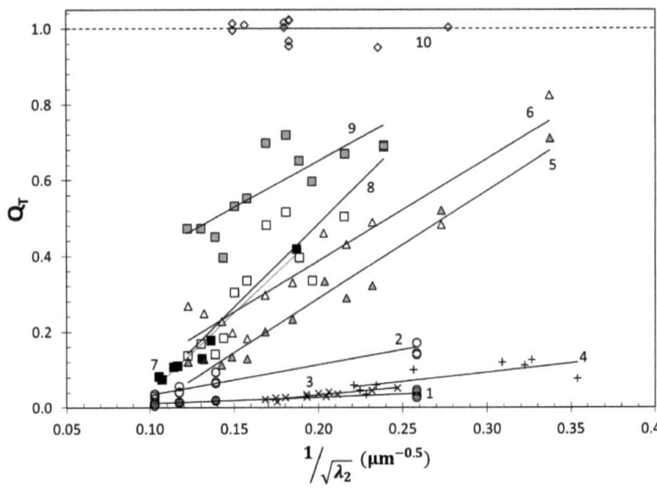

Fig. 1 Correlations between the inverse of square root of secondary dendrite arm spacing and Q_T

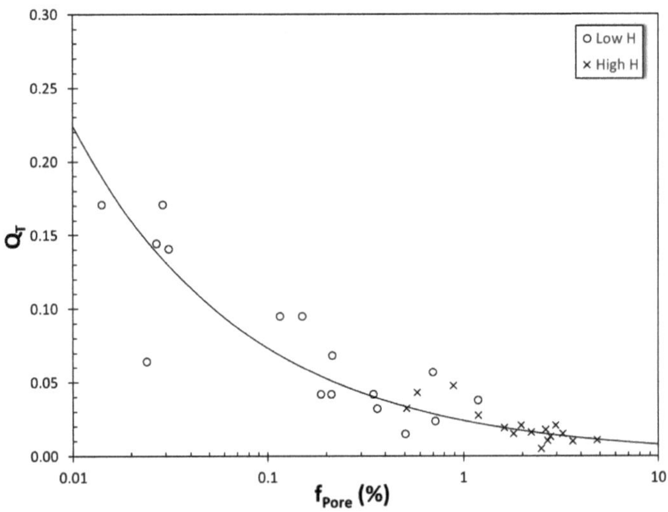

Fig. 2 Change in Q_T with volumetric pore percentage in 319 datasets (Data from Ref. [15])

even under reduced pressure. Therefore, there are usually more bifilms in the metal than needed to accommodate pores due to internal solidification shrinkage and/or hydrogen rejection. Consequently, a large number of bifilms remain inactive during solidification. These bifilms, however, would certainly contribute to the degradation of tensile properties, especially elongation. Evidence for these bifilms that remain inactive during solidification but open up during tensile testing can be seen in in situ deformation experiments in cast Al–Si alloys [21, 22]. With decreasing number of bifilms, there would be fewer bifilms that would remain inactive during solidification, but the tensile properties would be expected to improve at lower values of λ_2 but remain essentially the same at higher levels because of pores forming at long solidification times. Hence, the sensitivity of the metal to changing dendrite arm spacing would increase. This can be seen with maximum values in datasets 1–8 going up whereas minima remaining essentially constant, marked with a sharp increase in the slope of the correlations in Fig. 1. It is also remarkable that the datasets 1–8 estimate a $Q_T \approx 0.04$ at $\lambda_2 = 100$ μm ($1/\sqrt{\lambda_2} = 0.10$). The effect of further reduction in bifilms in the metal past this point leads to an increase in the elongation at high values of λ_2, leading to a decrease in the slope of the correlation, as evidenced by dataset 9 in Fig. 1. Further reductions in entrainment damage would lead to even more reduced slope and eventually a slope of 0, as seen in dataset 10, which are the maximum points on the yield strength elongation, i.e., ductility potential of cast Al–Si–Mg alloys [12]. Hence, the strength of the correlation between elongation (and therefore structural quality) and dendrite arm spacing is extrinsic, as it is determined by the bifilm content of the metal. As shown in Fig. 2, volumetric percentage of pores increases with λ_2 and hydrogen content. Zhang et al. [23] found that the length of the largest pores initiating fatigue fracture, $L_{Pore(max)}$, increased with λ_2 in A356 alloy castings, as presented in Fig. 3a. Moreover, the length of the β-Al_5FeSi platelets, generally thought to be weak and brittle, was also found [24, 25] to increase with λ_2, as shown in Fig. 3b. Because λ_2 changes with the 1/3 power of local solidification time [26], the degradation of elongation with increasing λ_2 can be attributed to growth of pores and β-platelets with time. We have already discussed the extrinsic nature of pores. Now, we will address the nucleation and growth of β-platelets, and whether they are intrinsically weak and brittle, as suggested repeatedly in the literature.

β-Al₅FeSi Platelets

One of the curious features of the microstructure of cast aluminum alloys is the presence of β-platelets. These platelets have the highest surface-to-volume ratio

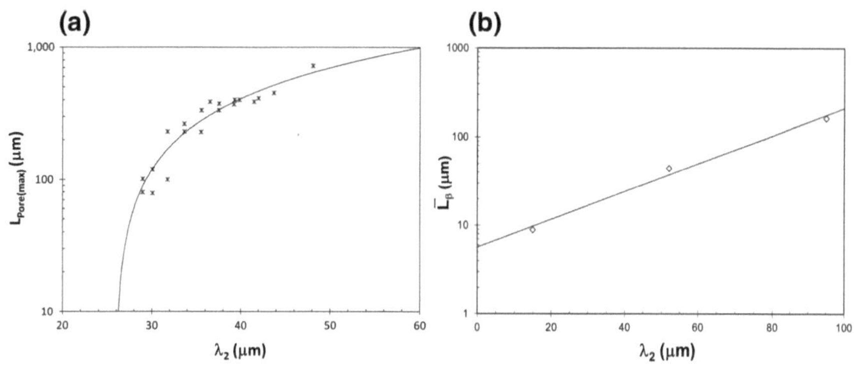

Fig. 3 Correlation between λ_2 and **a** maxi pore length and **b** length of β-Al_5FeSi platelets

among all microstructural features, making them almost a metallurgical mystery on how they overcome the surface energy barrier for nucleation and growth. Although iron is highly soluble in liquid aluminum, it has an extremely low equilibrium solubility, 0.03 at.% (0.06 wt%) in solid Al. As a result, the iron in solution in an aluminum melt eventually precipitates as coarse Fe-rich intermetallic phases, such as Al_6Fe, which precipitates at approximately 300 °C [27], as compared to β-Al_5FeSi platelets that nucleate above solidus in hypoeutectic Al–Si alloys. Under nonequilibrium conditions, up to 2 at.% (4 wt%) Fe was found to be soluble in aluminum [28], although no Fe-bearing constituents were found with a Fe content in solution up to 10 at.% [29]. Therefore, the precipitation of β-Al_5FeSi as platelets with enormous surface areas at temperatures above solidus strongly implies heterogeneous nucleation [30]. In situ observations of the precipitation of β-platelets in an A356 alloy [31] showed that they nucleate parallel to secondary dendrite arms, on the oxide skin of the metal on the surface or within pores, or on existing β-platelets. All three nucleation sites can be argued to be due to bifilms within the liquid metal. It was shown clearly in several studies [32–36] that β-platelets do indeed nucleate and grow on bifilms. Liu et al. [32] verified the presence of oxygen within β-platelets in A206 alloy castings. Hence, β-platelets, just like pores, act to make preexisting damage to the liquid metal visible and therefore are extrinsic. Because there is a bifilm inside or on one side of them, β-platelets are observed to debond and/or fracture early in plastic deformation. Fragments of bifilms would be expected to be present inside these platelets, causing the observed damage. Laz and Hillberry [37] even suggested that the size distribution of β-platelets is taken as the initial defect size distribution in fatigue studies of aluminum alloys.

The intrinsic properties of Fe-bearing constituents paint a picture that is very different from "weak and brittle." Li et al. [38, 39] showed in their ab initio study that the ideal strength of FeAl is between 13.9 and 18.6 GPa, accompanied by a 14% fracture strain in tension. Although there is no study in the literature on the Al_5FeSi phase, its ideal strength would be expected to be similar to FeAl. Seifeddine et al. [40] reported the modulus of elasticity (E) of β-Al_5FeSi as 196 GPa. Ideal tensile strength is approximately E/10 [41], and therefore, the results of Li et al. can be used as estimates for β-Al_5FeSi. Hence, β-platelets are strong and ductile; they are not weak, just extrinsically weakened.

Si Particle Debonding and Fracture in Al–Si Alloys

In cast hypoeutectic Al–Si–Mg alloys, fracture during tensile testing was observed [18] to start in the eutectic region with the appearance of microcracks. These microcracks have been attributed to the fracture and/or debonding of Si eutectic particles, which lead to the formation of voids and eventually to microcracks. The damage to the Si particles, i.e., debonding and fracture, starts at 1–2% plastic strain [42, 43] and increases linearly with plastic strain [18, 44]. Usually, a maximum of

10% of all Si particles is damaged in tensile or compression testing [45]. These tend to be the largest particles [46] with the highest aspect ratios.

Stresses developed in the Si eutectic particles during deformation have been measured in situ in several studies. Finlayson et al. [43] reported that Si particles in Sr-modified A356 alloy castings fracture at stresses between 200 and 300 MPa at 1% strain. Finlayson et al. [43] interpreted these values as the lower bound of Si particles; fracture stress of the "weakest particles." A fracture stress of 600 MPa was reported by Harris et al. [47] for a Si particle in an A319 alloy. Joseph et al. [48] investigated stress development in several Si particles in a cast Al–Si–Cu–Mg alloy with four different microstructures using the Raman technique. They found the fracture stress of the Si particles to be 500–1000 MPa, with fracture strains exceeding 10%. It is noteworthy that some of the particles tracked did not fracture at all even at 30% strain. Because Si is known to nucleate on bifilms [49], large Si particles with high aspect ratios can be expected to have nucleated heterogeneously on bifilms during solidification before necessary undercooling is reached for homogeneous nucleation of the Si eutectic. This explains why large particles with high aspect ratios fracture where others do not, as well as the discrepancies in fracture stresses.

Several ab initio molecular dynamics studies have been conducted to estimate the ideal strength of Si. Umeno et al. [50] found the ideal strength of Si to be 16 GPa and the fracture strain to be 30% in tension. Dubois et al. [51] determined the ideal tensile strength to be 16 GPa and fracture strain as 25% in the < 110 > direction. The results obtained in these two studies are remarkably similar. Moreover, they are much higher than the fracture stresses reported in the experimental literature.

Mueller et al. [52, 53] conducted in situ micromechanical tests on the Si particles in a cast Al–12.6 wt%Si alloy. They determined the stresses at which Si particles fracture to be as high as 13 GPa, which is in close agreement with the molecular dynamics studies. However, they also determined that Si particles were weakened by pinhole defects and interfaces visible on their surfaces. Mueller et al. [54] found Ti- and Fe-rich intermetallics at the bottom of the pinhole defects, embedded in the Si particles. Therefore, Si particles are also extrinsically weakened by intermetallics.

To evaluate the debonding of Si particles, the Si–Al interface has been investigated in several studies. Xia et al. [55] determined by nanoindentations and finite element modeling that the shear strength of the Al–Si interface is 240 MPa. Ward et al. [56, 57] found the interface strength in tension to be between 4 and 5 GPa via molecular dynamics simulations. Noreyan et al. [58] determined that the Si(111)/Al (111) interface had a tensile strength of 7.2 GPa but a shear strength 300 MPa, a value close to the one measured by Xia et al. In other orientations, shear strengths up to 1.2 GPa were obtained. Therefore, debonding of Si particles may be intrinsic, only when the weakest orientations occur. Because the rate of occurrence for the Si (111)/Al(111) alignment will be rare, it can be concluded that most of Si debonding events are also extrinsic.

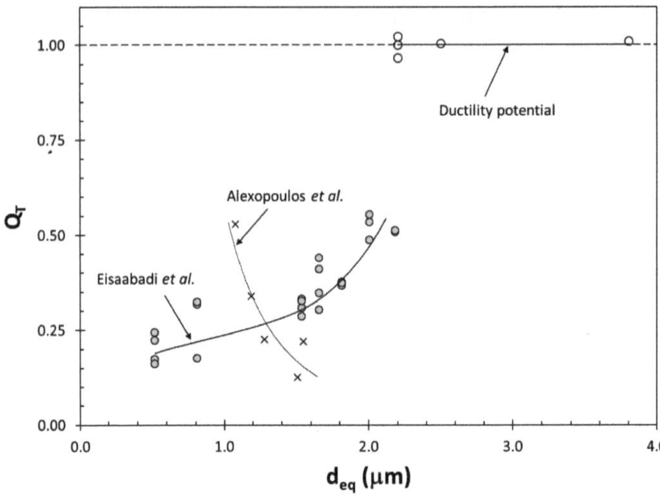

Fig. 4 Correlations between Si particle size and ductility

Turning our attention back to the effect of size of Si particles on the ductility of cast Al alloys, results from several studies are compared. Alexopoulos et al. [59] investigated the effect of alloying additions on microstructure and tensile properties in A357 and by Eisaabadi et al. [60] studied the evolution of Si particle sizes in A383 with solution treatment time and its effect on tensile properties. Data from these two studies are presented in Fig. 4, along with Si particle data from maximum ductility points for Al–7%Si–Mg alloys [12]. The reverse trends in particle size vs. quality index relationship are a clear indication that the effect of Si particle size on ductility is extrinsic, as also evidenced by the data from the ductility potential of cast Al–7%Si alloys.

Conclusions

Through analyses in this study and comparison with ideal properties, it has been determined that:

- The effect of secondary dendrite arm spacing on elongation is extrinsic. The strength of the effect is dictated by the bifilm content, i.e., structural quality, of the metal. Secondary dendrite arm spacing has no effect on ductility at very low and very high structural quality.
- The formation and subsequent fracture of β-Al_5FeSi platelets are extrinsic. Without bifilms, Fe would be retained in solution and precipitate as Al_6Fe in solid aluminum matrix. The β-Al_5FeSi phase is intrinsically strong and ductile, but extrinsically weakened by bifilms.

- Si particles are not expected to fracture during tensile deformation unless they have precipitated on bifilms or are weakened by intermetallics. Therefore, damage to Si particles during deformation is extrinsic.
- Intrinsic debonding of Si from aluminum matrix is possible only in very limited cases. The abundance of debonding observed in the literature suggests extrinsic factors weaken the Al–Si interface.

References

1. Campbell J (1967) PhD Thesis, University of Birmingham, Birmingham, UK
2. Campbell J (1968) Proceedings of the conference on the Solidication of Metals, Brighton, UK, Institute of Metals Publication, pp 18–26
3. Yousefian P, Tiryakioğlu M (2018) Metall Mater Trans A 49:563–575
4. Tiryakioglu M Metall Mater Trans A (in print)
5. Flemings MC (1974) McGraw-Hill, New York, pp 341
6. Reyes RV, Kakitani R, Costa TA, Spinelli JE, Cheung N, Garcia A (2016) Philos Mag Lett 96:228–237
7. Reyes RV, Bello TS, Kakitani R, Costa TA, Garcia A, Cheung N, Spinelli JE (2017) Mater Sci Eng A 685:235–243
8. Duarte RN, Faria JD, Brito C, Veríssimo NC, Cheung N, Garcia A (2016) Int J Mod Phys B 30:1550261
9. Canté MV, Spinelli JE, Cheung N, Garcia A (2010) Met Mater Int 16:39–49
10. Tiryakioğlu M, Campbell J (2009) Mater Sci Technol 25:784–789
11. Tiryakioğlu M, Campbell J, Alexopoulos ND (2009) Metall Mater Trans B 40:802–811
12. Tiryakioğlu M, Campbell J, Alexopoulos ND (2009) Metall Mater Trans A 40:1000–1007
13. Tiryakioğlu M, Campbell J, Alexopoulos ND (2009) Mater Sci Eng A 506:23–26
14. Tiryakioğlu M, Campbell J (2014) Int J Metalcast 8:39–42
15. Samuel A, Samuel F (1995) J Mater Sci 30:4823–4833
16. Grosselle F, Timelli G, Bonollo F, Molina R (2009) Metall Sci Technol 27:2–10
17. Grosselle F, Timelli G, Bonollo F, Tiziani A, Della E (2009) Corte, Metallurgia Italiana 101 25–32
18. Wang Q (2003) Metall Mater Trans A 34:2887–2899
19. Miguelucci E (1985) AFS Trans 93:913–916
20. Tiryakioğlu M, Yousefian P, Eason PD (2018) Metall Mater Trans A 49:5815–5822
21. Doglione R (2012) JOM 64:51–57
22. Ghassemali E, Riestra M, Bogdanoff T, Kumar BS, Seifeddine S (2017) Proc Eng 207:19–24
23. Zhang Chen (2000) Poirier. Fatigue Fract Eng Mater Struct 23:417–423
24. Vorren O, Evensen J, Pedersen T (1984) AFS Trans 92:459–466
25. Samuel A, Samuel F, Doty H (1996) J Mater Sci 31:5529–5539
26. Kirkwood D (1985) Mater Sci Eng 73:L1–L4
27. Saller BD, Hu T, Ma K, Kurmanaeva L, Lavernia EJ, Schoenung JM (2015) J Mater Sci 50:4683–4697
28. Saller BD, Sha G, Yang LM, Liu F, Ringer SP, Schoenung JM (2017) Philos Mag Lett 97:118–124
29. Nayak SS, Wollgarten M, Banhart J, Pabi SK, Murty BS (2010) Mater Sci Eng A 527:2370–2378
30. Shuey RT (2008) Personal communication
31. Puncreobutr C, Phillion AB, Fife JL, Rockett P, Horsfield AP, Lee PD (2014) Acta Mater 79:292–303

32. Liu K, Cao X, Chen XG (2010) Metall Mater Trans A 42:2004–2016
33. Cao X, Campbell J (2004) Metall Mater Trans A 35:1425–1435
34. Cao X, Campbell J (2003) Metall Mater Trans A 34:1409–1420
35. Cao X, Campbell J (2006) Mater Trans 47:1303–1312
36. Miller DN, Lu L, Dahle AK (2006) Metall Mater Trans B 37:873–878
37. Laz PJ, Hillberry BM (1998) Int J Fatigue 20:263–270
38. Li T, Morris JW, Chrzan DC (2006) Phys Rev B 73:024105
39. Li T, Morris J Jr, Chrzan D (2004) Phys Rev B 70:054107
40. Seifeddine S, Johansson S, Svensson IL (2008) Mater Sci Eng A 490:385–390
41. Zhu T, Li J, Ogata S, Yip S (2009) MRS Bull 34:167–172
42. Gangulee A, Gurland J (1967) AIME MET SOC TRANS 239:269–272
43. Finlayson T, Griffiths J, Viano D, Fitzpatrick M, Oliver E, Wang Q (2007) In: 2nd International symposium on shape casting, TMS, Warrendale, PA, pp 127–134
44. Caceres C, Griffiths J (1996) Acta Mater 44:25–33
45. Yeh J-W, Liu W-P (1996) Metall Mater Trans A 27:3558–3568
46. Poole WJ, Charras N (2005) Mater Sci Eng A 406:300–308
47. Harris SJ, O'Neill A, Boileau J, Donlon W, Su X, Majumdar B (2007) Acta Mater 55:1681–1693
48. Joseph S, Kumar S, Bhadram VS, Narayana C (2015) J Alloy Compd 625:296–308
49. Campbell J (2006) Mater Sci Technol 22:127–145
50. Umeno Y, Kushima A, Kitamura T, Gumbsch P, Li J (2005) Phys Rev B 72
51. Dubois SMM, Rignanese GM, Pardoen T, Charlier JC (2006) Phys Rev B 74:235203
52. Mueller MG, Žagar G, Mortensen A (2018) Acta Mater 143:67–76
53. Mueller MG, Fornabaio M, Žagar G, Mortensen A (2016) Acta Mater 105:165–175
54. Mueller MG, Fornabaio M, Mortensen A (2016) J Mater Sci 52:858–868
55. Xia S, Qi Y, Perry T, Kim K (2009) Acta Mater 57:695–707
56. Ward DK, Curtin WA, Qi Y (2006) Compos Sci Technol 66:1151–1161
57. Ward DK, Curtin WA, Qi Y (2006) Acta Mater 54:4441–4451
58. Noreyan A, Qi Y, Stoilov V (2008) Acta Mater 56:3461–3469
59. Alexopoulos ND, Tiryakioğlu M, Vasilakos AN, Kourkoulis SK (2014) Mater Sci Eng A 604:40–45
60. Eisaabadi GB, Yeom GY, Tiryakioğlu M, Netto N, Beygi R, Mehrizi MZ, Kim SK (2018) Mater Sci Eng A 722: 1–7

Measurement of Residual Strain in the Cylinder Bridge of High-Pressure Die Cast A383 Engine Blocks Using Neutron Diffraction

T. Liu, L. N. Brewer, Jeffrey R. Bunn, Chris M. Fancher, L. Nastac, V. Arvikar and I. Levin

Abstract This article describes the use of neutron diffraction to measure residual strains in high-pressure die cast aluminum alloy engine blocks. High-pressure die casting is widely used to manufacture aluminum alloy components for automobile applications such as engine blocks. Thermal gradients during this process can generate residual strain and stresses and may cause premature failure during service. Neutron diffraction can be used to measure the residual stress distributions deep inside large castings, as neutrons have much larger penetrating power than X-rays. In this work, neutron diffraction was used to measure the residual strain (in the axial and hoop directions) in the cylinder bridge region of high-pressure die cast A383 engine blocks in as-cast and heat-treated conditions. The results suggest that the residual strain in the cylinder bridge is tensile in both the axial and hoop components for both as-cast and heat-treated engines. The residual strain in hoop direction is higher than the strain in axial direction. Residual strain in axial and hoop directions is relaxed after heat treatment.

Keywords Residual stress · High pressure die casting · Distortion · Al-Si-Cu alloys

Introduction

High-pressure die casting produces extremely high solidification rates (up to 100 K/s) and is an important manufacturing process for fabrication of aluminum alloy engine blocks [1]. Aluminum has relatively low wear resistance, and therefore,

T. Liu · L. N. Brewer · L. Nastac
Department of Metallurgical and Materials Engineering, The University of Alabama, Box 870202, Tuscaloosa, AL 35487, USA

J. R. Bunn · C. M. Fancher
Oak Ridge National Laboratory, Oak Ridge, USA

V. Arvikar · I. Levin (✉)
Nemak Alabama, Sylacauga, AL, USA
e-mail: lnbrewer1@eng.ua.edu

M. Tiryakioğlu et al. (eds.), *Shape Casting*, The Minerals, Metals & Materials Series,
https://doi.org/10.1007/978-3-030-06034-3_29

protective liners, made of gray cast iron, are commonly used in particular locations within the engine blocks, such as within combustion chambers [2]. Significant residual stresses may be generated during the cooling process (after casting or heat treatment) in the cylinder bridge region due to thermal gradients and due to the large difference in thermal expansion coefficient between the gray cast iron and surrounding Al alloys ($\alpha_{Al} = 2.4 \times 10^{-5}\ K^{-1}$, $\alpha_{Fe} = 1.5 \times 10^{-5}\ K^{-1}$) [3]. Large residual stresses may cause permanent dimensional distortion or premature fracture [4]; therefore, measurement and control of the residual stresses in engine blocks are of vital importance for many engine block producers.

Neutron diffraction, as a nondestructive measurement, has the capability to penetrate much more material (200 mm for Al; 25 mm for Fe) than X-ray diffraction techniques. Neutron diffraction-based residual stress measurements can produce strain values with an accuracy of $\pm 50 \times 10^{-6}$ for strain with spatial resolution down to about 0.5 mm [5]. Lombardi first used neutron diffraction methods to measure the residual stresses throughout the whole cylinder bridge of sand cast A319 V6 engine blocks [6, 7]. Their results showed that tensile residual stresses in all three components (axial, hoop, and radial) were present and that the maximum residual stress magnitude for a service tested engine block was approximately +180 MPa (in the hoop direction). The large residual stress along the cylinder bridge was also found to trigger permanent dimensional distortion in the engine blocks [8]. Heat treatment has been found to reduce the magnitude of residual stresses in aluminum engine blocks [9–12]. Recently, Lombardi successfully observed the relaxation of residual strain along the cylinder bridge during solution heat treatment via in situ heating during neutron diffraction [3]. It was described in this work that the residual strain (in the axial direction) was relaxed gradually during the solution treatment process; however, tensile residual strains were regenerated during the cooling process due to thermal expansion mismatch. Combined with other ex situ neutron diffraction experiments, a measurable relaxation of residual stresses (in axial, hoop, and radial directions) was also observed after solution heat treatment [3].

The distribution and magnitude of residual stresses or strains in an engine block strongly depend on the composition and the manufacturing process (such as the particular casting process and subsequent heat treatment). As such, there is a need to analyze the residual strain profile in a high-pressure die cast aluminum engine block. In the present work, the residual strains and stresses along the cylinder bridge of high-pressure die cast A383 (Al-10.5Si-2.5Cu-1.5Zn-1.0Fe) engine blocks were measured by neutron diffraction. The effect of heat treatment (T5) on the residual strain profile along the cylinder bridge of high-pressure die cast A383 engine blocks was also investigated.

Experimental Procedures

The material used in this work consisted of two high-pressure die cast, V6 engine blocks (as-cast and heat-treated engine blocks) made of A383 aluminum alloys (the composition is shown in Table 1). These engine blocks were manufactured at

Table 1 Composition of HPDC A383 Al alloys, in wt%

Al	Si	Cu	Zn	Fe	Mn	Ni	Mg
Bal.	10.5	2.5	1.5(Max)	1.0(Max)	0.3	0.3	0.1

Nemak Alabama in the USA. Heat treatment (T5) was carried out at 215 °C for 3 h after casting process.

The neutron diffraction studies were carried out at the Neutron Residual Stress Facility (beam line HB-2B) at the High Flux Isotope Reactor (HFIR) at Oak Ridge National Laboratory [13], as shown in Fig. 1. In order to obtain a constant neutron wavelength, a double-focusing silicon (422) monochromator was applied. By calibrating with standards, the wavelength was corrected as 1.53344 Å. The aluminum {311} reflection was chosen for the neutron diffraction experiments in the present work. The residual strain, ε, in the cylinder bridge was determined in two directions (the axial and hoop directions) by using the "peak shift" method:

$$\varepsilon = \frac{d_{hkl} - d_{0,hkl}}{d_{0,hkl}}$$

where d_{hkl} was the measured d spacing value of Al {311} planes in the gauge volume *for a given component of strain*. The stress-free lattice parameter, $d_{0,hkl}$, was measured using stress-free samples on the same beam line. The stress-free samples were extracted directly from the cylinder bridge by removing the surrounding cast iron liner. To remove the remaining residual stress at the surface of the stress-free samples, they were machined by electrical discharge machining (EDM) into a "comb" geometry specimen. The residual stress for axial and hoop components was estimated by the equation: $\sigma = E \cdot \varepsilon$, where E is Young's modulus with a value 70 GPa used for this calculation.

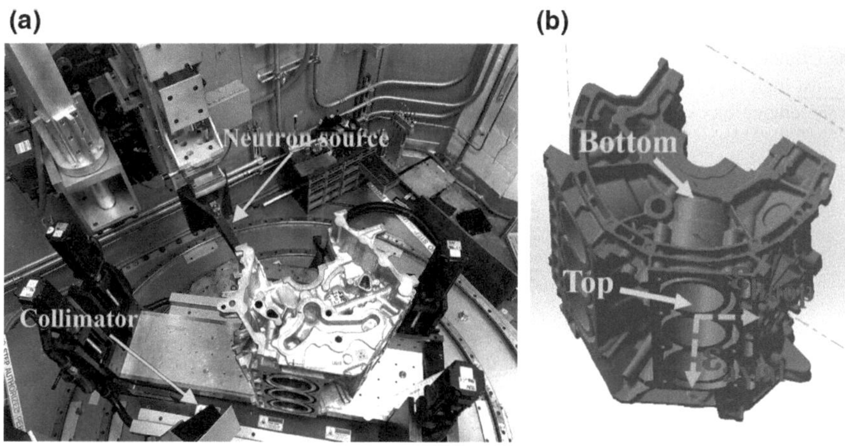

Fig. 1 **a** Experimental setup for neutron diffraction at NRSF-2. Note that in this figure, the hoop component of strain is being measured, **b** schematic that shows the geometry of the measurement

Table 2 Unstressed lattice spacing, d_0, values for as-cast and heat-treated strain-free samples

	Average d0 (Å)	Standard deviation (Å)
As-cast	1.222(4)	6.18×10^{-5}
Heat-treated	1.223(0)	6.41×10^{-5}

Experimental Results

Strain-Free Samples

The d spacing values for strain-free samples of as-cast and heat-treated samples were different, as shown in Table 2 (24 points were measured for each d_0 sample). The heat-treated sample had a higher d spacing value than the as-cast sample. As such, it was necessary to use different d_0 references for as-cast and heat-treated engines to determine residual strain.

Residual Strain Profile of As-Cast Engine Block

The measured residual strains along the cylinder bridge at both axial and hoop directions were tensile. The residual strain in the hoop direction was higher than the residual strain in the axial direction for the as-cast engine block, as shown in Fig. 2. The highest residual stresses were estimated as +45 MPa in the axial direction and +70 MPa in the hoop direction. The highest residual strain value was found at the top region of the cylinder bridge for the hoop direction, and the highest value for the axial direction was in the middle region of the cylinder bridge.

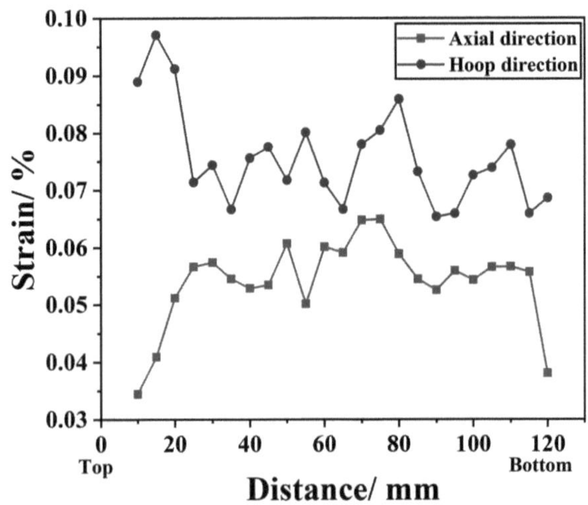

Fig. 2 Residual strains along the cylinder bridge profile of as-cast high-pressure die cast A383 engine block

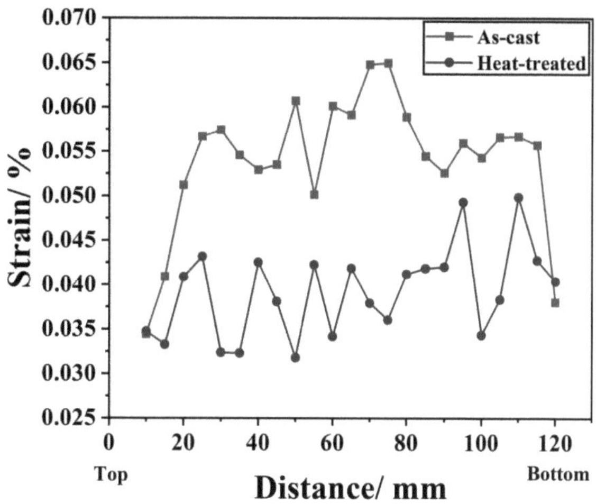

Fig. 3 Residual strain profile in the axial direction for the as-cast and heat-treated engine blocks

Effect of Heat Treatment on the Residual Strain Profile of High-Pressure Die Cast A383 Engine Blocks

Heat treatment affected the distribution and the magnitude of residual strains in the cylinder bridge region. The residual strains along the whole cylinder bridge of HPDC A383 engine blocks in the axial direction were significantly relaxed after heat treatment (but were still measurably tensile). As shown in Fig. 3, the profile of residual axial strains in the as-cast engine block was symmetric with depth about the mid-line of the cylinder: The residual axial strain value increased from the two ends (top and bottom) of the cylinder bridge toward the middle region. After heat treatment, the residual axial strain values decreased and its profile became more uniform than the as-cast engine block. The highest measured axial residual strain (+0.065%) was found in the middle region of the cylinder bridge for the as-cast engine block. After heat treatment, the residual axial strain value of the cylinder bridge oscillated over a very small range of +0.03 to +0.045%.

In contrast, the residual strain along the cylinder bridge in the hoop direction was only slightly relaxed after heat treatment. The profiles of hoop direction residual strain for both as-cast and heat-treated engine blocks were not uniform, as shown in Fig. 4. In the as-cast engine block, the residual strain value (in the hoop direction) was the highest at the top region of the cylinder bridge. The residual strain values decreased from the top region toward the middle region and then became stable and oscillated around +0.075%. For the heat-treated engine block, the highest hoop direction residual strain value was also found at the top region of the cylinder bridge, and it decreased from the top to the bottom of the cylinder bridge. In the region near the top of the cylinder bridge (0–40 mm), heat treatment had no effect on relaxing the hoop component of residual strain. Heat treatment only slightly relaxed the hoop component of residual strain in the middle region (40–70 mm). More residual strain was relaxed in the region (70–120 mm) near the bottom of the cylinder bridge after heat treatment.

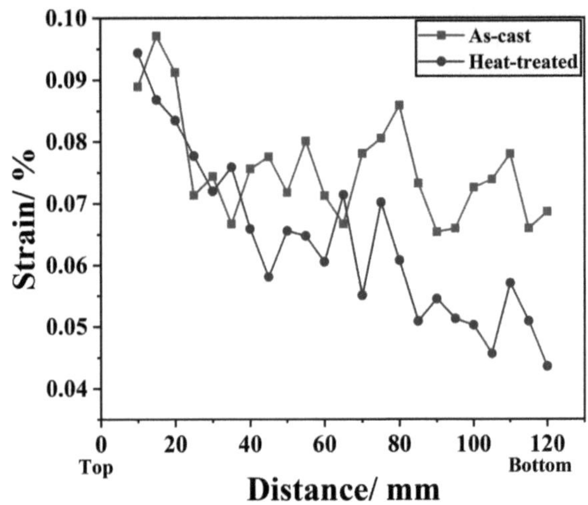

Fig. 4 Residual strain profile in the hoop direction for the as-cast and heat-treated engine blocks

Discussion

Residual Strain in High-Pressure Die Cast A383 Engine Blocks

For high-pressure die cast engine blocks, the heat treatment (T5) primarily relaxed the residual strain caused by thermal gradients during the solidification process. Upon cooling, thermal expansion mismatch between the Al cylinder bridge and the cast iron liners regenerated residual strains and stresses. These residual strains remained in both the axial and hoop directions, but were stronger in the hoop direction. After heat treatment, the residual strain in the axial direction was notably relaxed, but the residual strain in the hoop direction was only slightly relaxed (see Figs. 3 and 4). The difference between the strain in these components is likely due to the degree of mechanical constraint. Based on the geometry of this engine block, the hoop direction has greater mechanical constraint than the axial direction, thus generating larger residual strains upon cooling.

The residual strains measured in as-cast and high-pressure die cast A383 engine blocks are lower in magnitude than those measured in previous work on as-cast, sand cast A319 (Al-8Si-2.8Cu-0.6Zn-0.7Fe) engine blocks (Table 3). These A319

Table 3 Comparison between this work and previous neuron residual stress measurements on aluminum engine blocks. (comparing as-cast engine blocks only)

	Processing	Alloy	Max residual σ_{hoop} (MPa)	Max residual σ_{axial} (MPa)	Yield strength (MPa)
This work	HPDC	A383	+70	+46	195[a]
Lombardi [3]	Sand casting	A319	+220	+190	215

Note This value was measured from material taken from the bulkhead of the high-pressure die cast block. A yield strength number from the cylinder bridge is not currently available blocks were produced with precision sand casting with Cu chilling [3]. The estimated residual stresses in the high-pressure die cast blocks were 24–36% of the yield strength of this material. The residual strains reported by Lombardi et al. for the as-cast, sand cast A319 engine blocks can be used to estimate residual stresses in the cylinder bridge in the range of 88 -> 95% of the yield strength for that material. The reason for this difference is not completely known at this time. The thermal gradients during HPDC should be lower than those present during sand casting; however, differences in cylinder and cylinder bridge geometry could play a significant role as well.

Conclusions

The residual strains (in the axial and hoop directions) in the cylinder bridge of high-pressure die cast engine blocks (as-cast and heat-treated) were measured using neutron diffraction. The main conclusions are:

(1) The measured residual strains in the cylinder bridge of the high-pressure die cast A383 engine block were tensile. For the as-cast engine block, the residual strain in the hoop direction was higher than the residual strain in the axial direction. The corresponding highest value of residual stress is estimated to be +70 MPa in the hoop direction.
(2) Heat treatment relaxed the residual strains significantly in the axial direction; however, the residual strains in the hoop direction were relaxed only slightly after heat treatment. This difference is mostly likely due to increased thermal expansion mismatch and mechanical constraint in the hoop direction.
(3) Compared with Lombardi's work on sand cast A319 engine blocks, the residual strains in the cylinder bridge of the high-pressure die cast A383 engine blocks were considerably lower in magnitude.

Acknowledgements The authors are very grateful to the Alabama State Department of Commerce (Alabama Innovation Fund) and Nemak Inc. for financial support of this research. This research used resources at the High Flux Isotope Reactor, a DOE Office of Science User Facility operated by the Oak Ridge National Laboratory.

References

1. Bonollo F, Gramegna N, Timelli G (2015) High-pressure die-casting: contradictions and challenges. JOM 67(5):901–908
2. Darut G, Liao H, Coddet C, Bordes JM, Diaby M (2015) Steel coating application for engine block bores by plasma transferred wire arc spraying process. Surf Coat Technol 268:115–122

3. Lombardi A, Sediako D, Machin A, Ravindran C, MacKay R (2017) Effect of solution heat treatment on residual stress in Al alloy engine blocks using neutron diffraction. Mater Sci Eng A-Struct Mater Prop Microstruct Process 697:238–247
4. Patwardhan M, Paranjpe JM, Ramdasi SS, Karanth NV, Marathe NV, Bhat P (2015) Use of non linear analysis in powertrain design for prediction of cylinder bore distortion, design changes for reduction along with experimental validation, SAE Technical Paper
5. Withers PJ, Bhadeshia H (2001) Overview - residual stress part 1 - measurement techniques. Mater Sci Technol 17(4):355–365
6. Lombardi A, Sediako D, Machin A, Ravindran C, MacKay R (2015) Transient analysis of residual strain during heat treatment of multi-material engine blocks using in-situ neutron diffraction. Mater Lett 157:50–52
7. Lombardy A, Sediako D, D'Elia F, Ravindran CR, Mackay RI (2012) Neutron diffraction study on residual stress in aluminum engine blocks following machining and service testing, SAE International Journal of Materials and Manufacturing 5(2012-01-0188):115–121
8. Lombardi A, Ravindran C, Sediako D, MacKay R (2014) Determining the mechanism of in-service cylinder distortion in aluminum engine blocks with cast-in gray iron liners. Metall Mater Trans A 45(13):6291–6303
9. Wiesner D, Watkins T, Ely T, Spooner S, Hubbard C, Williams J (2005) Residual stress measurements of cast aluminum engine blocks using diffraction. Int Cent Diffr Data Adv X-Ray Anal 48:136–142
10. Su X, Jan J, Lasecki J, Allison J (2003) Thermal and residual stress analysis of an engine block with cast-in liners. In: Materials solutions conference 2003, vol. 2003
11. Carrera E, Rodriguez A, Talamantes J, Valtierra S, Colas R (2007) Measurement of residual stresses in cast aluminium engine blocks. J Mater Process Technol 189(1–3):206–210
12. Reihle M, Hofmann M, Wasmuth U, Volk W, Hoffmann H, Petry W (2014) In situ strain measurements during casting using neutron diffraction. In: Materials Science Forum 2014, pp 484–491. Trans Tech Publications
13. Cornwell P, Bunn J, Fancher C, Payzant E, Hubbard C (2018) Current capabilities of the residual stress diffractometer at the high flux isotope reactor. Rev Sci Instrum 89(9):092804

Relation Between Microstructure and Tensile Properties of V and B Added Al-7Si Alloy

Özkan Kesen, Ahmet Can Filiz, Selim Temel, Özen Gürsoy, Eray Erzi and Derya Dispinar

Abstract It is known that vanadium and boron elements are used as grain refiners. In this study, it was aimed to increase the mechanical properties of A356 by grain refinement. Boron and vanadium elements were added in an amount of 0.05 wt% and 0.1 wt%. Al–10 V and Al–3B master alloys were used. The melts were casted into sand and permanent molds at 730 °C. The cast samples were investigated by several tests: metallographic examination for microstructure and porosity measurements, tensile tests, and hardness tests for mechanical property characterization. Additionally, bifilm index was examined by reduced pressure test (RPT). The effect of alloying elements on the distribution, size, and shape of porosities has been determined. Weibull analysis was used. In both V and B additions, sand castings with larger grain size revealed higher and more reliable tensile values than permanent castings.

Keywords Aluminum · Vanadium · Boron · Casting · Grain refinement

Introduction

Al–Si alloys are one of the most popular alloys in the industries. The most important reasons for using these alloys are high mechanical strength despite their lightweight, high fluidity, weldability, ability of heat treatment and can be adapted to most of casting methods. The greatest effect of silicon on pure aluminum is to increase its casting ability.

The mechanical properties of aluminum alloys are believed to depend on its composition, but the mechanical properties are more dependent on distribution, size, and morphology of silicon. In order to change these properties of silicon, the cooling rate of the alloy and/or inoculant additions may take place. For example,

Ö. Kesen · A. C. Filiz · S. Temel · Ö. Gürsoy · E. Erzi · D. Dispinar (✉)
Faculty of Engineering, Department of Metallurgy and Materials Engineering, Istanbul University Cerrahpasa, 34320 Avcilar, Istanbul, Turkey
e-mail: deryad@istanbul.edu.tr

M. Tiryakioğlu et al. (eds.), *Shape Casting*, The Minerals, Metals & Materials Series,
https://doi.org/10.1007/978-3-030-06034-3_30

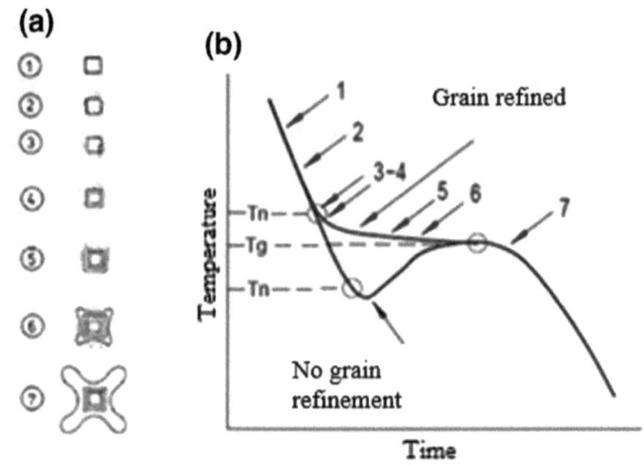

Fig. 1 **a** Illustration of the nucleation and growth of aluminum with "TiAl3" particles. **b** Cooling curve of aluminum alloy with and without grain refinement [1, 2]

fibrous silicones reduce ductility, impact resistance, and fatigue resistance despite increase in tensile strength. Coarser grained silicones increase wear resistance but cause lower tensile strength and premature fractures [1, 2].

One of the basic strength enhancement methods in aluminum silicon alloys is by grain refinement, microstructure homogeneity, increase in strength. All these effects make the grain refinement mechanism.

As can be seen in Fig. 1a, heterogeneous nucleation and growing begin from the surface of the grain refiner which is not dissolved in liquid aluminum. This continues until the liquid is fully solidified.

Figure 1b shows that when the cooling curve of casting without added grain refiner is examined, cooling to a nucleation temperature (Tn) below the temperature of the growth temperature (Tg) is required so that the nucleation can take place spontaneously. It is seen that in the cast with grain refiner, the nucleation occurs above the Tg temperature.

The effect of grain refinement on aluminum is known for many years. Grain refiners such as Al–Ti–B are commonly used in the industry. In summary, the grain refining mechanism of titanium is as follows: Al–Ti phase diagram shows a peritectic reaction at 1,2% wt. titanium at 665 °C. TiAl3 forms peritectically, and if there is enough titanium in the alloy (>0,15%wt.), first crystals of primary α-Al are formed in the liquid metal that nucleate on this phase. Al–Ti master alloys contain TiAl3 particles and when added to the liquid metal to be refined, disperse TiAl3 particles from the heterogeneous zones for nucleation. However, this effect disappears over time due to the settling of the TiAl3 particles in the liquid metal [3, 4].

Boron is used in aluminum and its alloys as a grain refiner and to improve conductivity by precipitating vanadium, titanium, chromium, and molybdenum (all of which are harmful to electrical conductivity at their usual impurity level in commercial grade aluminum). Boron can be used alone (at levels of 0.005–0.1%) as a grain refiner during solidification, but it becomes more effective when used with an excess of titanium. Commercial grain refiners commonly contain titanium and boron in a 5-to-1 ratio [5, 6].Töndel studied the efficiency of Ti-free B grain refinement and concluded that more globular grains were formed. Similar findings were reported by Dispinar [7–9].

According to the literature survey, the use of vanadium as grain refiners was initially thought that vanadium had a grain refining effect (Mondolfo 1976), and then this effect was confirmed at levels above 500 ppm. (Abdel- Hamid 1989). The addition of vanadium below 500 ppm showed no grain refining effect. (Abdel-Hamid 1989; Cook et al., 1997). According to the findings of Wang and his partners (2012), over 1000 ppm of vanadium, presumed to be due to the peritectic reaction in the Al–V phase diagram, has a certain ability to refining [10, 11].

In this study, it was aimed to observe grain refinement by adding Boron and Vanadium to A356 alloy with 0.05% and 0,1% wt.

Experimental Work

A356 alloy, Al–V10 and Al–B3 master alloys were used in this study. Step designed mold was used for microstructure analysis, reduced pressure test (RPT) samples were collected for bifilm index, and tensile bars were produced for mechanical characterization. Master alloys used in castings were added in such a way that the alloy element would be 0,05% and 0,1% wt (Table 1).

First of all, sand mold was prepared. For this process, resin-hardened sand was used. Permanent molds were also used which were heated to 140 °C.

The ingots were cut to the proper size, placed in the crucible, and charged to the furnace. The furnace temperature was set to 730 °C. After the melting process, master alloys of Al–V and Al–B were added at a ratio of 0.05% and 0.1% wt. After 10 min of holding, samples were produced.

Step and RPT samples were cut and grinded. Images were taken. Dendrite arm spacing (DAS) and secondary dendrite arm spacing (SDAS) were measured. In the scanned RPT samples, the number of bifilms was found, and bifilm index was calculated via the SIGMA SCAN PRO 5. Tensile tests were made according to ATSM E-80.

Results and Discussion

The microstructural analysis results are summarized in Figs. 2 and 3 for permanent and sand mold castings.

Table 1 Chemical composition of A356 alloy (% wt.)

Alloy	Si	Fe	Cu	Mn	Zn	Ti	Al
A356	6,8	0,35	0,02	0,03	0,04	0,04	Bal.

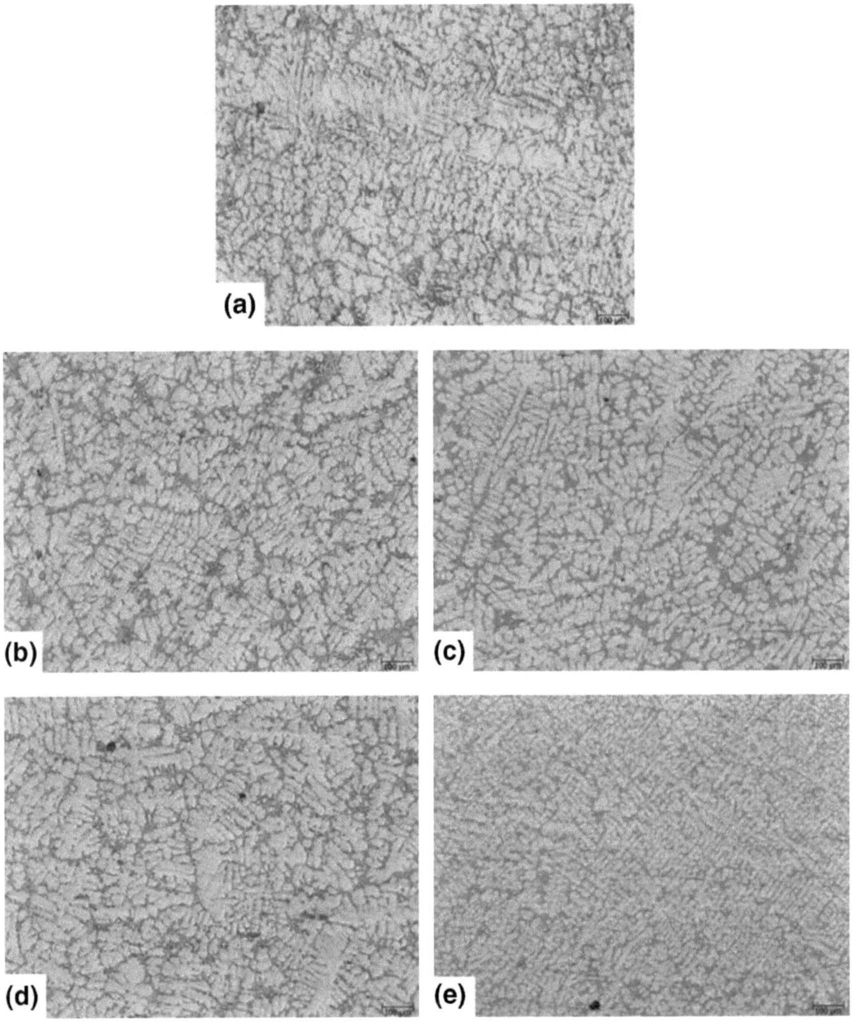

Fig. 2 **a** Permanent mold A356, **b** Permanent mold %0,05 B, **c** Permanent mold %0,05 V, **d** Permanent mold %0,1 B, **e** Permanent mold 0,1%V

DAS and SDAS measurements from the microstructure images from Figs. 2 and 3 are given in Figs. 4 and 5.

As shown in Fig. 4, boron and vanadium were compared in the grain refining process. Vanadium has been found to have little effect on DAS length. Boron has been observed to have reduced DAS length in contrast to vanadium.

Boron and vanadium were compared in the grain refining process. Vanadium has been found to decrease SDAS more than B addition. On the other hand, boron has been observed to have no effect on SDAS.

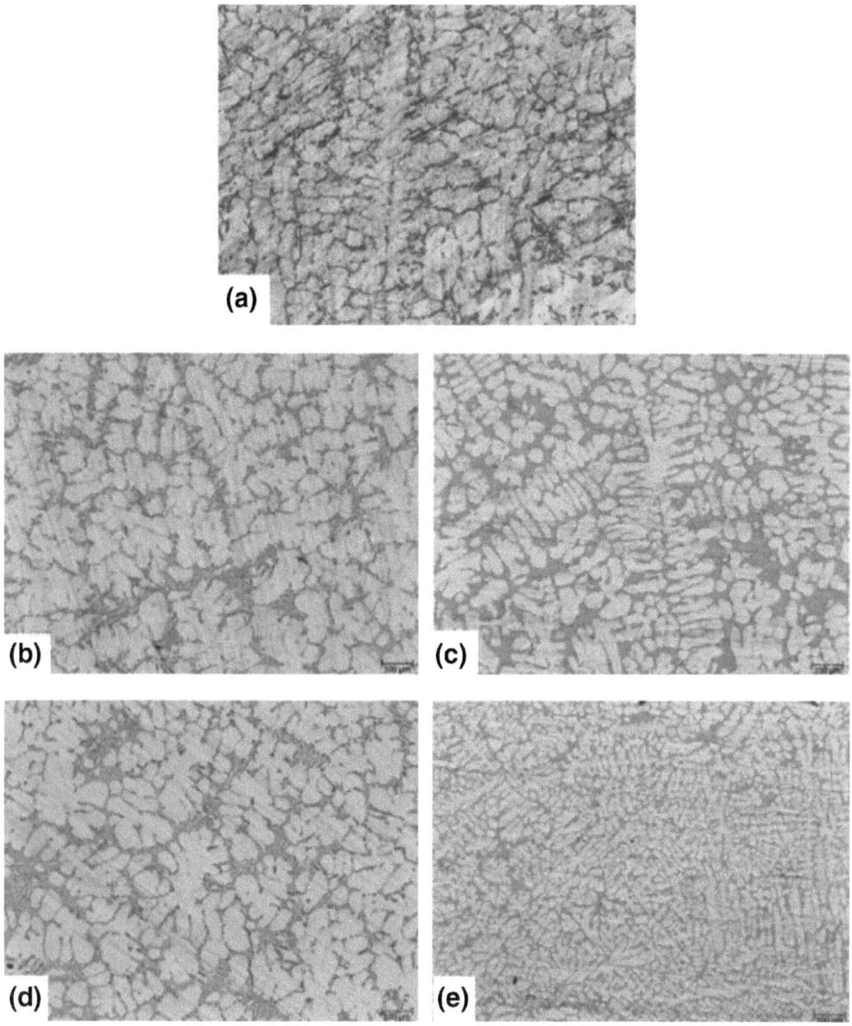

Fig. 3 Sand mold A356, **b** Sand mold %0,05 B, **c** Sand mold %0,05 V, **d** Sand mold %0,1 B, **e** Sand mold 0,1%V

The situations observed for Fig. 4 and the interpretations made for these situations are also valid for Fig. 5.

Figures 6 and 7 show the change in bifilm index and number of bifilms for all the melts that were studied in this work.

Tensile test results were analyzed by Weibull analysis. The results are as follows.

As shown in Fig. 8, the tensile test results of sand and permanent castings are examined. It can be seen that the sand castings gave better results than permanent

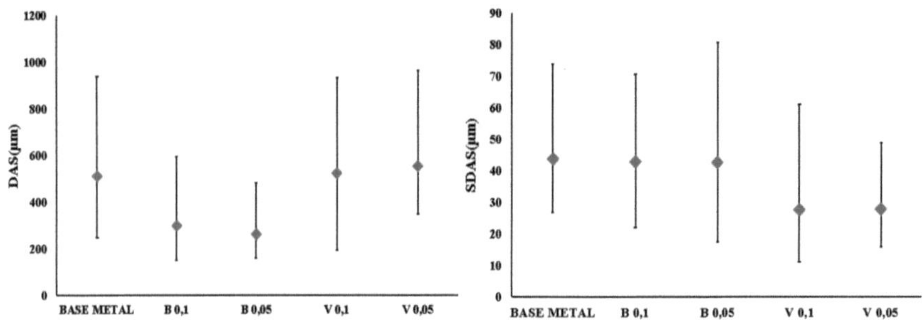

Fig. 4 DAS and SDAS sizes of the sand casting samples

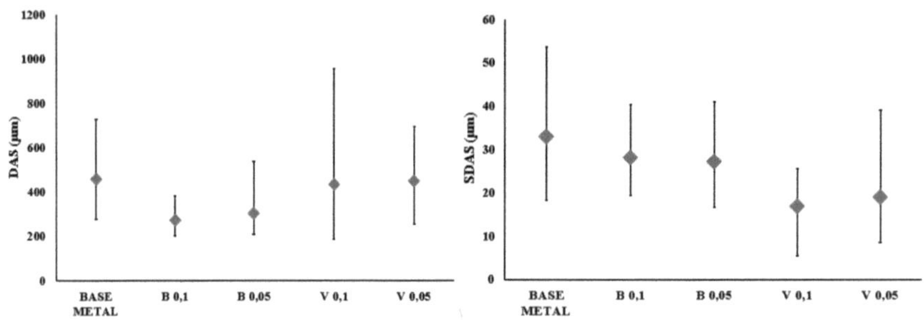

Fig. 5 DAS and SDAS sizes of the permanent casting samples

Fig. 6 Change in number of bifilms for all casting conditions

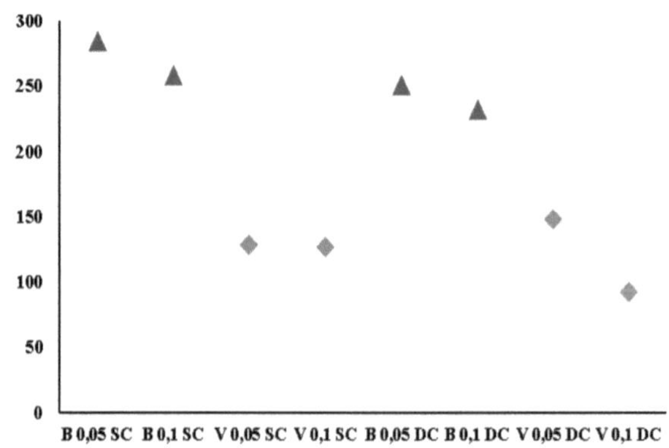

castings. It has been observed that in sand castings, vanadium gives better results when compared to boron.

As can be seen in Fig. 9, the elongation values of tensile samples were investigated. It has been observed that sand casting samples have much better elongation values than permanent casting samples. Compared with boron and vanadium, the elongation values of vanadium were better than boron.

Fig. 7 Bifilm index results for all conditions

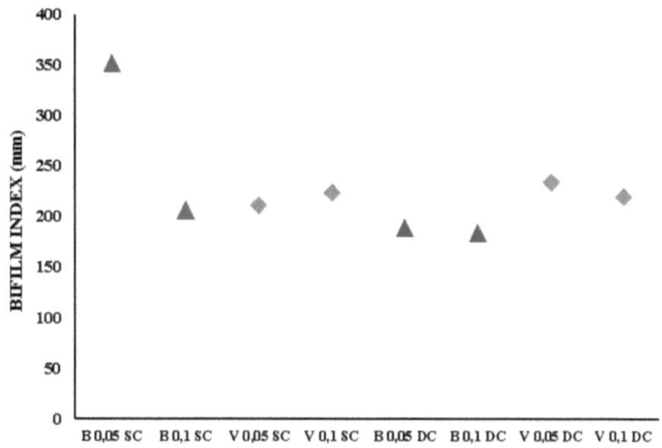

Fig. 8 Tensile test results of sand and permanent castings

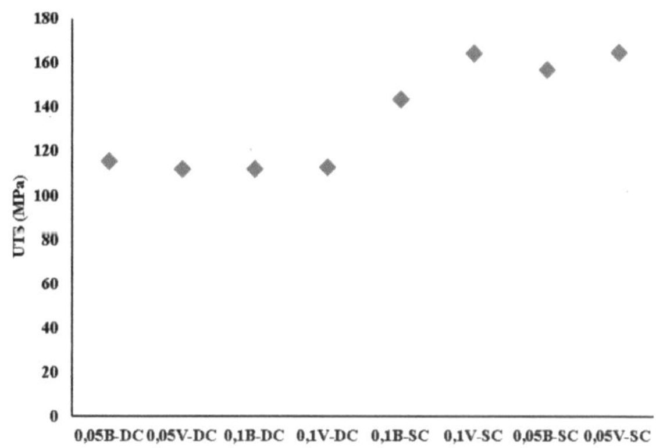

Fig. 9 Elongation values of sand and permanent casting samples

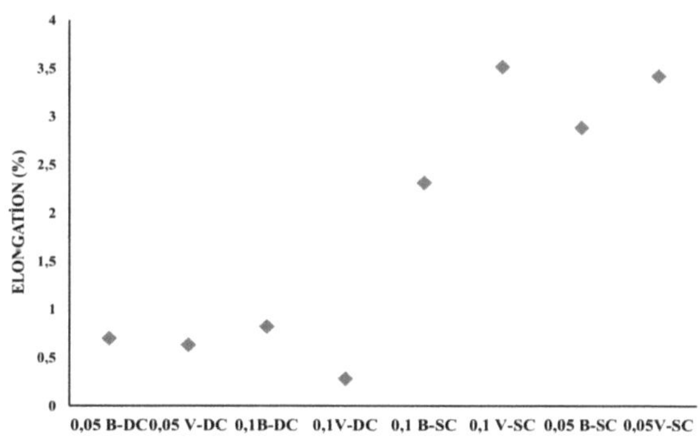

In the samples containing vanadium (Fig. 10), Weibull results show that sand casting samples give more reproducible and consistent results than permanent casting samples where Weibull modulus is 8 times higher which changes from 0.6 to 4.8 (as indicated in the legends of Fig. 10).

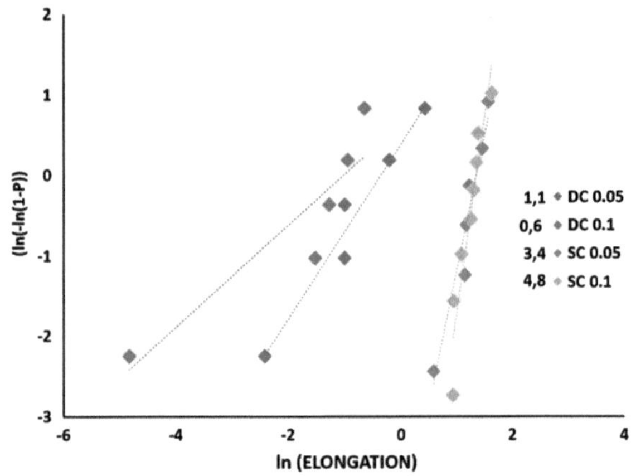

Fig. 10 Weibull analysis of elongation values of tensile samples contain vanadium

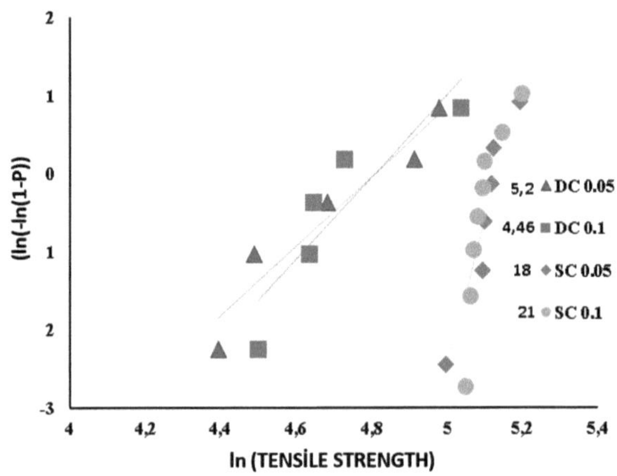

Fig. 11 Weibull analysis of tensile strength values of tensile samples contain vanadium

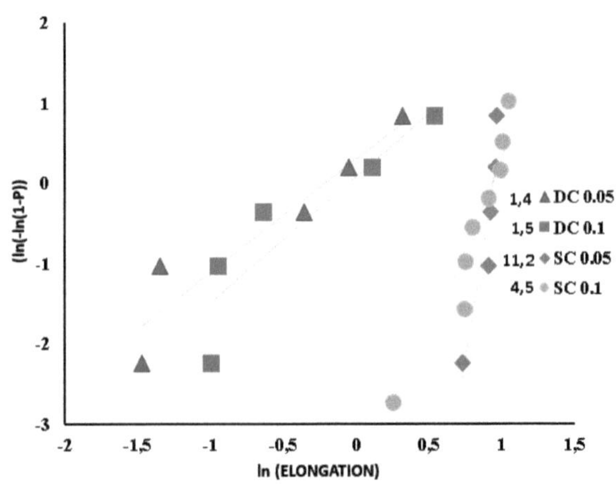

Fig. 12 Weibull analysis of elongation values of tensile samples contain boron

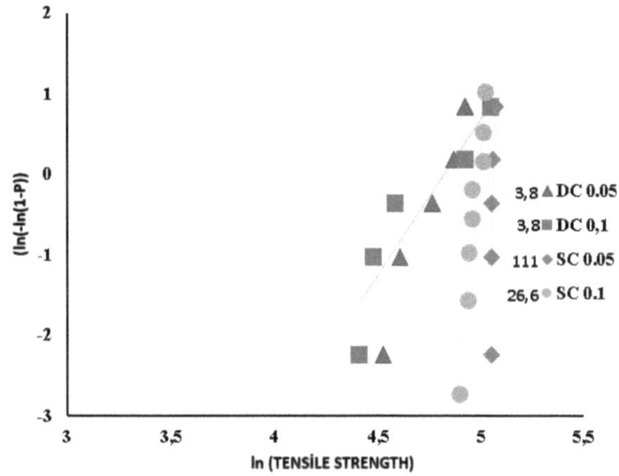

Fig. 13 Weibull analysis of tensile strength values of tensile samples contain boron

Table 2 Weibull parameters obtained from Figs. 10, 11, 12, 13

	Weibull modulus (m)		Characteristic value	
	UTS	%e	UTS	%e
Sand casting 0,05 V	18	3,4	169,6	3,8
Sand casting 0,1 V	21	4,8	168,3	1
Sand casting 0,05 B	111	11,2	157,5	2,5
Sand casting 0,1 B	26,6	4,5	146,3	2,5
Die casting 0,05 V	5,1	0,6	122,5	0,6
Die casting 0,1 V	4,46	0,6	122,6	0,3
Die casting 0,05 B	3,8	1,4	123	0,8
Die casting 0,1 V	3,8	1,5	124,1	0,9

Figure 11 shows that sand casting samples are more stable and repeatable than permanent casting samples for UTS values. The Weibull modulus change is from 5 to 18 which is almost 4 times as seen in Fig. 11 legends.

These figures also show that sand casting samples are more stable and repeatable than permanent casting samples too. Only Fig. 13 gave slightly more closer values than others. This shows that UTS of B grain refined castings are less affected by the cooling rate.

The general overview of Weibull results is summarized in Table 2. As can be seen, permanent casting has the lowest Weibull modulus and characteristic value. There is at least 4 times higher reproducibility in sand cast samples. When V and B are compared, B added samples have slightly lower UTS and elongation but higher reliability then V added samples.

Conclusions

When V is added as grain refiner to A356, DAS is significantly changed to a finer microstructure. On the other hand, SDAS is dominantly affected when B is used.

For mechanical properties, V gave higher UTS values (average of 140 MPa) whereas B revealed 130 MPa. However, Weibull statistics show that B grain refined A356 has more reproducible and reliable results than V added alloy.

The most important finding was found when permanent and sand casting samples were compared. It was concluded that sand casting revealed higher mechanical properties compare to permanent casting.

Both melts (V and B additions) had similar bifilm index values. Therefore, the comparison of mechanical properties were based on the fact that the effect of inclusions was the same for both castings. On the other hand, it is important to note that when V was added, there were more finely distributed smaller pores compared to large and less number of pores in B grain refined A356.

References

1. Lee YC, Dahle AK, StJohn DH, Hutt JEC (1999) The effect of grain refinement and silicon content on grain formation in hypoeutectic Al–Si alloys. Mater Sci Eng A 259:43–52
2. Baypınar F (2005) Grain refining of aluminium foundry alloys using Al5Ti1B. MSc thesis, Gazi University Institute of Science
3. Çolak M, Kayıkcı R (2009) Investigation of grain refining effect of AlTiB master alloy addition on ETIAL160 casting alloy. In: 5th International advanced technologies symposium (IATS'09), Karabük, Turkey, 13–15 May 2009
4. Schaffer PL, Dahle AK (2005) Settling behaviour of different grain refiners in aluminium. Mater Sci Eng A 413–414:373–378
5. Davis JR (2013) Light metals and alloys. ASM International, pp 372
6. Tøndel PA, Halvorsen G, Arnberg L (1993) Grain refinement of hypoeutectic Al–Si foundry alloys by addition of boron containing silicon metal. In: Light metals 1993. The Minerals, Metals & Materials Society, Warrendale, pp 783–783
7. Dişpinar D, Nordmark A, Voje J, Arnberg L (2009) Influence of hydrogen content and bifilm index on feeding behaviour of Al–7Si. In: 138th TMS annual meeting, pp 63–70, USA, 15–19 Feb 2009
8. Uludağ M, Çetin R, Dişpinar D, Tiryakioglu M (2018) Effect of degassing and grain refinement on hot tearing tendency in Al8Si3Cu alloy. Int J Metalcast 12:589–595
9. Uludağ M, Çetin R, Dişpinar D, Tiryakioglu M (2017) Characterization of the effect of melt treatments on melt quality in Al–7wt% Si–Mg alloys. Metals 7
10. Johanson A (2013) Effect of vanadium on grain refinement of aluminium. MS thesis, Institutt for materialteknologi
11. Küçük AO, Gürsoy Ö, Erzi E, Dişpinar D (2016) Characterisation of properties of V, B, Mo, Sr additions on HPDC of A380 alloy. In: High tech permanent casting, Venice, Italy, pp 1–10, 22–23 June 2016

The Effect of Friction Stir Processing on Bifilms and Structural Quality in A356 Alloy Castings

Nelson Netto, Murat Tiryakioğlu, Paul D. Eason, Boğaçhan Őndeş and Eray Erzi

Abstract The effect of friction stir processing (FSP) on the structural quality was tested in two A356 alloy ingots: conventional and continuously cast. Reduced pressure tests and microstructural analysis showed that the initial quality and microstructure in the two ingots were significantly different. The structural quality, as measured by the quality index Q_T, of the conventional ingot improved with each FSP pass, whereas that of the continuously cast ingot went up after the first pass and remained essentially constant with additional passes. The contribution of microstructural refinement as a result of FSP on the improvement of structural quality remains unknown. However, the largest contribution comes from bifilms breaking up during the friction stir process.

Keywords Bifilms · Microstructural refinement · Quality index

Introduction

Cast Al–Si–Mg alloys are usually used for high-strength components in aerospace and automotive industries [1]. However, structural defects, such as pores, oxide inclusions and intermetallics, reduce the mechanical properties of aluminum alloy castings, particularly the elongation and fatigue life. Several post-processing techniques, such as hot isostatic pressing (HIP) [2, 3], are available to reduce the effect of these structural defects. Nevertheless, structural defects remain prevalent in aluminum castings, to the point that some researchers consider them to be intrinsic

N. Netto · P. D. Eason · B. Őndeş
School of Engineering, University of North Florida, Jacksonville, FL 32224, USA

E. Erzi
Metallurgical and Materials Engineering Department, Istanbul University-Cerrahpaşa, Istanbul, Turkey

M. Tiryakioğlu (✉)
University of North Florida, Jacksonville, FL 32224, USA
e-mail: m.tiryakioglu@unf.edu

M. Tiryakioğlu et al. (eds.), *Shape Casting*, The Minerals, Metals & Materials Series,
https://doi.org/10.1007/978-3-030-06034-3_31

[4, 5] and therefore not possible to eliminate. In contrast, recent research [6, 7] has demonstrated that structural defects are extrinsic, and the true properties of cast aluminum alloys have yet to be determined.

Friction stir processing (FSP) is a technique, derived from friction stir welding [8, 9], where a rotating tool, consisting of a pin and shoulder, is plunged into the material, until the shoulder gets contacts the outside surface of the workpiece. Subsequently, the tool is forced along the plane of the surface of the material, while the shoulder remains in contact with the workpiece. The pin forces the material to undergo intense plastic deformation, resulting in a refined, homogenized, and recrystallized microstructure [8–11]. In cast Al–Si–Mg alloys, FSP was found to modify Si eutectic–dendritic structures and refine the oxide inclusions [12–14]. The elimination of these defects has important ramifications on the use of castings. The literature for cast Al–Si–Mg alloys attributes the refinement of Si particles and defect elimination by FSP as possible reasons for improvement in mechanical properties enhancement [15, 16]. If FSP is capable of eliminating/healing those defects, it becomes finally possible to determine intrinsic properties of these alloys. This study is motivated by the potential of FSP to heal casting defects and significantly improve the structural quality of aluminum alloys.

Experimental Details

Samples excised from conventional and continuously cast ingots of A356 were machined into 100 × 80 × 10 mm bars and also reduced pressure test (RPT) coupons. To determine the initial structural quality of the ingots, RPT coupons were melted at 700 °C and allowed to solidify under a pressure of 0.1 atm. FSP experiments were conducted with single and multiple passes of 100 mm on a smooth working surface. Two and three passes had 100% overlap over the previous zone. FSP was conducted on a Bridgeport vertical milling machine, with the FSP tool tilted 3° opposite to the processing direction. The tool rotation rate and transverse speed were kept constant at 700 rpm in clockwise direction and 50 mm/min, respectively. The FSP tool was made of H13 tool steel with a shoulder diameter of 18 mm. The cylindrical pin had a diameter of 5.9 mm, a length of 5 mm and M6 threads. Specimens were excised, parallel to the FSP direction, from FSPed materials at a CNC HAAS mini-mill following the E8-M ASTM standard for sub-size specimen, with the gage being completely within the stir zone. The specimens were then mechanically polished to have a smooth surface. Tensile tests were carried out on a Shimadzu tester with a strain rate of 10^{-3} s^{-1}. An extensometer with a gage length of 25 mm was attached to the specimen throughout the tensile tests. A Tescan Mira 3 field emission scanning electron microscope (FE-SEM) equipped with an Oxford X-Max 50 energy-dispersive spectrometer was used to evaluate microstructure on unetched specimens and fracture surfaces. CT scans of RPT specimens were obtained on a Shimadzu inspeXio SMX-225CT Microfocus X-Ray CT System.

Results and Discussion

The CT scans of RPT coupons of conventional and continuously cast ingots are presented in Fig. 1. Note that both ingots have sustained entrainment damage. Pores are larger in size and more spherical in the conventional ingot and the bifilms in the continuous ingot are also visible. Density measurements showed that volume percent of pores was 16.3% and 7.1% in conventional and continuously cast ingot, respectively.

The as-cast microstructure for the two ingots are presented in Fig. 2a and b. The conventional cast ingot microstructure contains discrete particles of silicon-rich phase, which possess a faceted, high-aspect-ratio morphology, while the continuously cast ingot exhibits an undulating, continuous, silicon-rich eutectic micro-constituent. After three passes, both microstructures are much more refined, as presented in Fig. 2c and d. Although the initial microstructures before FSP are quite different, the two final microstructures after three passes are similar with small Si particles distributed more homogeneously, also evidenced by increasing nearest neighbor distances for the continuously cast ingot reported previously [17].

The tensile engineering stress–strain curves are shown in Fig. 3, and the tensile properties for the as-cast and FSPed specimens are listed in Table 1, where σ_Y is yield strength S_T is tensile strength, e_F is elongation, and R_A is a reduction in area. Figure 3a shows that the as-cast specimen from the conventional ingot fractured shortly after yield strength. With every FSP pass, elongation increased, indicating a progressive increase in the structural quality of the metal [18–24]. In contrast, the as-cast specimen from the continuously cast ingot deformed extensively after yield strength before final fracture. A higher elongation than the conventional ingot is consistent with the structural quality assessment with RPT results in Fig. 1.

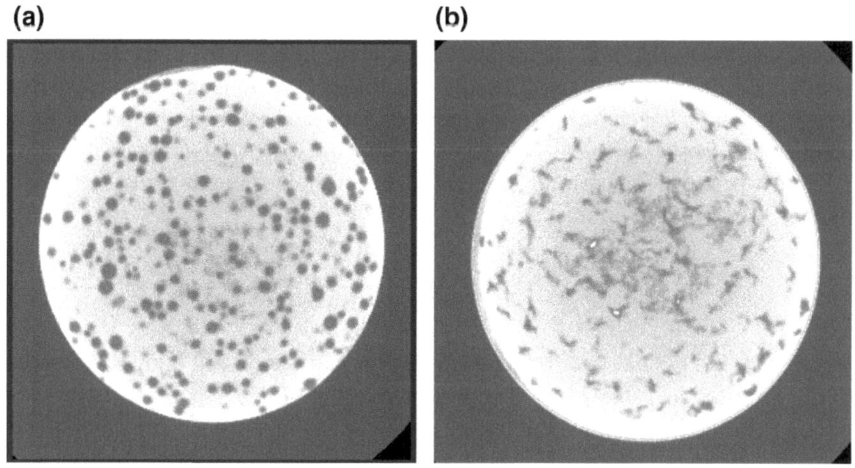

Fig. 1 CT scan images of RPT samples from **a** conventional and **b** continuously cast ingot

Fig. 2 Micrographs of the two ingots used in this study before and after FSP; **a** conventional, no FSP, **b** continuously cast, no FSP, **c** conventional after three passes of FSP, and **d** continuously cast after three passes of FSP

With FSP, elongation is increased with respect to the as-cast condition. However, tensile curves overlap and elongation values very close to each other are obtained with single or multiple passes, despite microstructural refinement in each pass. Hence, it can be concluded that one pass was sufficient to break the bifilms in the continuous ingot.

To normalize elongation for different levels of yield strength, structural quality index, Q_T [18–22] is used:

$$Q_T = \frac{e_F}{\beta_0 - \beta_1 \cdot \sigma_Y} \tag{1}$$

where $\beta_0 = 36.0\%$ and $\beta_1 = 0.064$ (%/MPa) for cast Al–Si–Mg alloys. The values of Q_T are listed in Table 1 and also presented in Fig. 4. Note that it takes two FSP passes to bring the conventional ingot to approximately the same structural quality

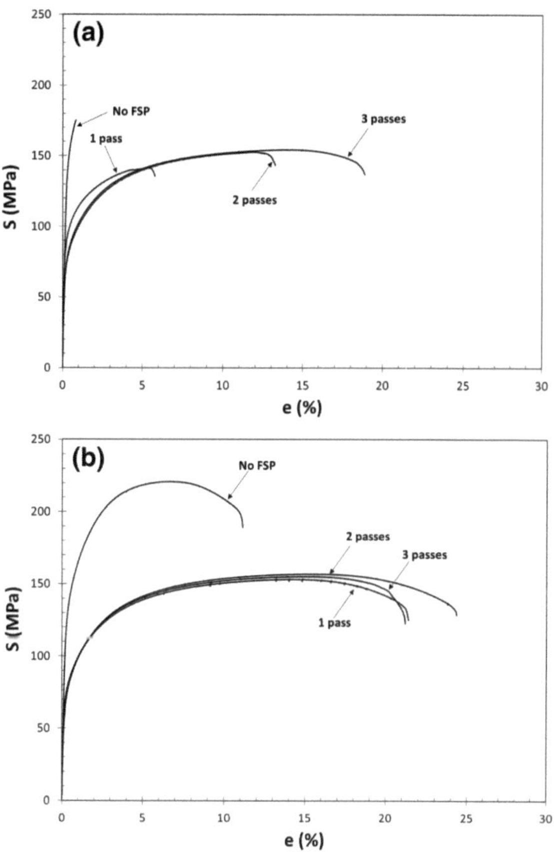

Fig. 3 Engineering stress–strain curves for specimens, no FSP and one, two, and three passes of FSP; **a** conventional and **b** continuously cast

Table 1 Summary of tensile data for the two ingots in as-cast condition and after FSP

Quality	FSP passes	σ_y (MPa)	S_T (MPa)	e_F (%)	R_A (%)	Q_T
Conventional ingot	No FSP	153.7	175.3	1.0	1.0	0.04
	1 pass	91.1	141.4	5.8	13.2	0.19
	2 passes	78.2	152.8	13.0	31.4	0.42
	3 passes	81.9	154.4	18.8	37.7	0.61
Continuously cast ingot	No FSP	121.0	221.0	10.9	24.7	0.39
	1 pass	79.5	153.7	21.1	39.5	0.68
	2 passes	75.5	157.0	24.0	43.3	0.78
	3 passes	77.9	155.2	20.9	39.3	0.65

level as the continuously cast ingot. The increase in Q_T with every FSP pass in the conventional ingot can be attributed to the bifilms being chopped to smaller and smaller sizes with further processing. Because the initial structural quality in the conventional ingot was lower than the continuously cast ingot, as evidenced by the

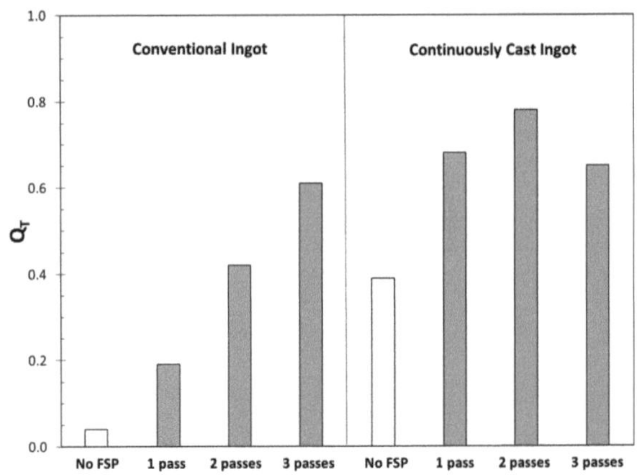

Fig. 4 Quality index for no FSP and one, two, and three passes of FSP for the two ingots

RPT results, it takes multiple passes to break up bifilms to a much less damaging size distribution. In the continuously cast ingot, however, it takes a single pass to reach the same bifilm size. Therefore, initial metal quality has a strong effect on FSP results.

It is noteworthy that Q_T does not reach the 1.0 level in spite of breaking up of bifilms during the friction stir process. Hence, the full potential of the FSP process could not be achieved in the present study. Nevertheless, the reduction in area levels that are 2 to 3 times those of elongation reported in Table 1 is a strong indication that the ductility of the specimens was changed significantly. It is not known at this time what the contribution of the change in the microstructure is to the enhance ductility. More research is needed to determine the microstructural contribution to this improvement. However, it is the authors' opinion that the benefit of microstructural refinement is greatly augmented by the simultaneous breaking up of bifilms.

Fracture surfaces in as-cast and after three passes of FSP for both ingots are presented in Fig. 5. Cleavage structure is prominent in the conventional ingot, which explains the low elongation in the as-cast condition. In continuously cast ingot, no cleavage was observed on the fracture surface. Large dimples covered the entire fracture surface. After FSP, fracture surface of the conventional ingot changed from cleavage to fine dimples. Even finer dimples were observed on the fracture surface of the FSPed continuously cast ingot, which was interpreted as the reason for the difference in the ductility (or Q_T) levels of the two specimens with three passes of FSP. Hence, bifilms in conventional ingot would probably take at least one more pass to be reduced to the size that results in the structural quality achieved after only single pass of FSP in the continuously cast ingot.

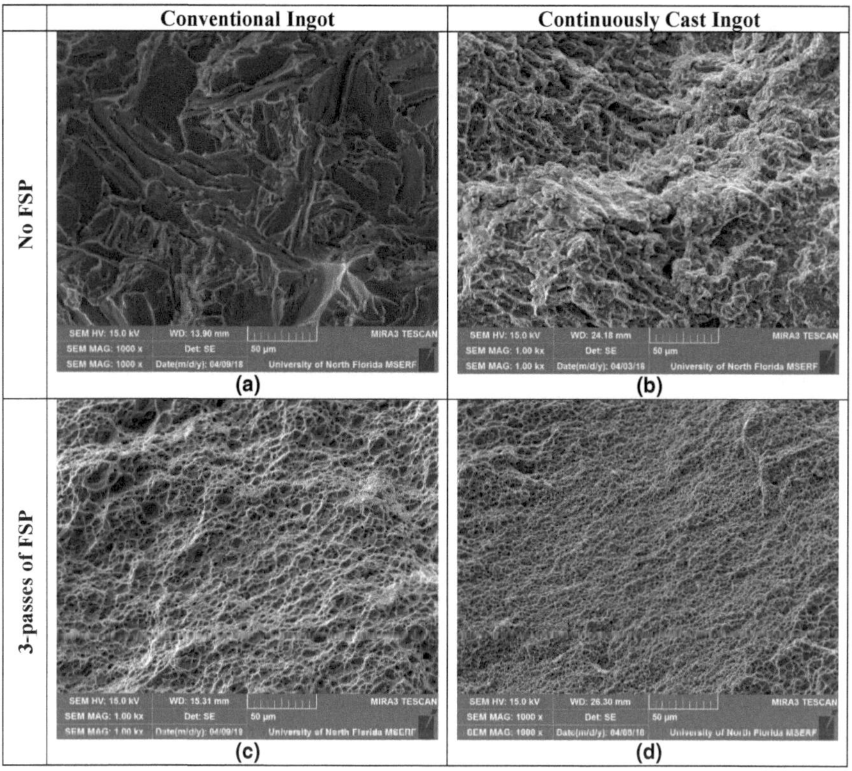

Fig. 5 Fracture surfaces of the two ingots used in this study; **a** conventional with no FSP, **b** continuously cast with no FSP, **c** conventional with three passes of FSP, and **d** continuously cast with three passes of FSP

Conclusions

- The reduced pressure test can be used to determine the structural quality of aluminium alloy ingots.
- The effect of friction stir processing on breaking up bifilms and thereby improving structural quality is strongly dependent on the initial quality of the metal. Lower-quality ingots take several FSP passes to achieve the same quality level that would be obtained after a single-FSP pass on a higher-quality ingot.
- The contribution of the change in microstructure to enhanced ductility, i.e., improved structural quality, remains unknown. The major contribution is thought to have come from the breaking up of bifilms during FSP.
- There is evidence that dimple size on the fracture specimen can be correlated to the ductility of FSPed Al–Si alloys.

References

1. John Baruch L, Raju R, Balasubramanian V, Rao AG, Dinaharan I (2016) Acta Metal Sin (English Letters) 29:431–440
2. Staley JT Jr, Tiryakioğlu M, Campbell J (2007) Mater Sci Eng A 465:136–145
3. Staley JT Jr, Tiryakioğlu M, Campbell J (2007) Mater Sci Eng A 460–461:324–334
4. Katgerman DGEL (2008) In search of prediction of hot cracking in aluminium alloys. In: Thomas Bollinghaus HH, Cross CE, Lippold JC (eds) Hot cracking phenomena in welds II. Springer, Berlin, pp 3–18
5. Nazarboland A, Elliott R (1997) Int J Cast Met Res 10:87–97
6. Yousefian P, Tiryakioğlu M (2018) Metall Mater Trans A 49:563–575
7. Tiryakioğlu M, Yousefian P, Eason PD (2018) Metall Mater Trans A 49:5815–5822
8. Mishra RS, Mahoney M, McFadden S, Mara N, Mukherjee A (1999) Scripta Mater 42:163–168
9. Mishra RS, Mahoney MW (2001) Mater Sci Forum. Trans Tech Publ, 507–514
10. Mishra RS, Ma ZY (2005) Mater Sci Eng R Rep 50:1–78
11. Jana S, Mishra RS, Grant G (2015) Friction stir casting modification for enhanced structural efficiency: a volume in the friction stir welding and processing book series. Butterworth-Heinemann
12. Alidokht SA, Abdollah-zadeh A, Soleymani S, Saeid T, Assadi H (2012) Mater Charact 63:90–97
13. Ma ZY, Sharma SR, Mishra RS (2006) Mater Sci Eng A 433:269–278
14. Ma ZY, Sharma SR, Mishra RS (2006) Metall Mater Trans A 37:3323–3336
15. Sun N, Apelian D (2012) TMS 2012 141st Annual meeting and exhibition, materials properties, characterization, and modeling Wiley, p 411
16. Sharma SR, Ma ZY, Mishra RS (2004) Scripta Mater 51:237–241
17. Netto N, Tiryakioğlu M, Eason PD (2018) Light metals. Springer, pp 417–422
18. Tiryakioğlu M, Campbell J (2014) Int J Metalcast 8:39–42
19. Tiryakioğlu M, Campbell J, Alexopoulos ND (2009) Mater Sci Eng A 506:23–26
20. Tiryakioğlu M, Campbell J, Alexopoulos ND (2009) Metall Mater Trans A 40:1000–1007
21. Tiryakioğlu M, Campbell J, Alexopoulos ND (2009) Metall Mater Trans B 40:802–811
22. Tiryakioğlu M, Campbell J (2009) Mater Sci Technol 25:784–789
23. Tiryakioğlu M, Staley JT, Campbell J (2008) Mater Sci Eng A 487:383–387
24. Tiryakioğlu M, Campbell J, Staley JT (2003) Scripta Mater 49:873–878

Effect of Copper and Nickel Addition on Mechanical Properties of A356 Alloy and Assessment of Mechanism of Pore Formation

Kerim Yildirim, Johannes Brachmann, Derya Dispinar, Andreas Bührig-Polaczek and Uwe Vroomen

Abstract In this study, changes in mechanical properties and hot tearing properties were investigated by adding 0.2 wt% Ni and 2 wt% Cu to Al–Si–Mg alloy. In addition, the porosity formation with regard to liquid metal cleanliness was investigated. Casting experiments have been carried out in certain compositions. The melt was degassed and up-gassed slowly. Micro100 alloy was added to the melt to investigate the effect of the distribution of porosity on the mechanical properties. A ring permanent mould was used to test the hot tearing characteristic and a tensile test bars were poured into permanent mould to test the mechanical properties.

Keywords Hot tearing · Tensile properties · Degassing · Porosity · Hydrogen

Introduction

Recently, light metal alloys have been increasingly used in a variety of applications such as automotive and aerospace due to their good pourability, lightweight and low strength-to-weight ratio. Since Al–Si–Mg (A356) cast alloys show excellent casting and mechanical properties, their use in these areas gives the best results [1].

A356 alloys contain 7% Si and 0.3% Mg. However, this type of alloy has problems with mechanical properties at high temperatures. Researchers are trying to improve mechanical properties by adding some alloying elements.

In this study, the addition of nickel and copper to the A356 alloy has been investigated as to how it affects the mechanical properties and hot tearing charac-

K. Yildirim (✉) · J. Brachmann · A. Bührig-Polaczek · U. Vroomen
RWTH Aachen University, Foundry Institute, Aachen, Germany
e-mail: kerim.yildirim@rwth-aachen.de

D. Dispinar (✉)
Istanbul University, Engineering Faculty, Metallurgical and Materials Engineering, Istanbul, Avcilar, Turkey
e-mail: deryad@istanbul.edu.tr

M. Tiryakioğlu et al. (eds.), *Shape Casting*, The Minerals, Metals & Materials Series,
https://doi.org/10.1007/978-3-030-06034-3_32

teristics of the alloy. Ni was added up to 0.2 wt% while Cu addition was increased up to 2 wt% [2]. On the other hand, with respect to temperature the amount of dissolved hydrogen in the liquid aluminum was evaluated by means of porosity formation [3].

During the casting of aluminum, the oxide layer formed on the surface of the liquid. This crack-like defects (bifilm), which can be mixed with the liquid, cause serious defects. In previous studies, it was observed that first bifilm is started to form and hydrogen in the liquid contributed to the formation of porosity which affects the mechanical properties [4–6].

Several different casting experiments were carried out with the given alloy concentrations. The melt was degassed and up-gassed slowly. To observe the relation between the distribution of porosity and its correlation on the mechanical properties, Micro100 alloy was added to the melt. To determine the hot tearing characteristic, a ring permanent mould was used to make a test. In addition, tensile test bars were poured into the standard permanent mould to test the mechanical properties.

Experimental Work

In this study, a A356 hypoeutectic Al–Si foundry alloy was used as the base alloy. Trace elements were added in concentrations of 1200 ppm and 1.5 wt% Ni and Cu, respectively. In order to understand the effect of trace elements, hydrogen and micro100 several different casting experiments were carried out with the given alloy concentrations. The melt was degassed and up-gassed slowly to obtain and increase the amount of hydrogen. Micro100 alloy was added to the melt to investigate relation between distribution of porosity and its correlation with mechanical properties. A ring permanent mould was used for measuring the hot tearing tendency of the alloys. Tensile test bars were poured into the standard permanent mould. The melt and pouring temperatures were kept constant at 720 C and 680 C, respectively.

It can be seen from Fig. 1 five different tensile test specimens were casted in accordance with DIN 50125 Form A to determine mechanical properties. In

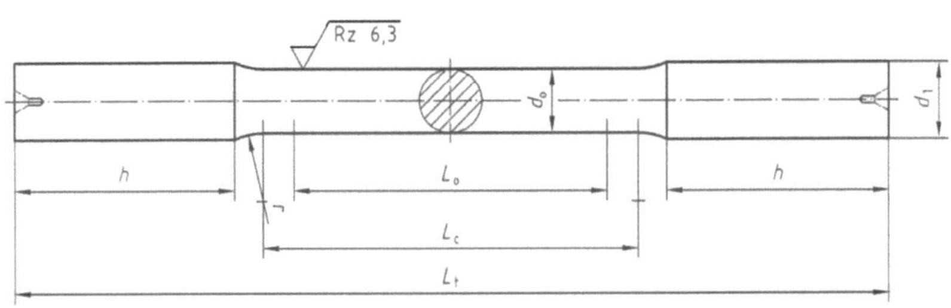

Key

d_o	test piece diameter	L_o	original gauge length ($L_o = 5\ d_o$)
d_1	diameter of gripped ends ($\geq 1{,}2\ d_o$)	L_c	parallel length ($L_c \geq L_o + d_o$)
h	length of gripped ends	L_t	total length of test piece

Fig. 1 Type A test piece, of circular cross-section with smooth and cylindrical ends [7]

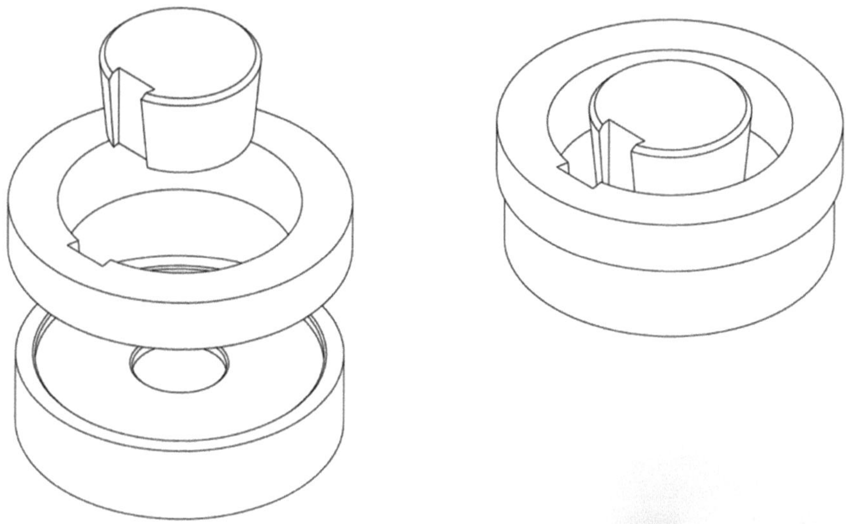

Fig. 2 Schematic drawing of the hot tearing permanent mould in an open and close position

addition, the spectroscopy sample and ring-hot tearing die cast were used to determine the chemical composition and hot tearing features before and after casting (Fig. 2).

Results

Chemical Results

In each casting, spectroscopy samples were poured and chemical analysis was carried out. Table 1, 2 and 3 show the average values of important elements in the different composition.

Table 1 Chemical composition of A356 with different combinations of Cu, hydrogen and Micro100

Compositions	Elements					
	Si	Cu	Mg	Ni	Fe	Mn
A356 + Cu	7,5	1,4	0,26	0,004	0,123	0,037
A356 + Cu + H	7,2	0,79	0,28	0,004	0,105	0,035
A356 + Cu + Micro100	6,65	1,51	0,25	0,007	0,076	0,011
A356 + Cu + H + Micro100	7,08	1,46	0,27	0,002	0,063	0,026

Table 2 Chemical composition of A356 with different combinations of Ni, hydrogen and Micro100

Compositions	Elements					
	Si	Cu	Mg	Ni	Fe	Mn
A356 + Cu + Ni	7,18	1,49	0,279	0,124	0,071	0,0239
A356 + Cu + Ni + H	7,28	1,61	0,273	0,115	0,061	0,0229
A356 + Cu + Ni + Micro100	7,01	1,42	0,274	0,116	0,064	0,0229
A356 + Cu + Ni + H+ Micro100	6,65	1,47	0,293	0,131	0,089	0,0006

Table 3 Chemical composition of A356 with different combinations of Cu/Ni, hydrogen and Micro100

Compositions	Elements					
	Si	Cu	Mg	Ni	Fe	Mn
A356 + Ni	7,39	0,06	0,28	0,114	0,054	0,024
A356 + Ni + H	7,46	0,04	0,27	0,123	0,057	0,025
A356 + Ni + Micro100	7,30	0,041	0,28	0,128	0,059	0,024
A356 + Ni + H + Micro100	6,69	0,014	0,281	0,132	0,078	0,009

Mechanical Results

A tensile test was performed to observe the effect of the addition of trace elements, oxide and hydrogen and the change in mechanical properties of porosity. The ultimate tensile strength (UTS) of the reference melt and the strength of this melt increase from 74 to 143 MPa with adding Cu and hydrogen to the A356, respectively, is shown in Fig. 3a. UTS increasing from 74 to 148 MPa via adding nickel, hydrogen and micro100 to the base alloy. In the last group of experiments, It can be seen that from Fig. 3c that the highest UTS is 139 with contamination of trace elements nickel and copper and hydrogen to the A356 (Fig. 4).

Metallographic and Fracture Results

It can be seen in Fig. 5 that the fracture surfaces of the tensile test sample were examined by scanning electron microscopy (SEM) (Fig. 6).

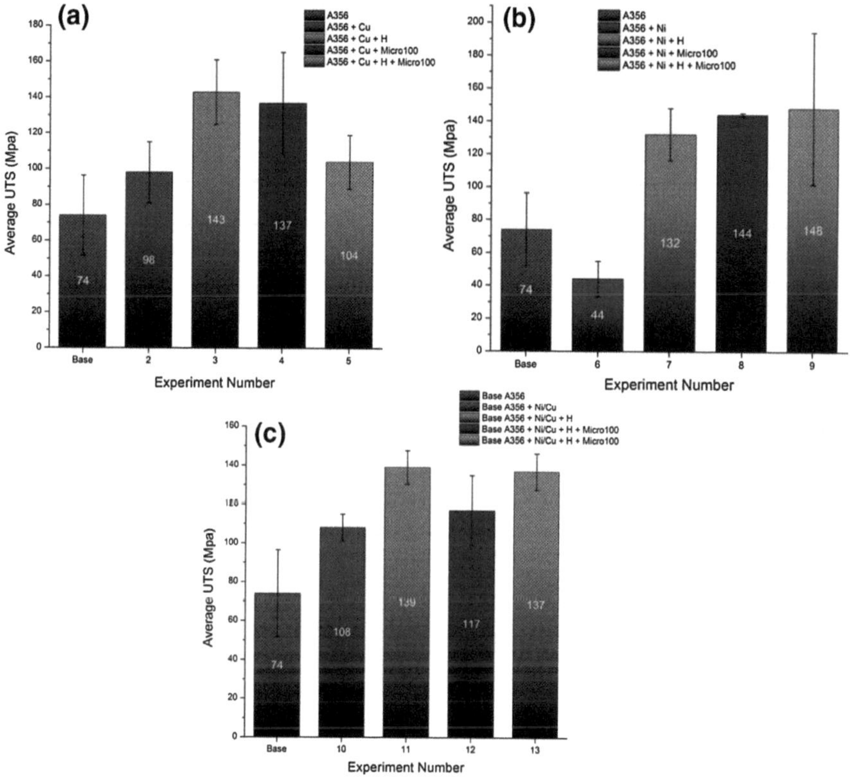

Fig. 3 Comparison of the relative UTS for the **a** Cu trace element, Hydrogen and micro100 **b** Ni trace element, hydrogen and micro100 **c** Cu and Ni trace element, hydrogen and micro100

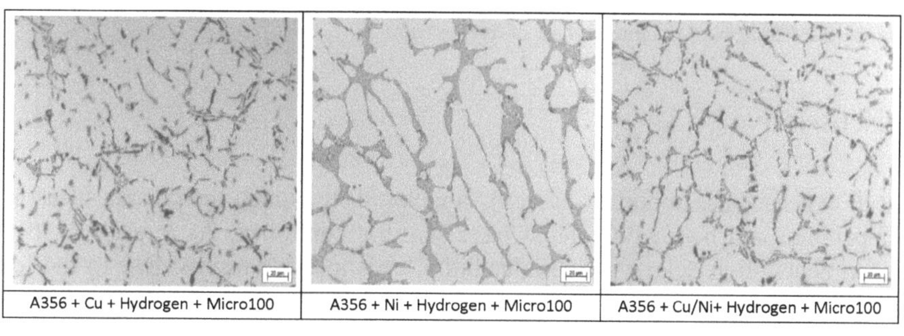

Fig. 4 Microstructure of A356 alloy obtained from different compositions

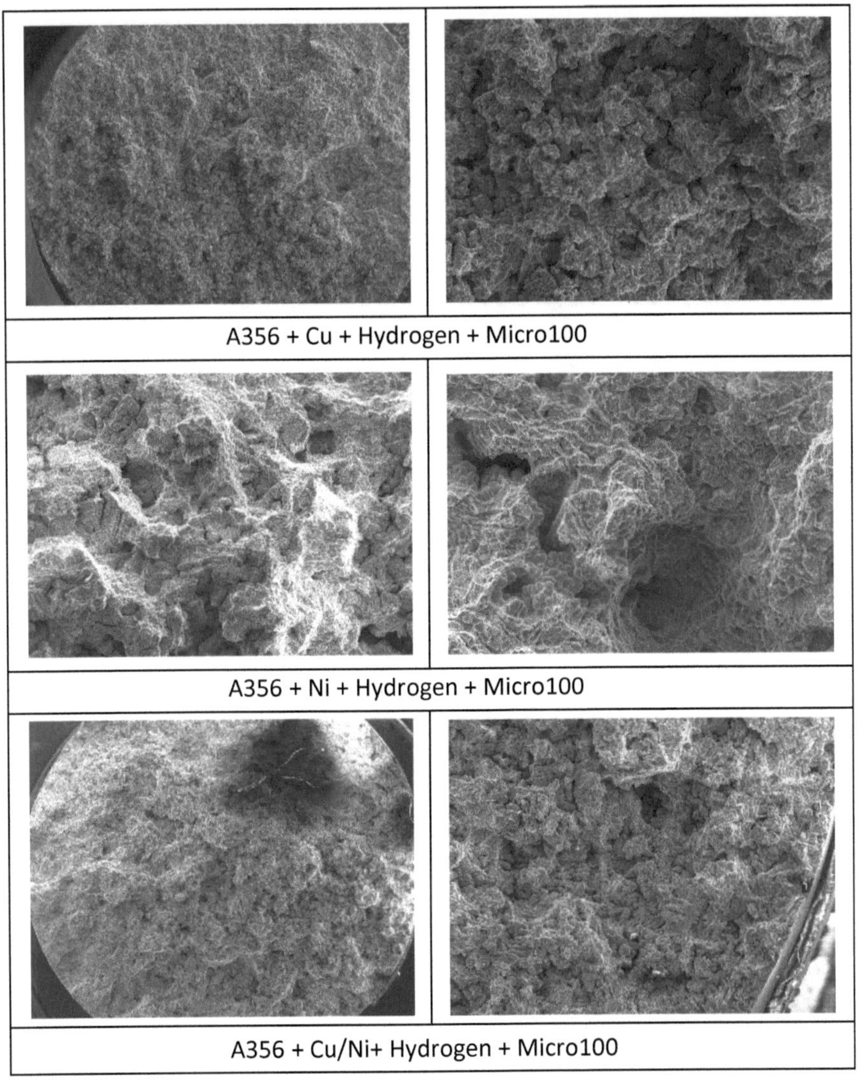

Fig. 5 SEM images of fracture surface of tensile bars

Discussion

In all experiments, important process variables such as melt temperature, lcasting temperature and hydrogen content were sensibly controlled. As a result of these experiments, high degree of reproducibility leads to close results. Due to the low hydrogen content in the experiments, the melt was up-gassed with wood to increase the hydrogen content and achieve high content.

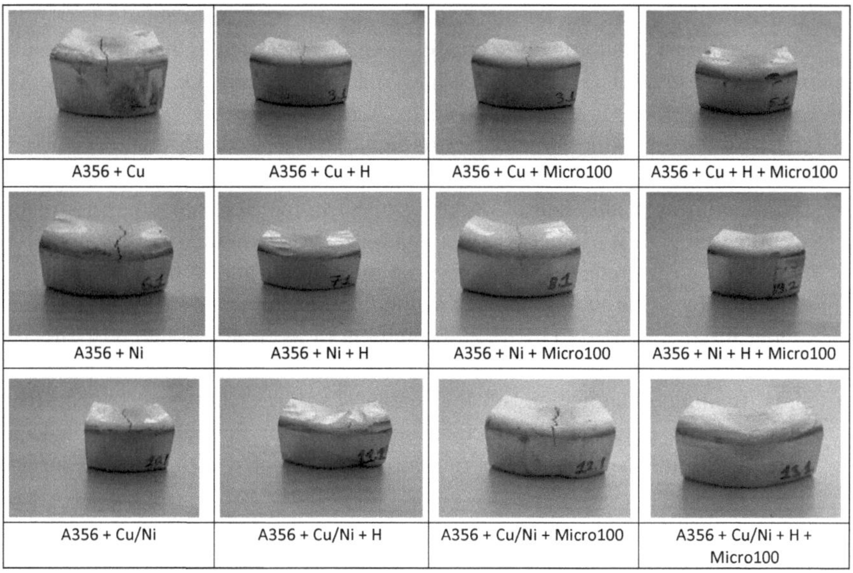

Fig. 6 Hot tearing of the different compositions in the hot spots of castings

The change in hydrogen content can be observed during the melt process. The excess of hydrogen concentration at the beginning indicates the presence of the moisture in the furnace liner and the porosity in the probe. The highest porosity was observed in specimens going upwards with the wood. The effect of the oxide additives on the mechanical properties is rather severe compared to an uncontaminated solution. This is in accordance with findings reported earlier by Dispinar [8, 9].

Hot tearing occurs completely randomly. Under the same conditions, there is no rule that hot tearing should occur in the same morphology in each alloy. Hot tearing is seen in Al–Cu alloy system. In this study, hot tearing was also observed in Al–Cu, Al–Ni and Al–Cu–Ni systems. However, hot tearing is not observed in the castings where the hydrogen ratio is increased.

Conclusion

The effect of Ni an Cu trace elements and also influence of hydrogen percentage and micro100 of as-cast foundry alloys in permanent mould casting process were investigated in the present work. Tensile test bars were performed to measure the mechanical properties. Moreover, microstructural and fractographic investigations were achieved to analyse the microstructures features involved in the fracture process.

- Ni extremely effects the tensile properties of the permanent mould cast A356 alloy in as-cast condition. It decreases both the yield strength ($R_{p0.2}$) and ultimate tensile strength (UTS) by 87% and 37%, respectively. On the contrary, The Cu addition increases $R_{p0.2}$ and UTS by 42% and 25%.
- The addition of oxide (Micro100) to liquid metal leads to the formation of smaller diameter porosities in a larger distribution.
- The cleanliness of the melt has a major effect in the development of porosities over the cast part. It has been shown that it is effective to degrade the amount of hydrogen and degrade it for the purpose of metal cleaning. It has been understood that by using the up-gassing method, either the wood or the argon-water vapour mixture is used to increase the amount of hydrogen in the most effective manner.
- SEM fracture examination of the fracture surface confirms the Si-driven quasi-cleavage nature of the fracture, as indicated by the number of cleavages planes and brittle Si-flakes. A lot of dimples are detected in the fracture surface.

Acknowledgements I am very grateful to Dr.-Ing. Uwe Vroomen for giving me the opportunity and great support to this study.

I would like to thank Johannes Brachmann for guidance and support through the course of my thesis, and for his excellent supervision during the entirety of this study.

I would like to express my gratitude to Associate Professor Dr. Derya Dispinar for his scientific opportunities, and for his enthusiastic support and advice.

References

1. Di Giovanni MT (2014)The Influence of Nİ and V trace elements on the high temperature tensile properties of A356 aluminum foundry. Norwegian University of Science and Technology
2. Casari D, Ludwig TH, Merlin M (2014) The effect of Ni and V trace elements on the mechanical properties of A356 aluminum foundry alloy in as-cast and T6 heat treated conditions. Mater Sci Eng A 610:414–426
3. Campbell J, Harding RA (1994) Solidification defects in casting. TALAT, European Aluminum Association
4. Dispinar D, Campbell J (2004) Critical assessment of reduced pressure test. Part 1: porosity phenomena. The University of Birmingham
5. Dispinar D, Campbell J (2011) Porosity, hydrogen and bifilm content in Al alloy casting. Mat Sci Eng A 528:3860–3865
6. Akhtar S, Arnberg L, Di Sabatino M, Dispinar D, Syvertsen M (2008) A comparative study of porosity and pore morphology in a directionally solidified A356 alloy. Int J Metalcasting. Winter 09
7. BEST Collection 08 (2018) Oberflächenbehandlung-2018, BEST Collection 21; DIN komplett-2018
8. Ludwig T, Dispinar D, Di Sabatino M, Arnberg L (2012) Influence of oxide additions on the porosity development and mechanical properties of A356 aluminium alloy castings. Int J Metalcasting 6(2):41–50
9. Dispinar D, Akhtar S, Nordmark A, Syvertsen F, Di Sabatino M, Arnberg L (2012) Tensile properties, porosity and melt quality relation of A356. Suppl Proc 2:201–208

Aluminum Alloy with High Mg Content: Casting Studies for Microstructural Evolution, Phase Formation and Thermophysical Properties with Different Alloying Elements

Armagan Gul, Özen Gürsoy, Eray Erzi, Derya Dispinar and Eyup Kayali

Abstract Global strategy for material science in the automotive industry is to use the lightweight material. In the present work, the casting of aluminum alloy with high magnesium content has been performed with the addition of B, Sr, Ti and Zr. Gravity sand casting and permanent mold casting methods were used. Microstructural evolution has been studied along with phase formations during solidification. Microstructural changes of each alloying element have been observed via optical and SEM microscopy. XRD analysis has been done for evaluation of phase formations. TGA analysis has been done to evaluate thermos physical properties. The reduced pressure test was carried out to quantify melt cleanliness and bi-film content. Indentation tests were carried out, and a linear relationship was found between mechanical test results and bi-film index.

Keywords Aluminum · Vanadium · Boron · Casting · Grain refinement

Introduction

High magnesium content with different alloying elements yields to pore and bi-film formations in case of gravitational casting without degassing or flux tablets. Because of the tendency of the Al–Mg alloy to bi-films and pores formations,

---

A. Gul
Oyak Renault, 16372 Bursa, Turkey

Ö. Gürsoy · E. Erzi · D. Dispinar (✉)
Faculty of Engineering, Department of Metallurgy and Materials Engineering, Istanbul University Cerrahpasa, 34320 Istanbul, Avcılar, Turkey
e-mail: deryad@istanbul.edu.tr

E. Kayali
Faculty of Chemical and Metallurgical Engineering, Istanbul Technical University, Istanbul, Maslak, Turkey

M. Tiryakioğlu et al. (eds.), *Shape Casting*, The Minerals, Metals & Materials Series,
https://doi.org/10.1007/978-3-030-06034-3_33

reduced pressure test (RPT) and microstructural evolutions have been evaluated. Campbell studied and showed that oxide films within the metal melt are the source of pore formation and property degradation, which cause post-failure under service conditions [1]. In order to evaluate final properties, bi-film index studies and pore formation must be studied and assessed for a given alloy system [2]. This study can be done via RPT tests [3]. Therefore, low mechanical properties can be identified if it is related to melt quality or alloying type. Therefore it will be possible to assess if the alloy requires degassing or another treatment.

Experimental

AlMg5 and AlMg 5083 have been used as base alloys. Strontium, boron and titanium have been added to the base alloy to study alloying effects. Gravity molds have been used. According to Fig. 1, step size mold, which serves to study solidification upon different cooling rates, and U-type single column specimen molds have been used. Gravitational permanent mold casting has been done. Electrical resistance melting furnace has been set to 750 °C as smelting temperature level. A3 type casting crucibles have been used to smelt and cast the alloys. Despite the positive effect of degassing [2], no treatment has been done to investigate alloying elements effect on pore formation and bi-film counts. After a stage of alloying and 45 min holding time, casting has been done to preheated molds. Castability studies have been done using a spiral permanent mold. Three main compositions have been studied. Different preheating has been applied.

Bi-film Index Studies

For bi-film study and counts cupcake geometry RPT mold has been used to perform casting under vacuum condition. As a standard, 5 min have been set for holding time in the RPT machine after the casting of the alloy into the cupcake crucible.

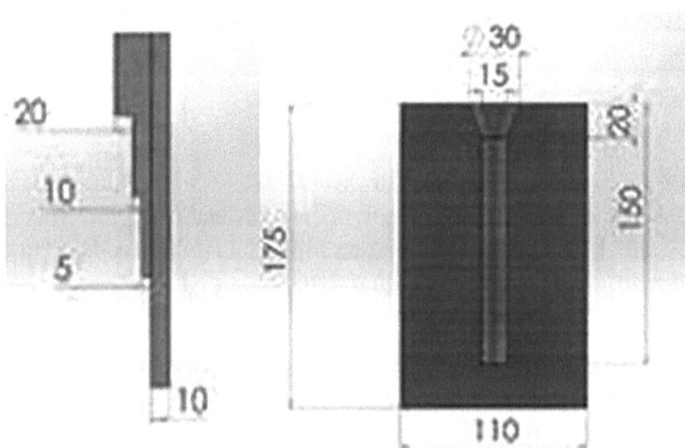

Fig. 1 Molds that have been used for tensile tests and microstructural evaluation

Microstructural Studies

Cold resin molding has been selected to investigate microstructure. SiC grinding papers have been used to prepare the specimens in Metkon grinder/polisher machine. 400/800/1200 particle grades have been selected.

Mechanical Tests

Tensile tests have been done in Instron testing machine with 10^{-2} strain rate. Specimens have been prepared according to ASTM B 557M specification. Each test has been repeated with 3 specimens, and average results have been taken. Vickers hardness tests have been done under 0.2 kg load in Zwick micro hardness machine. Each microhardness test has been repeated 3 times, and average results have been taken into account.

Results and Discussion

In Table 1, the major alloying scheme has been given for present studies. Alloying derivation of titanium boron strontium zirconium has been studied. AlMg5 has been evolved to AlMg4 as a result of alloying elements.

According to Table 2, from spiral mold fluidity studies, the addition of titanium increase fluidity of AlMg5 alloys for the same preheating condition when compared with boron strontium alloyed counterpart. Zirconium addition did not give better results in terms of fluidity. It must be taken into account that fluidity study has been conducted with a permanent steel mold without any fluidity agent or mold surface modifier such as special paints.

In Table 3, evolution of bi-film index has been studied. As a result, antimony addition has been found without any effect on bi-film formation. SrTi vs SrZr and SrTiZr have yielded similar results. Boron and boron strontium have caused higher index numbers which are negative for final properties.

Table 1 List of alloy complexity of the castings

Alloys	Composition
AlMg5	Standard composition
AlMg5	5083 H111 grade
AlMgXY	X: Boron and Y: Strontium
AlMgX	
AlMgXT	T: Titanium
AlMgXZ	Z: Zirconium
AlMgXYZTW	W: Antimony

Table 2 Spiral mold fluidity values

	Distance reached	Preheat T °C
AlMg–BorSr	23	100
AlMg–BorSr	24,5	150
AlMg–BorSr	28,5	200
AlMg–TiBorSr	24	100
AlMg TiBorSrZr	24	100

Table 3 Bi-Film index counts and numbers

AlMg5+	Bi-Film index	Number
Bor	416	496
Bor	322	349
BorSr	355	172
BorSr	205	252
BSrTi	44	69
BSrTi	162	244
BSrTi	65	84
BSrTi	107	150
BSrSnTi	230,9	159
BSrSnZr	268	182
BSrSnTiZr	182,8	183

Effect of Ti and Zr on AlMgB microstructure has been given in Fig. 2. Ti and Zr addition have positive effect on microstructure by preventing sharp dendrite formation (Fig. 2b and c).

Bi-films have neutral buoyancy, they are dispersed in the metal melt. As we have not used degassing tablets or degassing system and the permanent mold we used has a single line of metal fill without a filling system. Thus upon pouring into the die cavity, existing bi-films have been in the melt, and they have been randomly dispersed in the final cylindrical specimen.

According to tensile tests in Table 4, bi-films and porosities have a dominant effect on mechanical properties as the values are lower than expected levels. Thus, alloying elements' effect is not visible in the results. Inspected fracture surfaces indicate that the specimens have failed at porosity and stress intensity locations.

This is attributed to bi-films that are entrapped in the structure that acted as defect spots which caused failure. Thus treatment of metal melt is very important for stable mechanical properties and lower bi-film index values according to Dispinar [2].

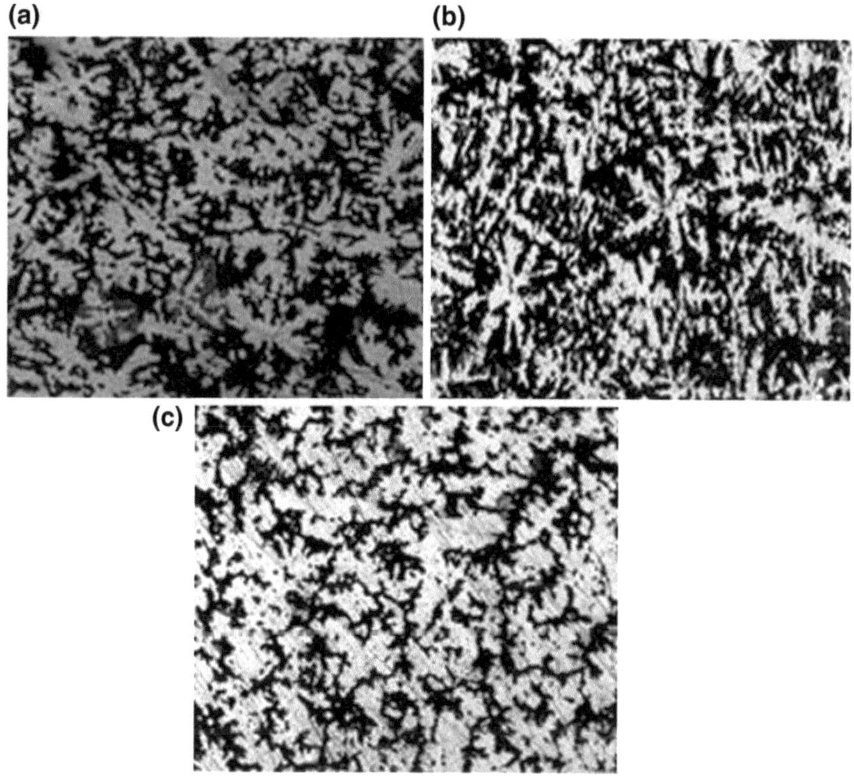

Fig. 2 **a** AlMgB Microstructure **b** AlMgBTiZr Microstructure **c** AlMgBorTiSr

Table 4 Tensile tests results (in MPa, % elongation)

	Yield strength MPa	Tensile strength MPa	%El. at break
AlMg4B	82	151	6,7
AlMg4BSr	84,5	140,5	6.35
AlMg4BZr	75	149,5	10,5
AlMg4BSrTi	81,3	157,6	13,45
AlMg4BSrZr	80	147	8,25

Conclusions

Bi-film count is decreased with alloying elements to Al-Mg5 as shown in Table 3. Titanium and zirconium are successful alloying options even with non-degassed systems. Even with higher Ti–Zr contents, bi-film counts are similar to the low alloying composition.

Melt quality is an important factor when high content master alloying is desired. Tensile tests show lower experimental values than predictions. As specimens have been fabricated via conventional gravity casting into the steel molds, these low values have been attributed to faster cooling rates, porosity formation and casting defects. Bi-films entrapped in the structure have the dominant effect of porosity and defect formation as no special melt quality treatment have been applied such as a mixer, degassing by Argon or degassing tablets. On the other hand, filling system and mold that have been used have no special arrangement to modify bi-film entrapment by changing melt flow and filtering bi-films.

It has been concluded that higher magnesium content for aluminum alloys requires melt quality treatment methods and more uniform/slow cooling rates. Otherwise, porosities and metal films will stay in the microstructure causing abrupt failure. For high magnesium and high master alloy content casting studies, sand casting, degassing, sustaining nonturbulent metal flow, low-pressure casting must be taken into account if good casting quality is desired.

For future work, different types of mold will be used with higher preheating conditions and with different filling types to prevent bi-film effect and shrinkage upon solidification.

Acknowledgements A. Gul is supported by TUBITAK- BIDEB. 2211-/A national fellowship program for Ph.D. studies.

Doc. Dr Murat Baydoğan from ITU Metallurgy and Material Department is highly acknowledged for his support in mechanical tests setup.

References

1. Campbell J (2006) Entrainment defects. Mater Sci Technol 22:127–145
2. Uludag M, Çetin R, Dispinar D, Tiryakioglu M (2017) Characterization of the effect of melt treatments on melt quality in Al-7wt %Si-Mg alloys. Metals (5):157
3. Dispinar D, Campbell J (2006) Use of bifilm index as an assessment of liquid metal quality. Int J Cast Met Res 19:5–17

Correlation Between Melt Quality and Machinability of Al9Si3Cu HPDC Alloy

Ibrahim Halil Kalkan, Omer Vardar, Ibrahim Göksel Hızlı, Eray Erzi, Ozen Gürsoy and Derya Dışpınar

Abstract High-pressure die casting is the most suitable method for mass production of Al–Si alloys. The cast parts are typically subjected to machining for the final finish. In this work, the cutting tool life was investigated with regard to the melt quality. The melt cleanliness is defined by reduced pressure test and bifilm index was measured. Since bifilms are oxides, the tool life can significantly be affected by the presence of bifilms in the cast part. The study was carried out with four different melt cleanliness. Sixteen cutting tools were changed after 10000 parts for the castings with bifilm index higher than 200 mm. When bifilm index was decreased to 20 mm, the tool life was increased to an average of 4500 parts and totally four cutting tools were changed in 12000 parts.

Keywords Bifilm · Melt quality · High-pressure die casting · Cutting tool · PCD tool

Introduction

One of the most important expense items in the machining of aluminum cast parts is cutting tools. A cutting tool's cost depends on production methods, material variation or customer requests [1]. Regardless of the quality of the cutting tool, the most important feature request in these types of tools is their long life and their ability to make the measurements as stable as possible. Very expensive coatings and materials are made to supply that stability and long life requests for cutting tools [2, 3]. The machinability of

I. H. Kalkan
Daimler Mercedes-Benz Turk, Istanbul, Turkey

O. Vardar
Can Metal, Yeşilova Holding, Bursa, Turkey

I. G. Hızlı · E. Erzi · O. Gürsoy · D. Dışpınar (✉)
Metallurgical and Materials Engineering Department, Istanbul University, 34320 Istanbul, Avcilar, Turkey
e-mail: deryad@istanbul.edu.tr

M. Tiryakioğlu et al. (eds.), *Shape Casting*, The Minerals, Metals & Materials Series,
https://doi.org/10.1007/978-3-030-06034-3_34

aluminum parts according to their type of alloy. Due to the pure aluminum inherent softness, it is very easy to machining, but the chip ability is very low. Therefore, the aluminum is alloyed with Si element to increase the machinability of the aluminum. However, as the Si element increases the machinability of the aluminum, the life of the tool is reduced because of its hardness (>400 HV). For this reason, PCD (Polycrystalline Diamond) tools are used instead of traditional carbide tools to increase the life of tool and stability of machining. As is known, some dimensions may be within very challenging tolerances on the working draw; therefore, the role played by the tools is becoming very important to produce these tolerances. Broken or abraded tool has to be changed in during production, so it can be very expensive because basically, it means: stopping production, setting up new parameters, making new parts and measuring to ensure with related dimensions (sometime with caliper, sometimes coordinate measurement machines). The worst of the part is loss of stability for production.

A basic machining process and detailed cutting tool areas can be seen in Figs. 1 and 2 [1]. Wearing or breaking of tool is related with shear plane angle, rake angle

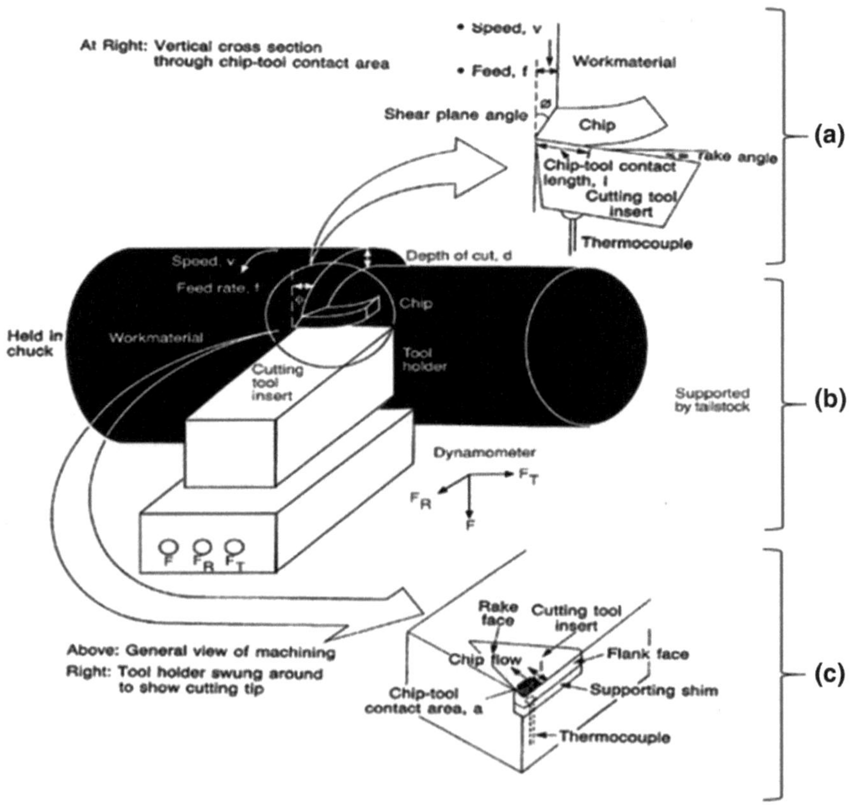

Fig. 1 Basic operation of turning is showed in this figure. Thermocouple is used for laboratory types [1]. **a** Shear plane is area for contact surface on the tool. **b** Typical turning machine diagram. **c** Basic schematic drawing for cutting tool and its areas

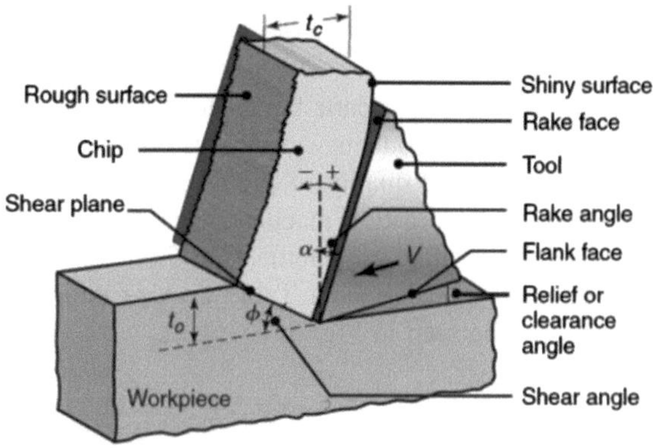

Fig. 2 Detailed cutting tool areas for basic machining

and speed of tool. It is an open secret to increase shear stress force is useful to increase the tool life, but for aluminum parts, quality of the melt must to be taken into consideration and adapted into these parameters.

As noted above, for Al–Si alloys, tool life is binding with percentage of Silicon base on material. PCD tools with the diamond particles show remarkable resistance to wear or friction [4, 5]. Theoretically, the stability of PCD tools should be more effective during Al–Si alloys. But in practice, we see that the quantity of tools used is not the same and differ with each production, even though the percentage of silicon is the same and same alloy. The stability and quantity of the tool life should not be directly related with the silicon percentage. Because this amount of differences of the tool quantities will be directly related to the quality of the melting aluminum.

During the production of aluminum ingots and castings, the surface oxide on the liquid may be folded into the bulk liquid to produce crack-like defects (bifilms) that are extremely thin, but can be extensive, and so constitute seriously detrimental defects. Campbell and Dispinar have done an extensive work in the field of casting quality [6]. Yang and Campbell proved the optimized pouring [7]. Rezvani et al. [8] and Nyahumwa et al. [9] showed the importance of filling systems over the reliability of the mechanical properties of castings [10]. Basic bifilm formation is shown in Fig. 3 [11].

In this study, parts, which are produced with AlSi9Cu3 (ENAC 46000), are investigated according to their bifilm ratio for tool life.

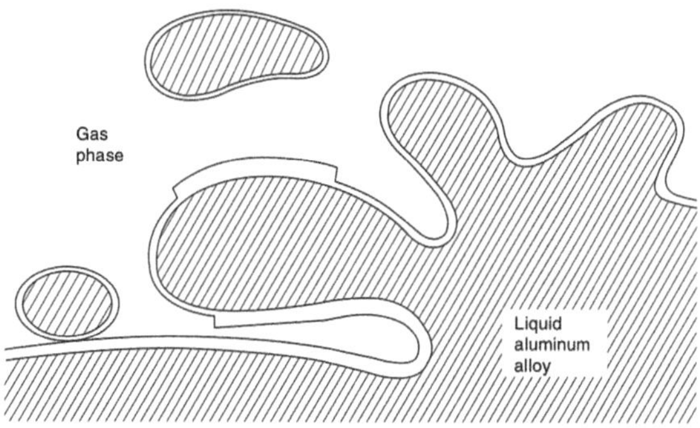

Fig. 3 Surface turbulence and formation of bifilm

Experimental Study

Two different die cast parts have been produced with AlSi9Cu3 alloy. The difference between the products was that the bifilm index ratio, in other words, quality of melting. Ratio and quality of ingots are shown in Fig. 4. Casting has been done with 430 T Zitai casting machine. RPT analyses were done in 30 mbar vacuum pressure and casting temperature was 715 °C. Production quantities and usage of tool numbers are shown in Table 1. Also unfinished parts with runners and air pockets can be seen in Fig. 5.

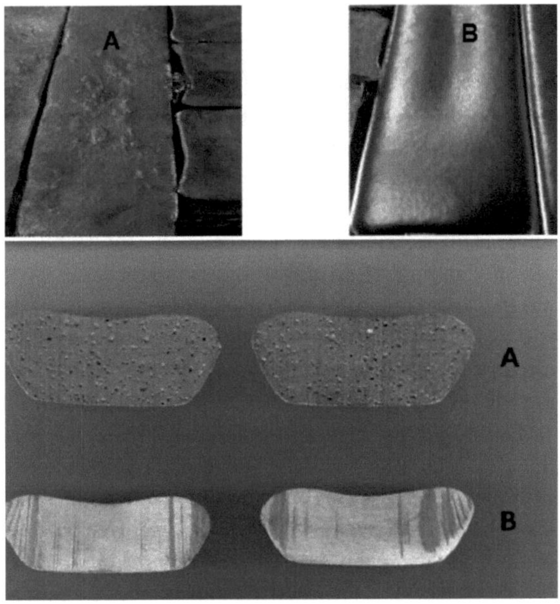

Fig. 4 Quality of ingots is showed in this picture. There is clearly difference terms of quality between A and B both for experimental and visually. A: Lack of degassing and turbuleted ingots B: Effective degassing, no-turbuleted ingots

Table 1 Summary of production and scrapped PCD tools quantity

Quality of ingots	Total machined parts	Number of usage PCD tool
Bifilm index > 200 mm (A)	10.000	16
Bifilm index < 20 mm (B)	12.000	4

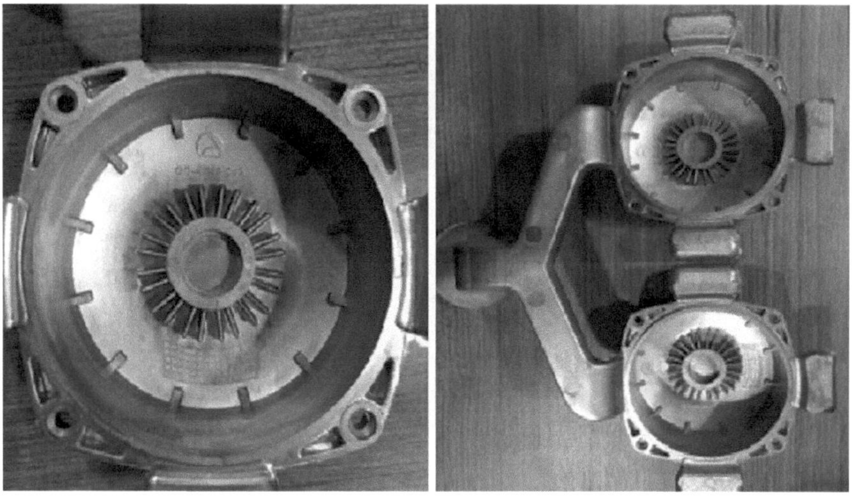

Fig. 5 Unfinished part is with runner and air pockets. The inside of part is machined in turning machine

Results and Discussion

After constant die casting and machining parameters, the difference of scrapped PCD quantities is shown in Table 1. It is clear that there is a significant decreasing, four times less, for usage of PCD tools with minimum bifilm index casting.

Table 2 is examined production which is cast with B ingots (Bifilm index < 20 mm) for PCD tools and their machined quantities.

In light of Table 2, one tool can machine 5500 parts, so it is nearly half of production quantity. If it is spoken on quality of B ingots, 5500 machined parts are a remarkable performance when compared to quality of A ingots.

In below, Figs. 6, 7, 8, 9, 10 and 11, 1. and 2. PCD tools are photographed according to their visual changes.

Table 2 Examination of production which is produced with B ingot

PCD tool number	Machined quantity	The reason for change	Amount of intermetallics parts
1. PCD tool	5500	Abraded	13
2. PCD tool	4250	Abraded Broken	51
3. PCD tool	*95*	*Broken*	*3*
4. PCD tool	2100	Not abraded, production finish	5

Fig. 6 1. PCD tool in 500th part

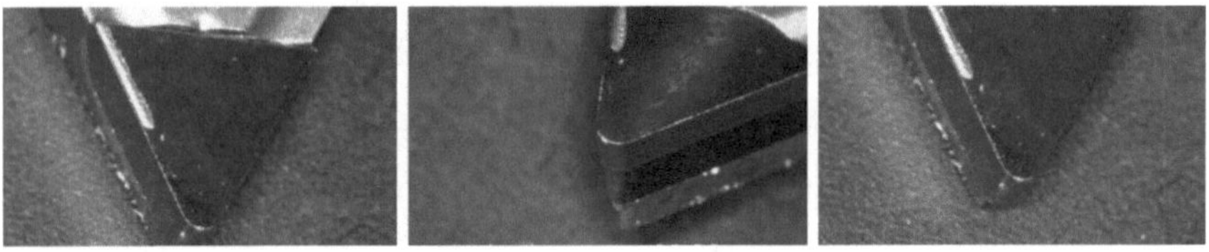

Fig. 7 1. PCD tool in 1000th part

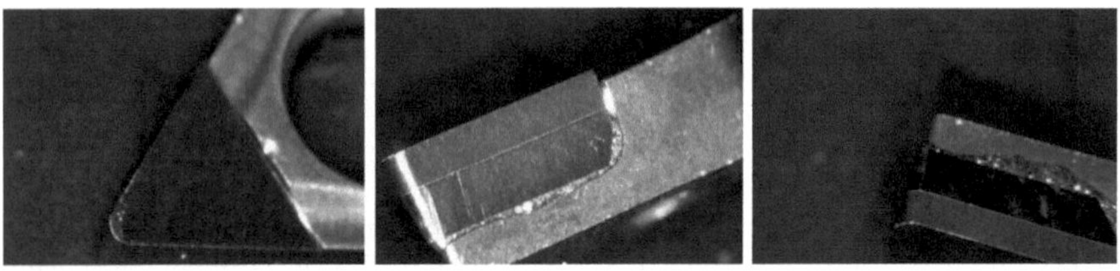

Fig. 8 1. PCD tool in 2500th part

Fig. 9 1. PCD tool in 350th part

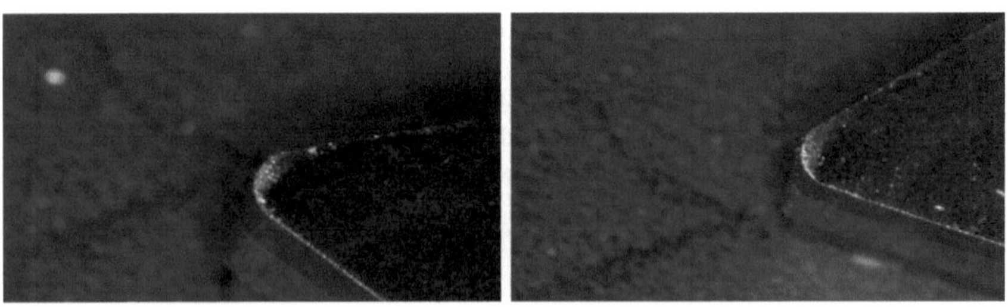

Fig. 10 1. PCD tool in 5500th part

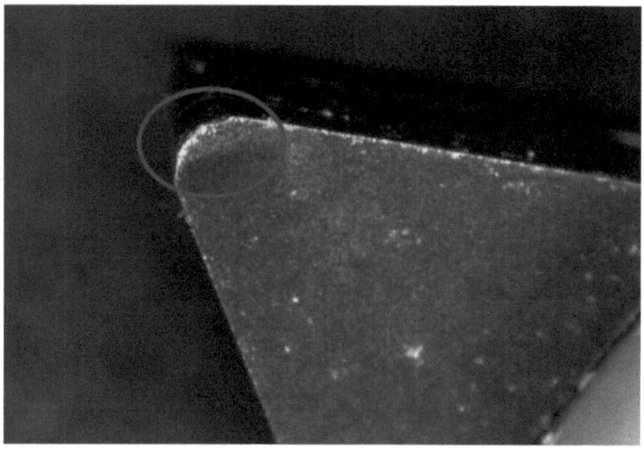

Fig. 11 2. PCD tool in 4250th part. Shear plane area is not just abraded, it is also damaged mechanically

1. PCD tool is machined 13 intermetallic parts before its abrasion. Figure 9 shows that clearly the shear plane area is abraded and changing were inescapable for stability of tolerances.

Contrary to 1. PCD tool, 2. PCD tool is exposed 51 intermetallic parts and the characterization of abrasion is different as seen in Fig. 11. Shear plane area is abraded but in addition to this it's also damaged mechanically. But anyway, this is better result than ingots A.

The part which is causing the change of the 2. PCD tool and the intermetallics, it contains, are shown in Fig. 12.

Figure 13 shows the 3. PCD tool. It has surprisingly broken in 95th part and it was exposed 3 intermetallic part. However, when on got focused that PCD tool has damaged completely mechanically and it was on the rake angle area. Related failed part can be seen in Fig. 14.

Failed part, which is shown in Fig. 14, is examined under scanning electron Microscope in Fig. 15 and also EDS analysis performed on this area in Fig. 16.

It is known that the life of the cutting tools depends on the type of machining material, machine speed, rake angle and shear angle. Apart from these, the most important point to look at is the quality of liquid aluminum. Especially if more

Fig. 12 Failed part which is causing the change of tool. Intermetallics can be seen clearly

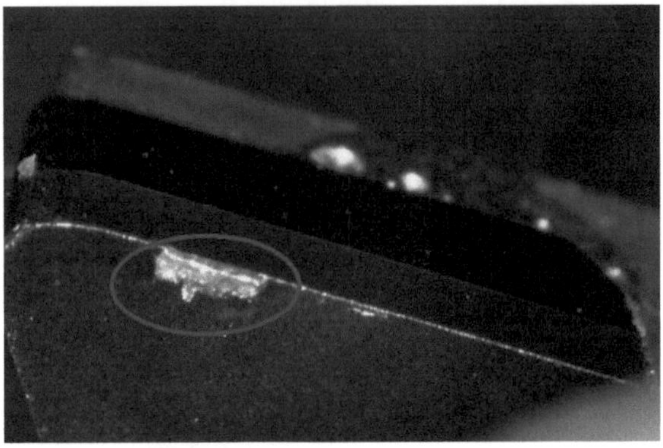

Fig. 13 3. PCD tool is in 95th part. As seen is the damaged area is clearly broken and it is not shear plane area; it's on rake angle area

Fig. 14 Failed part which is causing the change of 3. PCD tool. Intermetallic can be seen clearly

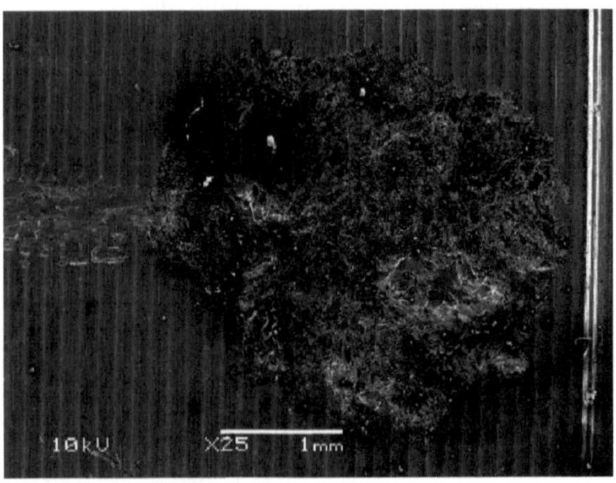

Fig. 15 Failed area is showed under SEM

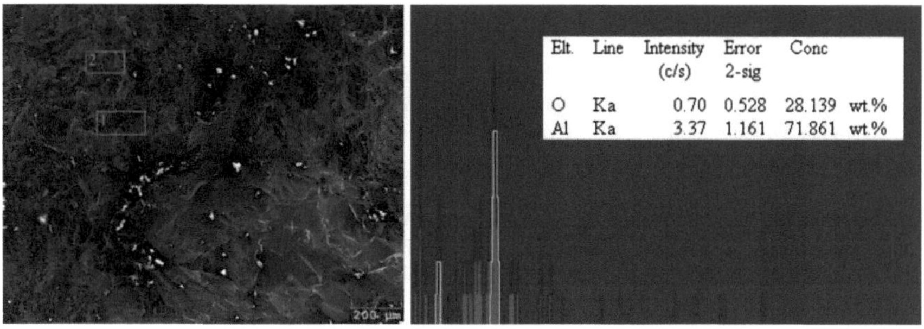

Fig. 16 EDS result is showed for failed area

stable production is desired, the performance of the cutting tools starts from quality of liquid aluminum. As seen in this paper, the performance of the cutting tool is also improved when the quality of the liquid metal is numerically increased during ingot production. Reduced pressure test is generally used for quality of melting aluminum to clarify numerically. It has been already proved that effective degassing and without turbulent casting is mechanically improved the aluminum parts. It is also an important finding that bifilm index affects the performance of the tool in addition to all these. Also both of SEM and EDS results can be shown as an important evidence for effect of melt quality on tool life. As is seen the related area is exact root cause to breaking of tool. Finally, it should not be forgotten that performance of cutting tools are also affecting to efficiency of production. Because changing of cutting tools mean uncalculated quality assurance costs; extra measuring time and risking for stability in CNC machining production.

Conclusion

Reduced pressure test is generally used for quality of melting aluminum to clarify numerically. The performance of the cutting tool is improved with liquid metal quality. Effective degassing and non-turbulent casting is mechanically improved the aluminum parts. The service life of the PCD tool tips is directly proportional to the quality of the liquid metal.

Acknowledgements This work was supported by Arslan Makina Dokum A.S, Istanbul, Turkey.

References

1. Trent EM, Wright P (2000) Metal cutting, 4th edition
2. Kim J (2005) Cutting performance of Ti–Al–Si–N-coated tool by a hybrid-coating system for high-hardened materials. Surf Coat Technol 193(1):249–254

3. Klocke F (1999) Coated tools for metal cutting–features and applications. CIRP Ann Manuf Technol 48(2):515–525
4. Davim JP (2008) Machining: fundamentals and recent advances. Springer, London, p 3
5. Davim JP (2011) Machining of hard materials. Springer, London
6. Dispinar D, Campbell J (2004) Critical assessment of reduced pressure test. Part 1: porosity phenomena. Int J Cast Met Res 17(5):280–286
7. Yang X, Campbell J (1998) Int J Cast Met Res 10(5):239–299
8. Rezvani M, Yang X, Campbell J (1999) Trans Am Foundrym Soc 107:181–188
9. Nyahumwa C, Green NR, Campbell J (1998) AFS Trans 106:215–224
10. Campbell J (2003) Castings, 2nd edn. Butterworth

Change in Sr Modification by Duration and Its Effect on Mechanical Properties of A360 and A413 Alloy

Inal Kaan Duygun, Ozen Gursoy, Eray Erzi and Derya Dispinar

Abstract Strontium modification has many advantages for the mechanical properties of Al–Si alloys. Holding time is an important factor in Sr modification. Its high affinity to oxygen may result in forming of $SrO.Al_2O_3$ spinel oxides. Therefore, Sr quantity in the melt can decrease by time due to fading. In this work, this phenomenon was investigated in A360 and A413 alloys under different casting conditions. The melt was held at 700 °C for 4 and 17 h after the addition of 300 ppm Sr. Metallographic examination was carried out to observe the change in eutectic morphology. Melt quality change was measured by means of reduced pressure test. In addition, tensile test samples were produced. It was found that there is a clear correlation between mechanical properties and bifilm index of Al–Si alloys.

Keywords A360 · A413 · Sr modification · Melt quality · Mechanical properties

Introduction

Aluminum–silicon-based alloys are important materials for many different areas, especially in automotive and aircraft industries. Their good mechanical properties, castability and corrosion resistance make them very useful for use in cylinder blocks and heads, plain bearings and internal combustion engine pistons [1–3]. Eutectic silicon morphology is one of the most important factors in castability and mechanical behaviors of hypoeutectic and eutectic Al–Si alloys. It is often possible to control mechanical properties in these alloys depending on shape and density of eutectic silicon phase. Modification is an effective melt treatment in order to obtain good quality casting parts. Improved properties of Al–Si alloys can be obtained by

I. K. Duygun · O. Gursoy · E. Erzi · D. Dispinar (✉)
Faculty of Engineering, Department of Metallurgical and Materials Engineering, Istanbul University-Cerrahpasa, Istanbul, Avcılar, Turkey
e-mail: deryad@istanbul.edu.tr

M. Tiryakioğlu et al. (eds.), *Shape Casting*, The Minerals, Metals & Materials Series,
https://doi.org/10.1007/978-3-030-06034-3_35

modification in two different ways, either by rapid cooling rate in metal mold or by addition of a modifier element, such as Sr or Na. Sr is one of the most used modifier elements for this purpose [2, 3].

Eutectic morphology is changed from brittle to coarse flakes which result in poor mechanical properties to fine and fibrous structure which affects the mechanical properties in a positive way [4]. Microstructure evolution of hypoeutectic Al–Si alloys during solidification occurs in two steps: primary dendrite Al phase formation and the subsequent eutectic transformation [5]. The hypoeutectic Al–Si alloys show eutectic phase located between the α-Al dendrites. Unmodified eutectic Al–Si alloys exhibit an acicular or lamellar eutectic silicon well dispersed throughout the aluminum matrix [6]. The eutectic silicon phase crystallizes into a coarse, plate-like morphology during eutectic formation. Coarse morphology of flaky eutectic silicon is detrimental for the performance of Al–Si casting parts. Sharp corners of this structure are the possible sites for increment of stress concentration which may cause fracture during use. Sr modification may reduce stress concentration in these regions by altering morphology and improve the mechanical properties of the alloy [7–9].

In this work, the effect of holding time of A360 and A413 alloys was investigated by comparing modified and unmodified morphologies of alloys at 700 °C casting temperature. Reduced pressure test was also used to evaluate changing metal quality by means of bifilm index. Mechanical properties of samples were also investigated by comparing modified and unmodified alloys.

Experimental Work

The A360 and A413 ingots were cut into pieces weighing 500 g and charged into resistance furnace. The melts were modified with addition of Sr by using Al–15%Sr master alloy, and for all samples, it was chosen 300 ppm Sr for eutectic modification. In order to assess the effect of temperature on the modification, the melts were held at 700 °C. For each temperature, four samples were poured into metal mold at 4th and 17th hours. The chemical composition of two different alloys used in this work is summarized as in Tables 1 and 2.

Table 1 Chemical composition of A360 alloy

Fe	Si	Cu	Mn	Mg	Zn	Ni	Ti	Pb
0,50	9,00–10,00	0,10	0,40–0,60	0,30–0,45	0,10	0,10	0,15	0,05

Table 2 Chemical composition of A413 alloy

Fe	Si	Cu	Mn	Mg	Zn	Ni	Ti	Pb
0,60	11,50–13,50	0,10	0,20	0,10	0,10	0,10	0,15	0,05

For the investigation of eutectic Si morphology of modified and unmodified samples, optic microscope and Clemex vision software was used. In order to measure Bifilm index of specimens, reduced pressure test was used. Sigma-Scan image analysis software was used to investigate bifilms of samples which were cast and applied reduced pressure. Tensile test specimens were also prepared and their tensile strength and % elongation determined in order to observe the modification and different casting conditions on the mechanical properties of A360 alloy.

Results and Discussion

The microstructures of modified and unmodified A360 and A413 alloys are compared in Figs. 1 and 2, respectively. The effect of Sr modification can be seen in finer eutectic microstructure of samples in Fig. 1b and d. The eutectic Si phase of modified alloys shows a fine fibrous morphology. The unmodified alloys in Fig. 1a and c show a coarse primary Al and needles of Si phase. When the holding time is increased to 17 h at furnace, the eutectic silicon area of modified alloy decreases and becomes coarser compared to the modified alloy which was held for 4 h in furnace.

The same conditions were applied to the samples of A413 alloy at 700 °C. The microstructures of samples indicate that the eutectic Si phase of modified alloys

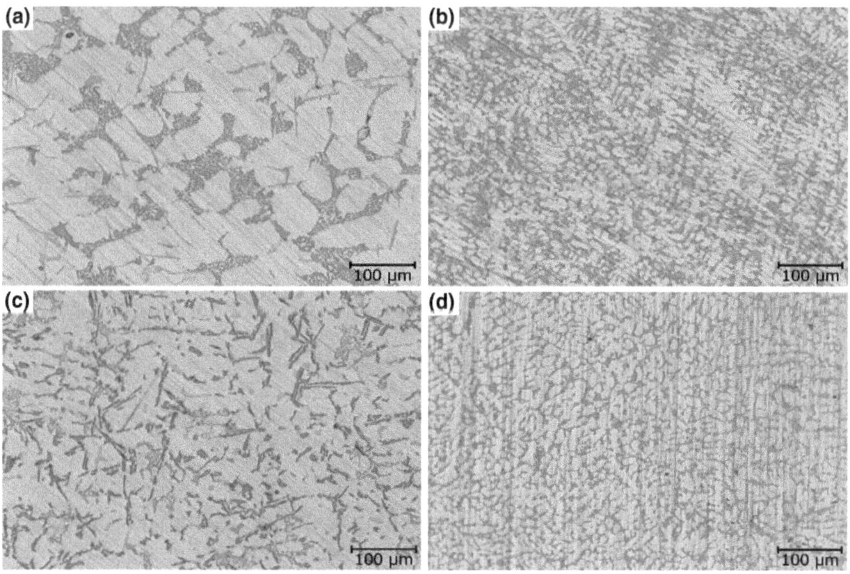

Fig. 1 Microstructures of samples which was poured into permanent (metal) mold at 700 °C; **a** 4 h, unmodified, **b** 4 h, Sr modified (300 ppm), **c** 17 h, unmodified, **d**17 h, Sr modified (300 ppm)

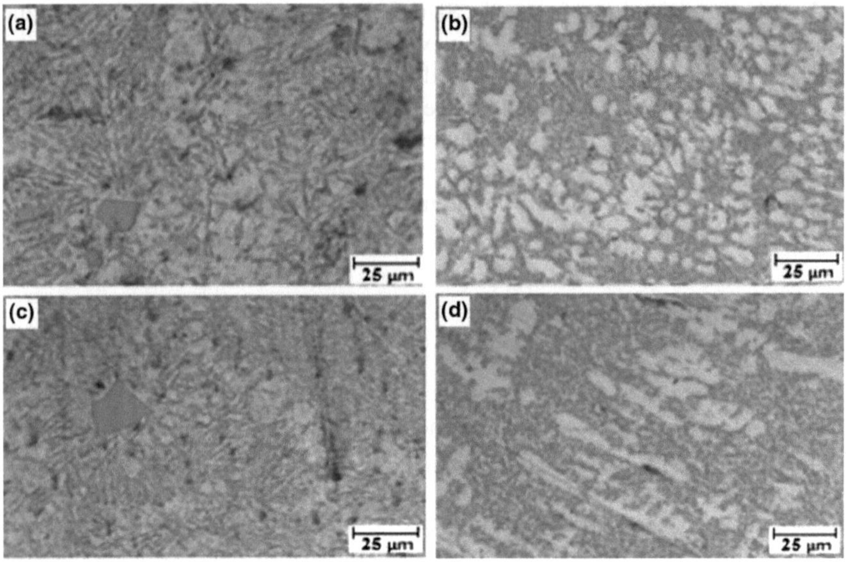

Fig. 2 Microstructures of samples which was poured into permanent mold at 700 °C; **a** 4 h, unmodified, **b** 4 h, Sr modified (300 ppm), **c** 17 h, unmodified, **d** 17 h, Sr modified (300 ppm)

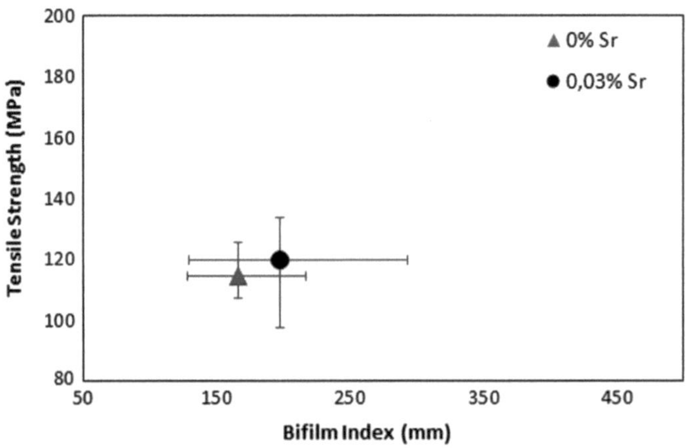

Fig. 3 Relation between average tensile strength and bifilm index of modified and unmodified A360 alloys which were melted at 700 °C, with 4 h of holding time

shows a fine and fibrous structure, in addition, the dendritic structure was formed because of Sr modification, while the unmodified alloys in Fig. 2a and c show coarse primary Al flakes and needle-shaped silicon particles.

In Figs. 3 and 4, average tensile strength and average bifilm index of A360 alloy samples after 4 hours of holding time indicates that Sr addition results in bifilm

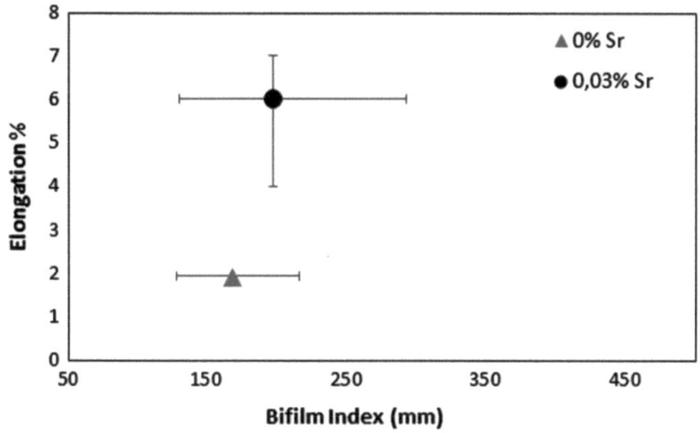

Fig. 4 Elongation % of modified and unmodified A360 alloys which were melted at 700 °C, with 4 h of holding time

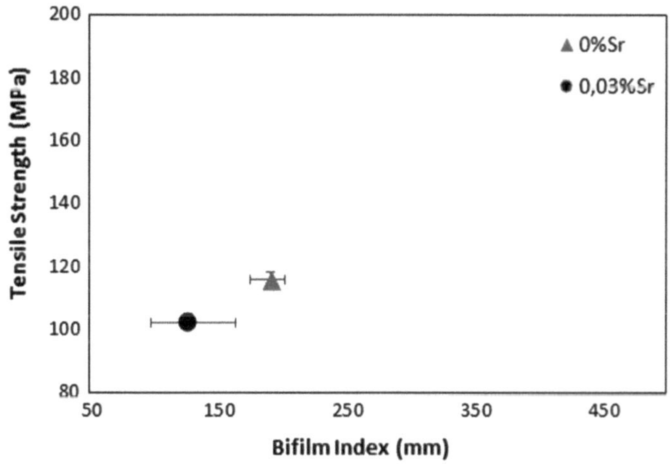

Fig. 5 Relation between average tensile strength and bifilm index of modified and unmodified A360 alloys which were melted at 700 °C, with 17 h of holding time

index of modified A360 samples to increase compared to the unmodified samples at same holding time. In addition, there is not an obvious increment in tensile strength of samples under these conditions. It can also be seen that there is not a significant difference between modified and unmodified samples in terms of tensile strength. On the other hand, in both holding time, Sr modification has a positive effect on elongation % of samples compared to unmodified alloys. Besides that, it can be seen from Figs. 5 and 6 that mechanical properties tend to decrease with holding time possibly because of Sr fading.

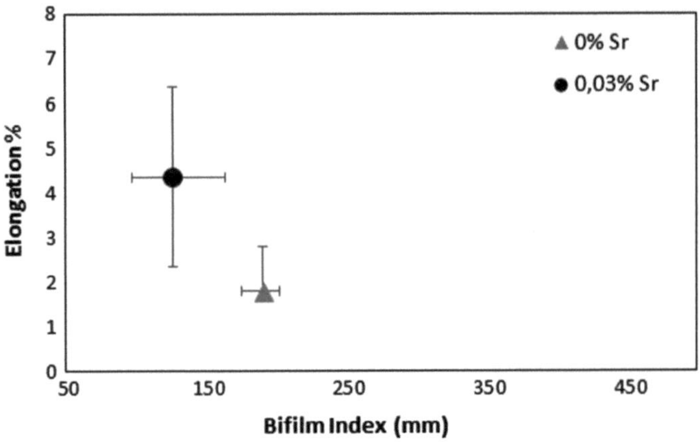

Fig. 6 Relation between % elongation and bifilm index of modified and unmodified A360 alloys which were melted at 700 °C, with 17 h of holding time

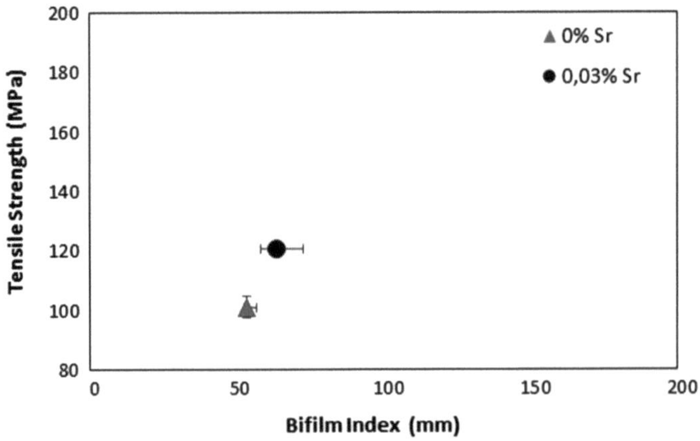

Fig. 7 Relation between average tensile strength and bifilm index of modified and unmodified A413 alloys which were melted at 700 °C, with 4 h of holding time

The relation between tensile strength and bifilm index of modified and unmodified A413 samples is given in Figs. 7 and 9, while elongation % and bifilm index relationship are given in Figs. 8 and 10. The holding time results in different changes in bifilm index of modified and unmodified A413 alloys. There is a little increment in tensile strength of samples by Sr modification, and the unmodified alloys have lower bifilm index. Besides that, the modified alloys at 700 °C do not show an important change with increasing holding time. It is also evident that increasing holding time affects negatively elongation % of A413 samples despite Sr modification. On the other hand, higher elongation % was observed in modified samples after 4 hours of holding time.

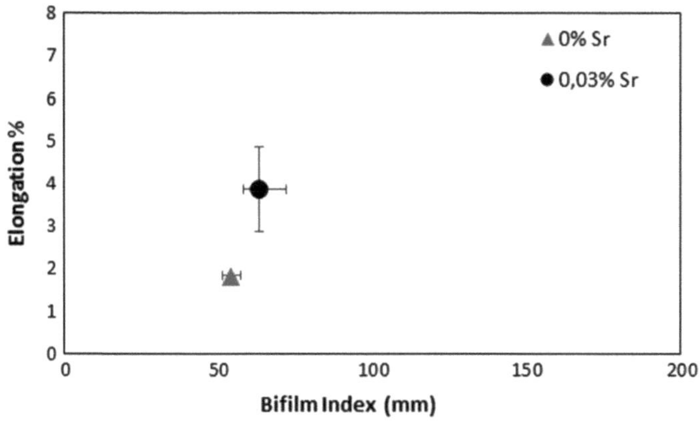

Fig. 8 Relation between % elongation and bifilm index of modified and unmodified A413 alloys which were melted at 700 °C, with 4 h of holding time

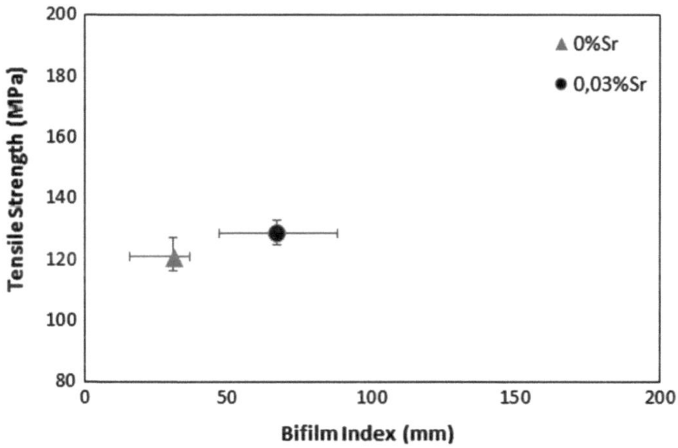

Fig. 9 Relation between average tensile strength and bifilm index of modified and unmodified A413 alloys which were melted at 700 °C, with 17 h of holding time

Although there is not an important improvement in tensile strength of alloys in this work, there are different reports in the literature. Sui et al. [10] investigated the mechanical properties of modified and unmodified Al–12Si–4Cu–2Ni–0,8Mg alloys and demonstrated that the addition of Sr increased from 0 to 0,02% results in ultimate tensile strength increased from 196 MPa to 249 MPa and the elongation increase from 0,74 to 0,95% as an indication of better ductility. Samuel et al. [11] stated that although the porosity increases with Sr addition, Sr modification improves the ductility of Al–Si alloys. There is also a significant correlation between bifilm index and mechanical properties of the alloy. Uludag et al. [12] showed that as bifilm index of A356 alloy was decreased and because of that an

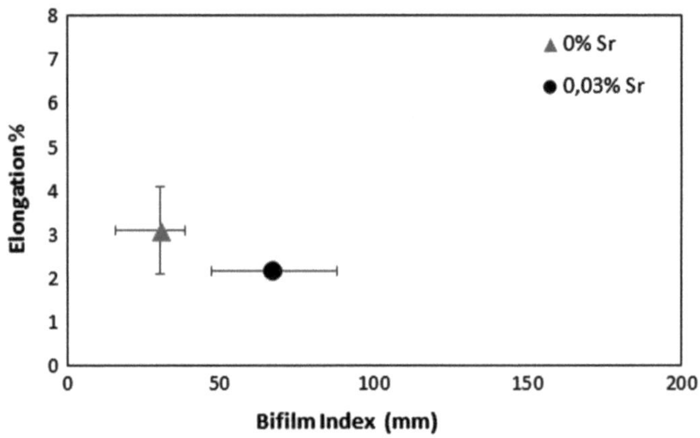

Fig. 10 Relation between % elongation and bifilm index of modified and unmodified A413 alloys which were melted at 700 °C, with 17 h of holding time

improvement was obtained in mechanical properties. The results in this study show that as bifilm index was decreased by Sr modification, the mechanical properties, especially ductility increase significantly.

Conclusion

The following conclusions can be drawn from the experimental work:

1. Sr modification results in finer eutectic silicon in A360 alloy and forming dendrites in A413 alloy because of changing crystallization mechanism. When the holding time is increased from 4 to 17 h the eutectic silicon exhibits a coarser shape at same Sr contents.
2. Mechanical properties of A360 and A413 samples were improved by Sr modification, especially ductility of alloys were increased, regardless of holding time.
3. It was found that there is a strong correlation between bifilm index and mechanical properties of A360 and A413 alloys. In addition, when holding time is increased to 17 h average bifilm index of samples decreases, possibly due to fading Sr.

References

1. Eiken J, Apel M, Liang SM, Fetzer RS (2015) Impact of P and Sr on solidification sequence and morphology of hypoeutectic Al–Si alloys: combined thermodynamic computation and phase-field simulation. Acta Mater 98:152–163
2. Xiang C, Yanxiang L (2010) Effects of strontium and boron interactions in Al–7Si alloys. China Foundry 7:328–330
3. Kahraman F, Kulekci MK (2007) Microstructure and Eutectic Morphology of Al–12.5% Si alloy refined with antimony. SAlj Fen Bilünleri Dergisi 1:10–14
4. Sarada BN, Srinivasamurthy PL (2013) Swetha, microstructural characteristics of Sr and Na modified Al–Mg–Si alloy. Intl J Innov Res Sci Eng Technol 2:3975–3983
5. Yu S, Shao-Ping P, Xue-Ran L, Zi-Run Y, Guo-Xiong S (2011) Nucleation and growth of eutectic cell in hypoeutectic Al−Si alloy. Trans Nonferrous Met Soc China 21:2186–2191
6. Gruzleski John E, Closset Bernard M (1990) The treatment of liquid aluminum–silicon alloys. American Foundrymen's Society, Inc., Des Plaines, Illinois
7. Cross CE, Olson DL (1982) Modification of eutectic weld metal microstructure. Weld Res. Supplement, 381–387
8. Kósa A, Gácsi Z, Dúl J (2012) Effects of strontium on the microstructure of Al–Si casting alloys. Mater Sci Eng 37(2):43–50
9. Haquea MM, Ismail AF (2005) Effect of superheating temperatures on microstructure and properties of strontium modified aluminium-silicon eutectic alloy. In: 13th international scientific conference on achievements in mechanical and materials engineering, Gliwice-Wisla, pp 267–270
10. Sui Y, Wang Q, Wang G, Liu T (2015) Effect of Sr content on the microstructure and mechanical properties of cast Al–12Si–4Cu–2Ni–0,8Mg alloys. J Alloy Compd 622:572–579
11. Samuel AM, Alkahtani SA, Doty HW, Samuel FH (2016) Porosity formation in Al–Si–Cu alloys. Int J Metall Mater Sci Eng 6:1–14
12. Uludağ M, Kocabaş M, Dışpınar D, Çetin R, Cansever N (2017) Effect of Sr and Ti addition on corrosion behaviour of Al–7Si–0,3Mg alloy. Arch Foundry Eng 17:125–130

Part V
Poster Session

Aluminum Matrix Graphene-Reinforced Composite Materials

Okan Aydın, Aziz Kocaveli, Özen Gürsoy, Eray Erzi and Derya Dışpınar

Abstract In this study, due to the enhanced properties of graphene, it was aimed to produce aluminum matrix composite by graphene reinforcement. Several different methods were applied such as lost foam casting, sand casting, die casting, ceramic mold casting and squeeze casting. The incorporation of the graphene was established by only squeeze casting method. Graphene was added by two methods: as layers and in aluminum foils. The produced samples were subjected to bending tests, and it was found that 25% increase in bending strength was achieved by 0.1 wt% graphene addition to A356 alloy.

Keywords Aluminum · Graphene · Composite · Casting

Introduction

The composite term is broadly defined as "composite material" where materials in the same or different groups of two or more materials are assembled at the macro-level to combine the best features together or to create a new feature [1].

Aluminum is an element in the group IIIA of the periodic table with an atomic number of 13, +3 valence with an atomic weight of 26.89, a density of 2,7 gr/cm^3 at 20 °C, a melting point of 659.8 °C and a boiling point of 2450 °C.

Aluminum is typically alloyed with copper, zinc, silicon, magnesium, manganese, iron, nickel and titanium. Some features are lightweight, improved mechanical properties, good heat and electrical conductivity, suitable for heat treatments and resistant to certain chemical effects. Aluminum alloys are suitable for casting. At the same time, forging, drawing and rolling are also applied [2, 3].

Graphene is the name given to a single layer of graphite, an allotropy of carbon element capable of hybridizing sp, sp^2 and sp^3. It was first introduced in 1994 by

O. Aydın · A. Kocaveli · Ö. Gürsoy · E. Erzi · D. Dışpınar (✉)
Faculty of Engineering, Metallurgical and Materials Engineering Department, Istanbul University-Cerrahpasa, 34320 Istanbul, Avcılar, Turkey
e-mail: deryad@istanbul.edu.tr

M. Tiryakioğlu et al. (eds.), *Shape Casting*, The Minerals, Metals & Materials Series,
https://doi.org/10.1007/978-3-030-06034-3_36

Table 1 General properties of single layer graphene [4]

Properties	Value
Hybrid shape	sp^2
Number of layers	Monolayer
Crystal structure	Hexagonal
Dimension	Two
Degree of purity (%)	99
Mass (bulk) density (g/cm^3)	~0.3
True density (g/cm^3)	2.25
Thickness (nm)	~1–2
Surface area (m^2/g)	2600
High-temperature resistance	−75 +200 °C
Thermal conductivity (WK^{-1}/m)	4840–5300
Electro mobility [$cm^2/(V.s)$]	$\sim 2.5 \times 10^5$
Elastic modulus (Tpa)	~1

Boehm [4] (Table 1). In most of the studies [5–7], powder metallurgy was used to produce composites.

Experimental Work

A356 aluminum alloys were melted at 750 °C in resistance furnace and lost foam casting method was used for production of the composite. Layers of foams were created and graphene was placed in between the layers, then the charge was poured. Metallographic and EDS analyzes were carried out by scanning electron microscopy (SEM) for the microstructural characterization.

In another trial, squeeze casting method was used. The mold cavity was 150 × 15 × 15 mm. After the pouring of the melt into cavity together with graphene, the alloy was solidified under pressure at 150 bar.

In another trial, semisolid casting was applied. The samples were heated in a resistance furnace at 630 °C for 20 min with graphene in between the alloy. The samples were squeezed under 220 bar.

In the other trial, two set of tests were carried out. In one of the tests, graphene was placed in aluminum foil which was placed vertically in a sand mold cavity and the charge was poured. In the second test, the same method was applied, but the cast part was thixoforged afterward under 220 bar.

Bending test samples were produced from each experimental work.

Results

In the results of the SEM studies, due to the difference in density of the powder graphene (density value in the literature is between 0.8 and 1.6 g/cm^3) with aluminum (2.7 g/cm^3) and the wetting properties of the graphene, it was observed that

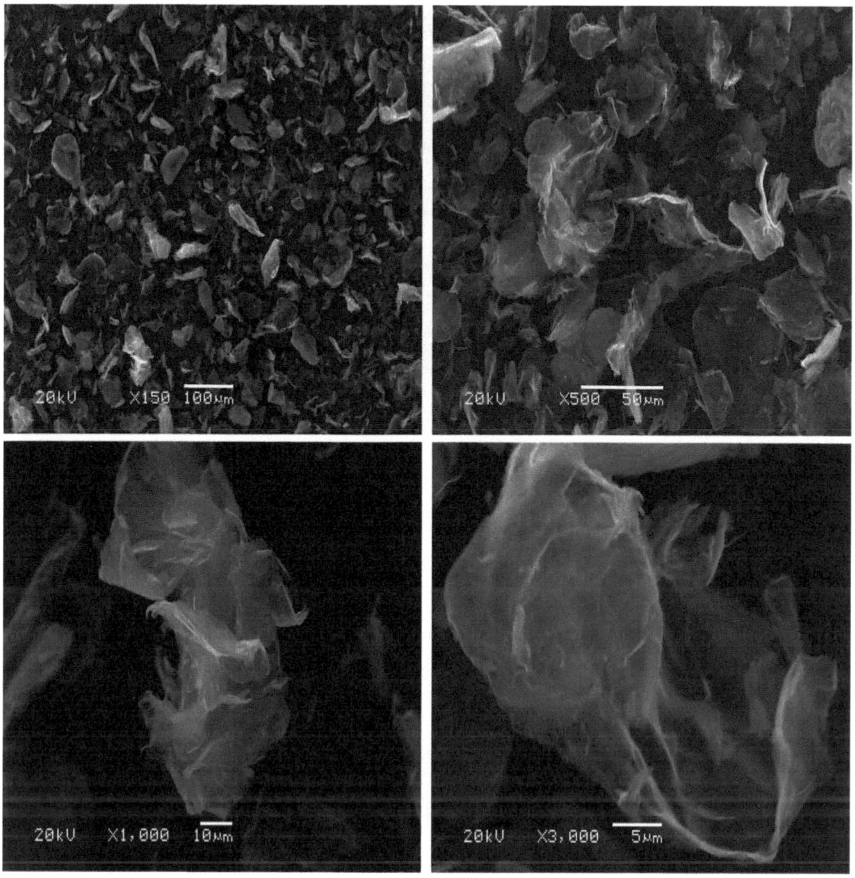

Fig. 1 SEM images of lost foam cast process

the graphene and aluminum were not homogeneously mixed and collected heterogeneously between the layers on the casting surface.

SEM examinations of the casting method with carving of cubic foam showed the same results (Fig. 1).

In the SEM examinations made on samples which are held for 1, 2 and 3 h in the resistance furnace in the squeeze casting method, it was found that after 1 h, it was seen that the layers were not wetted by the aluminum. At the end of 2 h, the aluminum covered the surface of the graphene, however in the metallographic examination, graphene layers of the samples were found to be scattered. At the end of 3 h of holding, graphene was wetted by aluminum and SEM examinations show the carbon peaks (Fig. 2).

The results of squeeze casting showed that the contact of the graphene with the aluminum did not result in homogenous mixing of the graphene due to the wettability. It was not successful.

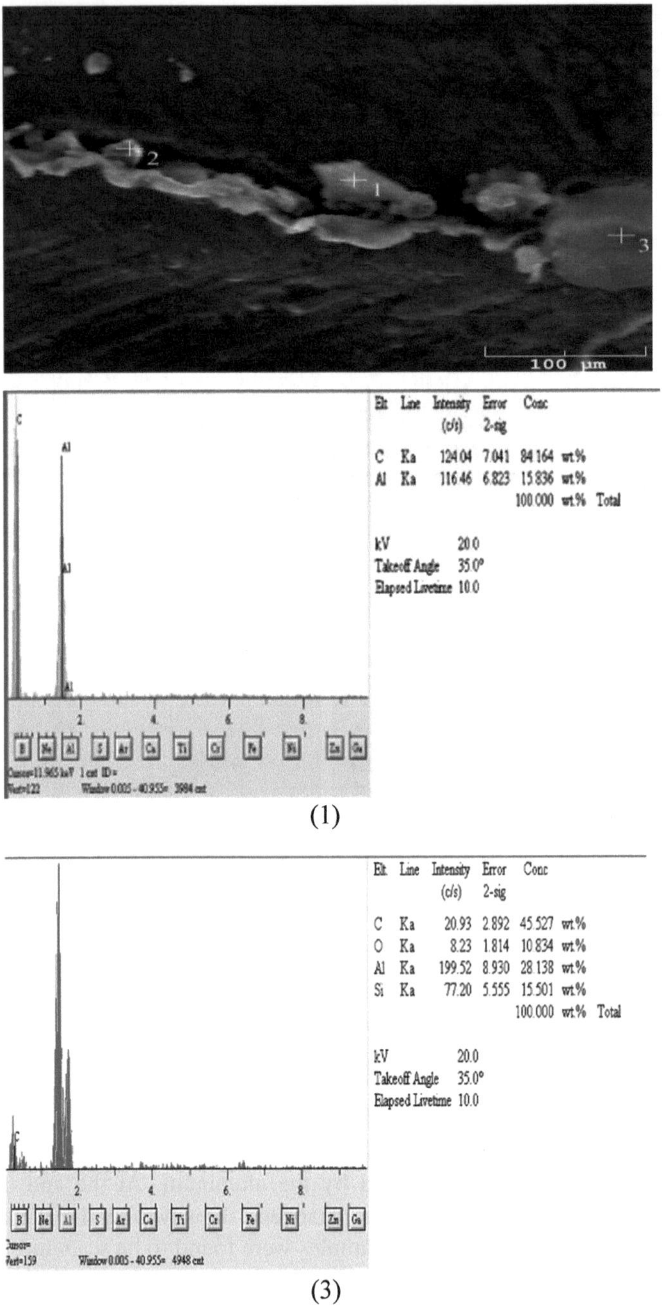

Fig. 2 SEM images of squeeze cast composite

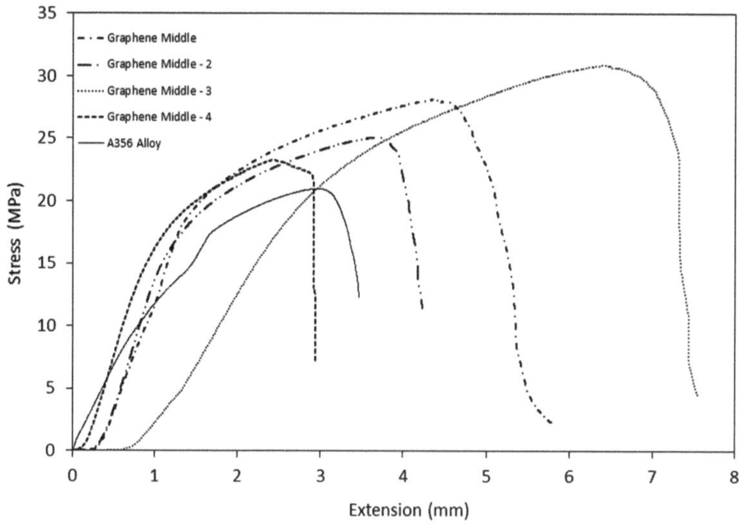

Fig. 3 Graphene powder added medium of A356 alloy—A356 aluminum alloy three-point bending test results

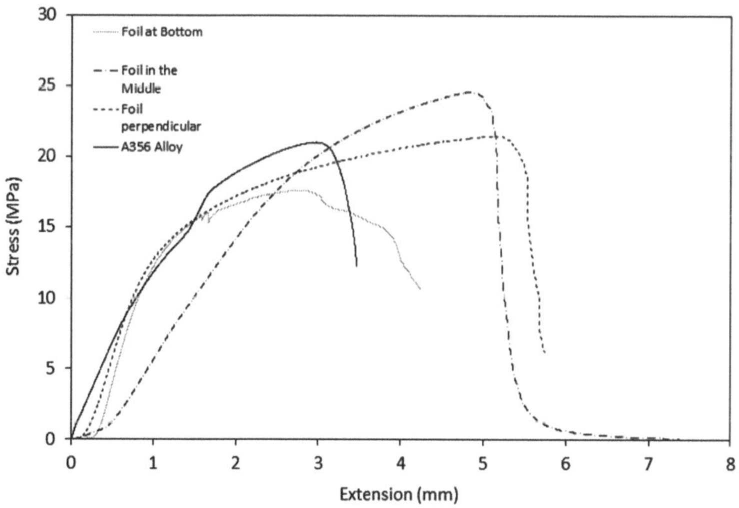

Fig. 4 Graphene powder in aluminum foil—aluminum alloy without graphene three-point bending test results

For the sand cast samples, the graphene again did not wet and no incorporation of the graphene was achieved. Due to the slow cooling, graphene was floated to the surface.

The best composite production was achieved for the castings where graphene was placed in sand mold in aluminum foil and thixoforged. Three-point bending tests were carried out and results are given in Figs. 3, 4, 5 and 6.

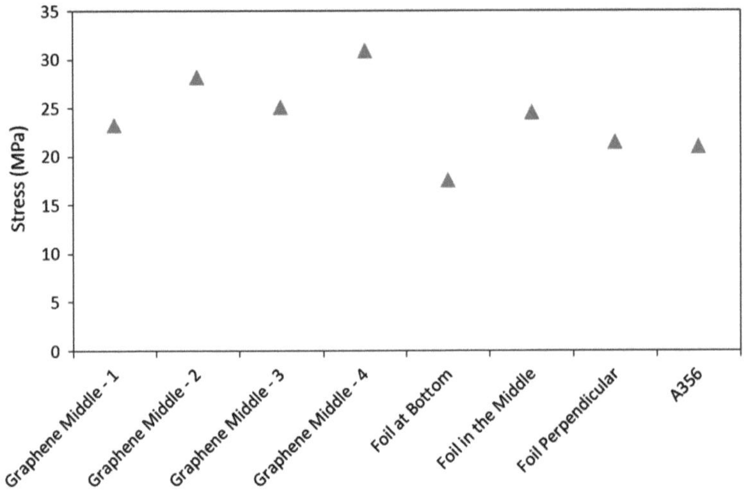

Fig. 5 Strain values according to three-point bending test results

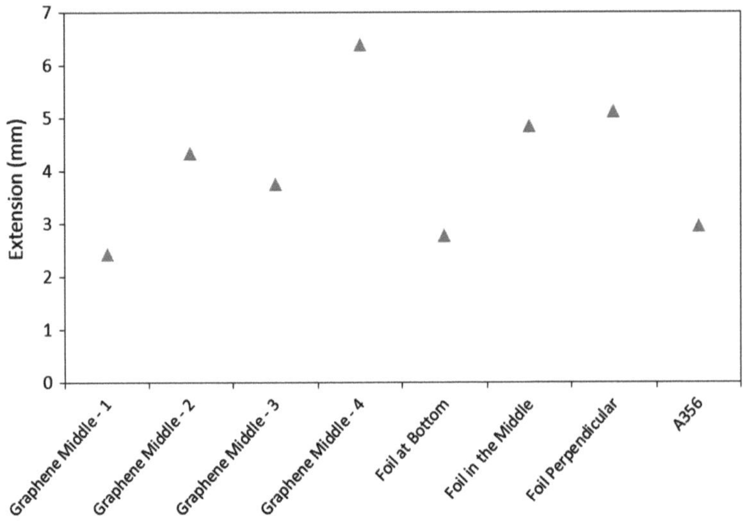

Fig. 6 Extension values according to three-point bending test results

In Fig. 3, it can be seen that the bending strength of the A356 aluminum alloy was 20 MPa, while the powder graphene added in squeeze casting were found to have 25 and 30 MPa.

In Fig. 4, the bending strength of the composites produced by thixoforging was found to change between 20 and 25 MPa.

Conclusion

Production of graphene-reinforced A356 alloy was quite challenging mainly due to the density difference and wettability of graphene. Several different casting methods were applied. The best solution was found by squeeze casting and thixoforging where the graphene was added in the form of foils placed in mold cavity and then forged under pressure.

According to the bending test results, it was observed that the bending strength of A356 was 20 MPa, while graphene added alloy had bending strengths of 25–30 MPa. As a result of this study, it was observed that bending strength of graphene matrix aluminum-reinforced composites increased by 25–50% compared to the base alloy.

References

1. Campbell J (2015) Complete casting handbook: metal casting processes, metallurgy, techniques and design. Elsevier Science
2. Totten GE, MacKenzie DS (2003) Handbook of aluminum: volume 2: alloy production and materials manufacturing. Handbook of aluminum. CRC Press
3. Polmear I, StJohn D, Nie JF, Qian M (2017) Light alloys: metallurgy of the light metals. Elsevier Science
4. Boehm HP (1994) Some aspects of the surface chemistry of carbon blacks and other carbons. Carbon 32(5):759–769
5. Aybarc U, Yavuz H, Dispinar D, Seydibeyoglu MO (2018) The use of stirring methods for the production of SiC-reinforced aluminum matrix composite and validation via simulation studies. Int J Met, 1–11
6. Aybarc U, Dispinar D, Seydibeyoglu M (2018) Aluminum metal matrix composites with SiC, Al_2O_3 and graphene–review. Arch Foundry Eng
7. Akçamlı N, Gökçe H, Uzunsoy D (2016) Processing and characterization of graphene nano-platelet (GNP) reinforced aluminum matrix composites. Mater Test 58(11 12):946 952

Influence of Melt Quality on the Fluidity of AlSi12Fe

İbrahim Göksel Hızlı, Meltem Salkır, İbrahim Halil Kalkan and Derya Dışpınar

Abstract The usage of secondary aluminum has a large volume in the industry due to the economic benefits. The effect of degassing on the melt quality, microstructure, and mechanical properties of secondary alloys was studied in this article. Fluidity and melt quality are having important roles on mold filling. Fluidity was examined with new design mold, and melt quality was determined by bifilm index (BI) via using reduces press test (RPT). In summary, the relationship between melt quality, mechanical properties, and fluidity has been investigated. Inverse ratios were observed between bifilm index and fluidity distance. The aspect ratio is constant at all parameters. Degassing was sufficient to increase the fluidity.

Keywords Bifilm index · Melt quality · Fluidity · Degassing · Scrap

Introduction

Since secondary aluminum production is 95% cheaper than the primary, secondary aluminum is preferred in all casting methods. Another advantage is that it provides recycling before wasted, and the content of the melted scrap is also very important. 8 kg bauxite, 4 kg of chemical product, and 14 kWh of electricity could save up due to recycling of 1 kg aluminum [1, 2]. Recovery from scrap is often used in decoration scraps and car parts. In the first production of these parts, the inclusions and bifilms added to the structure continue to exist in the structure.

After recovery, the liquid metal must be cleaned from these strangers to obtain quality metal. Otherwise, if there is existence in the structure, the unconverted surfaces in the produced parts may be opened by the gas that is diffused into the

İ. G. Hızlı (✉) · M. Salkır · İ. H. Kalkan · D. Dışpınar
Metallurgical and Metallurgy Engineering Department, Istanbul University-Cerrahpasa, Istanbul, Turkey
e-mail: gokselhizli@gmail.com

İ. H. Kalkan
Daimler Mercedes-Benz Turk, Istanbul, Turkey

M. Tiryakioğlu et al. (eds.), *Shape Casting*, The Minerals, Metals & Materials Series,
https://doi.org/10.1007/978-3-030-06034-3_37

structure and create porosity, may cause hot rupture during shrinkage, and may decrease the ductility and mechanical properties after solidification. Since the bifilms are not visible, they are visualized in the RPT and evaluated numerically [3–5].

The fluidity of pure aluminum decreases with the addition of alloying elements, until it reaches the minimum level. This level is close to maximum freezing range composition.

Fluidity increases in eutectic composition classically and again decreases where the freezing range is wider in hypereutectic area. However, Al–Si phase system has been shown different behavior like maximum fluidity on hypereutectic area. Alloys with planar solidification behaviors have higher fluidity. There is no consensus about the effect of the fine-grained structure on the fluidity of the aluminum alloys [6].

The advantages of the sand mold casting method have become noticeable in the aluminum industry in recent years. Especially in the automobile, aviation, and aerospace industries, the use of sand molds for critical parts castings has begun to increase. Therefore, every improvement in the production of secondary ingots will increase the yield of production of critical parts and reduce their costs and make humankind beings a step closer to their dreams.

In this study, the relationships between duration of degassing, liquid metal quality, and fluidity distance of secondary ingots were investigated.

Experimental Study

Firstly, the secondary alloy made of scrap in the experiments was used in the 500 kg oven working with natural gas. After the alloying, ingots as samples were poured at 750 °C when 10 and 40 min degassed by purging N_2 through graphite lance. The sample castings to be used in the analyses were melted from the ingots. After remelting, the amount of BI in the liquid metal does not change. Also, holding the metal did not cause a significant effect of quality of melt [2].

Table 1 Chemical composition of alloy

Elements	(wt%)
Si	11,90
Fe	0,77
Cu	0,35
Mn	0,65
Mg	0,21
Zn	0,12
Ti	0,02
Al	Bal.

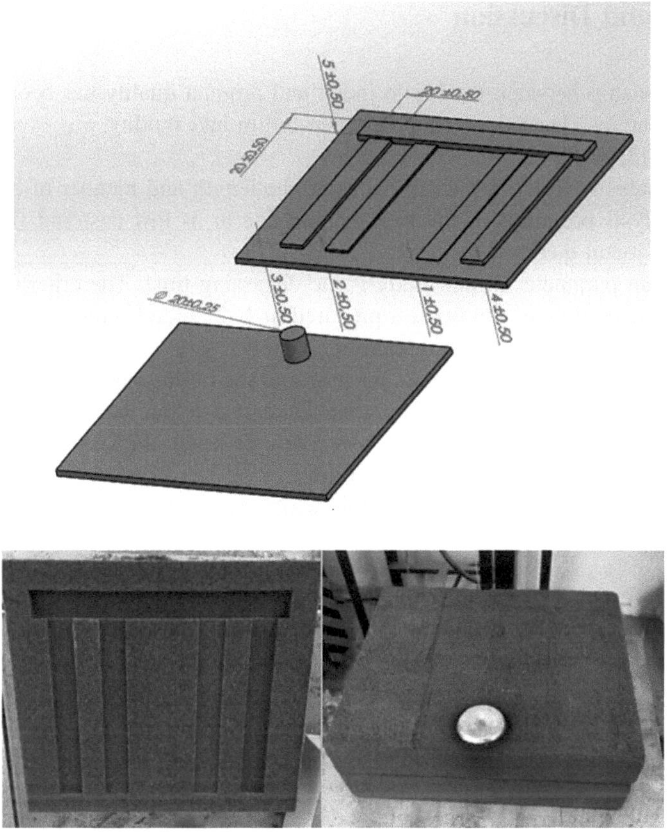

Fig. 1 Geometry of the fluidity mold consisted of different thicknesses

A resistance furnace was used to melt 2, 5 kg of the alloy in a SiC crucible at 750 °C. The chemical composition of alloy is given in Table 1 which is provided by ALGISS MT, Istanbul.

The sand mold shown in Fig. 1 was used to measure the filling distance in different thicknesses. Five samples were cast for each degassing time. Aspect ratio, bifilm index, and filling distance were examined.

The geometry (Fig. 1) consisted of four different thickness planes as 1, 2, 3, and 4 mm, respectively. All planes have 20 mm width and 150 mm length.

Reduces pressure test (RPT) samples were collected from the melt at same time on each pouring of fluidity mold to measure the melt quality via bifilm index [3].

Results and Discussion

The relationship between the bifilm index and product quality has been proven in previous studies. The relationship between bifilm and fluidity was investigated in this study [3, 4, 7].

RPT analysis, which can easily measure the length and number of bifilms with image analysis programs, is the best method due to its low cost and simplicity in measuring metal quality [7].

The main parameter in this study is the degassing time. The criterion which is important in sand mold fluidity test prepared in 1, 2, 3, and 4 mm thickness is full filling of test sample in 1 mm thickness.

The relation between the degassing time and the bifilm index is given in Fig. 2. The BI decreases as the degassing time increases. It has been investigated that changes in filling distance and aspect ratio of each thickness are related to increasing degassing and decreasing bifilms.

The fluidity test results of the ingots which are poured at different degassing times are given in Fig. 3. 4 mm thickness is full filled (150 mm) even without degassing.

In different degassing times, the filling distance of the liquid metal in sand mold and their BI are given in Fig. 4. A clear relationship was observed between the degassing time and the filling of the sections at thicknesses of 1, 2, and 3 mm. The effect of liquid metal cleaning on mold filling is thought to be due to different bifilm ratios as shown in Fig. 5. Depending on the size of the bifilm and the thickness of the mold, the bifilms may limit the flow of liquid metal.

In order to increase the filling distance in parts smaller than 4 mm, the amount of BI must be reduced. This changes in proportion to the duration of degassing.

About filling the thinnest section, 1 mm, the shortest mold fill amount was observed at 30 mm on the undegassing sample and the longest mold filling amount

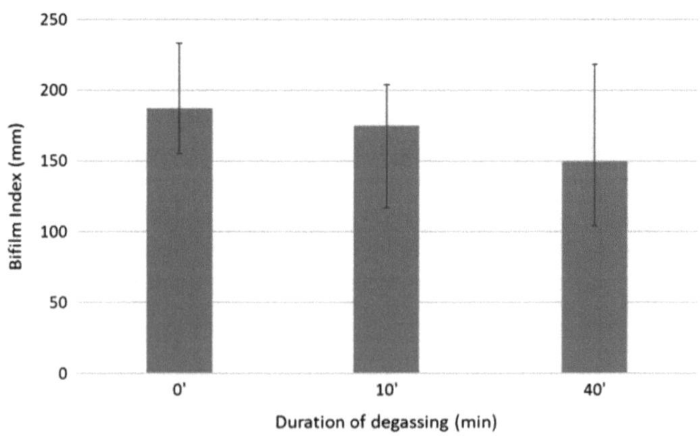

Fig. 2 Relation between bifilm index and degassing time

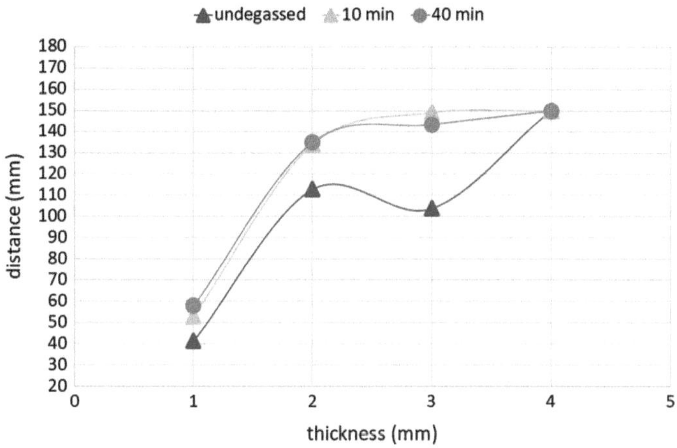

Fig. 3 Filling distance in different thicknesses

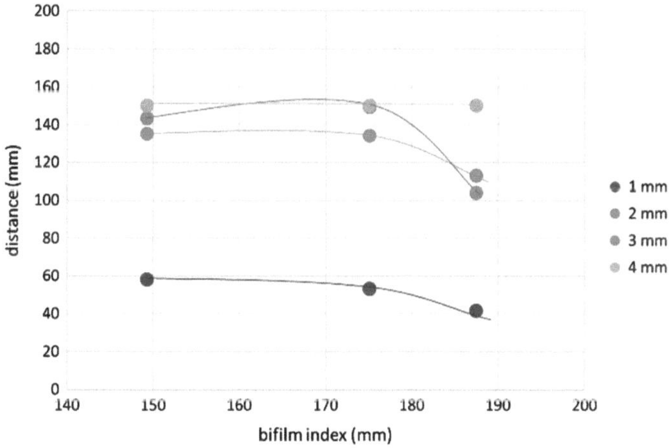

Fig. 4 Filling distance–BI relation in different thicknesses

Fig. 5 Fluidity behavior of liquid metal with and without bifilms

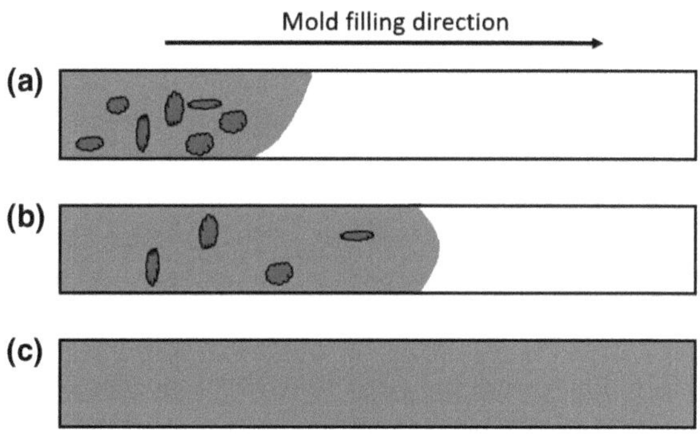

was provided with 82 mm at 40 min degassing time. In liquid metal quality measurement, the lowest bifilm index was observed as 104 at 40 min degassing sample and the highest bifilm index 233 was observed at undegassing sample.

On average (Table 2), the bifilm index has reduced from 187 to 149 mm after 40 min degassing time. After the degassing operation, the fluidity was increased from 41 to 58 mm for 1 mm section thickness.

In 2 mm thickness, filling distance of sand mold has increased to 135 mm from 113 mm after 40 min degassing time. In 3 mm thickness, filling distance of sand mold has increased to 143 mm from 104 mm after 40 min degassing time.

The aspect ratio is almost constant regardless of parameters such as degassing time, BI, and thickness. Figure 6 shows that the aspect ratio is not an effect on liquid metal quality and fluidity.

Regardless of percentage of scrap content in main liquid, fluidity decreases when the contents and oxide content increase [8]. The use of dirty metal also increases inclusions and oxide content. But the percentage of recycled aluminum alloys does not significantly influence the fluidity for the same inclusion and oxide content.

Table 2 Average bifilm index and fluidity length

	0′	10′	40′
1 mm	41	53	58
2 mm	113	134	135
3 mm	104	149	143
4 mm	150	150	150
BI	187	175	149

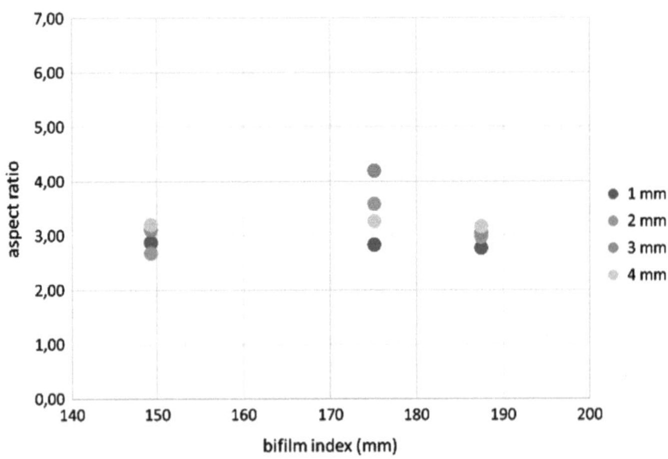

Fig. 6 Bifilm index and aspect ratio relation on different sand mold thicknesses

Conclusions

- The bifilm index has decreased with the degassing time.
- Bifilm index reduction increases the filling distance of all thicknesses.
- Aspect ratio does not affect fluidity.

Acknowledgements This work was supported by the generous management of Algiss Metal Technologies, Istanbul, Turkey.

References

1. Dispinar D, Kvithyld A, Nordmark A (2011) Quality assessment of recycled aluminium. In: Light metals 2011. Springer, Cham, pp 731–735
2. Kvithyld A, et al (2012) Quality comparison between molten metal from remelted sheets; Mill finish and coated. In: Light metals 2012. Springer, Cham, pp 1031–1035
3. Dispinar D et al (2010) Degassing, hydrogen and porosity phenomena in A356. Mater Sci Eng A 527(16–17):3719–3725
4. Dispinar D, Campbell J (2011) Porosity, hydrogen and bifilm content in Al alloy castings. Mater Sci Eng A 528(10–11):3860–3865
5. Campbell J (2015) Chapter 3 - Flow. In: Campbell J (ed) Complete casting handbook, 2nd edn. Butterworth-Heinemann, Boston, pp 91–134
6. Ravi K et al (2008) Fluidity of aluminum alloys and composites: a review. J Alloy Compd 456(1–2):201–210
7. Dispinar D, Campbell J (2006) Use of bifilm index as an assessment of liquid metal quality. Int J Cast Met Res 19(1):5–17
8. Di Sabatino M, et al (2005) The influence of oxide inclusions on the fluidity of Al–7 wt% Si alloy. Mater Sci Eng A 413:272–276

Author Index

M. Tiryakioğlu et al. (eds.), *Shape Casting*, The Minerals, Metals & Materials Series,
https://doi.org/10.1007/978-3-030-06034-3

Subject Index

M. Tiryakioğlu et al. (eds.), *Shape Casting*, The Minerals, Metals & Materials Series,
https://doi.org/10.1007/978-3-030-06034-3

MIX
Papier aus verantwortungsvollen Quellen
Paper from responsible sources
FSC® C105338

If you have any concerns about our products,
you can contact us on
ProductSafety@springernature.com

In case Publisher is established outside the EU,
the EU authorized representative is:
Springer Nature Customer Service Center GmbH
Europaplatz 3, 69115 Heidelberg, Germany

Printed by Libri Plureos GmbH
in Hamburg, Germany